SENTENCING REFORM *in* OVERCROWDED TIMES

SENTENCING REFORM *in* OVERCROWDED TIMES

A Comparative Perspective

Edited by
Michael Tonry
& Kathleen Hatlestad

New York   Oxford
Oxford University Press
1997

Oxford University Press

Oxford   New York
Athens   Auckland   Bangkok   Bogota   Bombay   Buenos Aires
Calcutta   Cape Town   Dar es Salaam   Delhi   Florence   Hong Kong
Istanbul   Karachi   Kuala Lumpur   Madras   Madrid   Melbourne
Mexico City   Nairobi   Paris   Singapore   Taipei   Tokyo   Toronto

and associated companies in
Berlin   Ibadan

Copyright © 1997 by Oxford University Press, Inc.

Published by Oxford University Press, Inc.
198 Madison Avenue, New York, New York 10016

Oxford is a registered trademark of Oxford University Press

Library of Congress Cataloging–in–Publication Data
Sentencing reform in overcrowded times : a comparative perspective /
Michael Tonry and Kathleen Hatlestad, editors.
p.   cm.
Includes bibliographical reference.
ISBN 0–19–510786–1;   ISBN 0–19–510787–X
1. Prison sentences.   2. Prison sentences—United States.
3. Alternatives to imprisonment—United States.
4. Sentences (Criminal procedure)—United States.
5. Discrimination in criminal justice administration—United States.
6. Prisons—Overcrowding—United States.   I. Tonry, Michael H.   II. Hatlestad, Kathleen.
HV8708.S44   1996
364.6'0973—DC20   96–8403

9 8 7 6 5 4 3 2 1

Printed in the United States of America
on acid-free paper

Preface

Sentencing Reform in Overcrowded Times is a play on words. American sentencing policies have produced badly overcrowded prisons. On January 1, 1996, American prisons operated at 125 percent of their capacity, and prisons or jails in upward of forty states were subject to federal court orders related to overcrowding. American federal and state prisons on December 31, 1980, held 328,695 inmates, a number that more than tripled to 1,127,132 on December 31, 1995. Another 500,000 people were confined in American jails on that date. During much of the 1990s, corrections spending grew faster than any other category of state government expenditure. In addition, all of the articles in this book first appeared in Overcrowded Times, a bimonthly journal that specializes in sentencing and corrections policy and research, and in initiatives aimed at reducing America's overreliance on imprisonment as a penal sanction.

This book has two siblings. The first, Michael Tonry's Sentencing Matters (1996), also published by Oxford University Press, is an analysis of sentencing policy and practice in the United States and other Western countries and comments on many of the research and program developments discussed in this book. The second, Intermediate Sanctions in Overcrowded Times (1995), edited by Michael Tonry and Kate Hamilton, was published by Northeastern University Press, and, like this book, it combines thematic groupings of articles first published in Overcrowded Times with new introductions.

From its inception, Overcrowded Times has been supported by the Edna McConnell Clark Foundation, the only major national foundation that has shown continuing interest in criminal justice and corrections problems. The aim is to make reliable, authoritative information on important research and program developments available to policy makers, practitioners, and researchers. Articles are written in nontechnical language for a nonspecialist audience, but the writers are generally among the best-informed researchers on their subjects or are public officials who have been actively involved in the developments they discuss.

Half of the articles in this book concern sentencing reform in the United States. Nearly a quarter concern developments elsewhere. Countries covered include most of the major English-speaking countries—Australia, Canada, England, New Zealand, and South Africa—and most of the Western European countries that have empirical research traditions—Finland, Germany, the Netherlands, Sweden, and Switzerland. Sentencing and corrections issues are much the same in all Western countries, and increasingly, efforts are being made to import policies and practices that have succeeded elsewhere, particularly in Europe.

So far, the United States has not been notably successful as an importer; efforts to develop community service programs patterned on English experience and day fines patterned on German and Swedish experience can charitably be described as halting and incomplete. Likewise, distinctively American innovations have seldom been replicated elsewhere. No other country has as yet been persuaded to adopt U.S.-style numerical sentencing guidelines and, while some countries have been attracted to boot camps, "three-strikes" laws, and electronic monitoring, none has as yet adopted them on a significant scale. Nonetheless, understanding of American dilemmas and the search for better solutions can only be enhanced by learning more about other countries' approaches to common problems.

The balance of the book consists of articles on racial disparities caused by modern American crime control policies and on public opinion about crime. Racial disparities in the justice system have long been extreme, but since 1980 they have steadily worsened. The adult incarceration rate for American blacks is eight times that for whites. Disparities are worse in the juvenile system. American drug policies, especially laws requiring mandatory minimum penalties and distinguishing between crack and powder cocaine, are the principal reason for worsening disparities.

Many harsh laws and policies recently have been adopted in the name of public opinion. The best evidence, however, is that ordinary people's views are commonly misunderstood by public officials. Ordinary people's considered views are not unqualifiedly harsh. Most people hold conflicted views that parallel those of most practitioners—wrongful actions should be punished and vigorous efforts should be made to help offenders become law-abiding, self-supporting citizens.

Except for minor stylistic editing, articles appear here as they first appeared in *Overcrowded Times*. Quotations from named individuals without additional citations of sources appear in several articles; in these cases the quotations are from comments made to the articles' writers during the course of their inquiries or research. We are grateful to writers for their cooperation, to Yale Law School's Daniel J. Freed and the University of Toronto's Anthony Doob for their encouragement, and to the Justice Program of the Edna McConnell Clark Foundation and its director, Kenneth Schoen, for their support of *Overcrowded Times*.

So much for our aims, method, and scope. Readers will decide for themselves whether the effort to publish short, timely, readable articles on sentencing reform in overcrowded times has been worthwhile.

Castine, Maine   M. T. & K. H.
November 1996

Contents

Contributors

Hans-Jörg Albrecht is professor of law at Technische Universität, Dresden, Germany.

Andrew Ashworth is Edmund-Davies Professor of Criminal Law and Criminal Justice at King's College, London.

Jen Kiko Begasse is a research associate at Doble Research Associates in Englewood Cliffs, New Jersey.

David Boerner is professor of law at the Seattle University School of Law, Tacoma, Washington.

Kathleen Bogan, a criminal justice consultant in Portland, Oregon, was formerly executive director of the Oregon Criminal Justice Council.

Kenneth E. Carlson is a research scientist at Abt Associates in Cambridge, Massachusetts.

Stevens H. Clarke is professor of public law and government, Institute of Government, University of North Carolina at Chapel Hill.

Debra Dailey is executive director of the Minnesota Sentencing Guidelines Commission.

David Diroll is executive director of the Ohio Criminal Sentencing Commission.

John Doble is director of Doble Research Associates in Englewood Cliffs, New Jersey.

Anthony N. Doob is professor of criminology at the University of Toronto.

David Factor was formerly executive director of the Oregon Criminal Justice Council in Salem.

Steve Farkas is senior research associate at the Public Agenda Foundation, New York.

Richard S. Frase is professor of law at the University of Minnesota Law School.

Arie Freiberg is professor of criminology at the University of Melbourne, Victoria, Australia.

Richard Gebelein is a superior court judge in Wilmington, Delaware, and chairman of Delaware's Sentencing Accountability Commission.

Angela Gorta is research manager of the Independent Commission Against Corruption, Sydney, Australia.

David Gottlieb is professor of law at the University of Kansas School of Law.

Judy Greene is State-Centered Program coordinator for the Edna McConnell Clark Foundation.

Tracy Huling is a criminal justice consultant based in Freehold, New York.

Stephen Immerwahr is a research associate for the Public Agenda Foundation in New York.

Cynthia Kempinen is senior associate director of the Pennsylvania Commission on Sentencing.

Martin Killias is professor of criminology at the University of Lausanne, Switzerland.

John Kramer is executive director of the Pennsylvania Commission on Sentencing.

André Kuhn is assistant professor of corrections at the University of Lausanne, Switzerland.

Roxanne Lieb is assistant director of the Washington State Institute for Public Policy, Evergreen State College.

Robin Lubitz is executive director of the North Carolina Sentencing and Policy Advisory Commission.

Marc Mauer is assistant director of the Sentencing Project in Washington, D.C.

Douglas McDonald is a research scientist at Abt Associates in Cambridge, Massachusetts.

Jerome Miller is president of the National Center on Institutions and Alternatives in Alexandria, Virginia.

Rod Morgan is dean of the Faculty of Law at the University of Bristol, England.

David Moxon is principal research officer at the Home Office Research and Planning Unit in London.

Richard J. Oldroyd is director of research for the Commission on Criminal and Juvenile Justice in Salt Lake City, Utah.

Carl Pope is professor of criminal justice at the University of Wisconsin in Milwaukee.

Stan C. Proband is a contributing editor for *Overcrowded Times*.

Thomas J. Quinn is visiting fellow at the U.S. Department of Justice, National Institute of Justice.

Curtis R. Reitz is professor of law at the University of Pennsylvania.

Kevin R. Reitz is professor of law at the University of Colorado.

Julian V. Roberts is professor of criminology at the University of Ottawa.

Simone Rônez is a statistician at the Office fédéral de la statistique, Bern, Switzerland.

Peter J.P. Tak is professor of law at the Catholic University of Nijmegen, the Netherlands.

Stephan Terblanche is professor of law at the University of South Africa in Pretoria.

T. M. Thorp is a justice of the High Court of New Zealand, Auckland.

Patrik Törnudd is director of the National Research Institute of Legal Policy in Helsinki, Finland.

Andrew von Hirsch is professorial fellow in the Institute of Criminology, Cambridge University.

Martin Wasik is professor of law at the University of Manchester, England.

Thomas Weigend is professor of law at the University of Cologne, Germany.

Marianne Wesson is professor of law at the University of Colorado at Boulder.

Ronald F. Wright is professor of law at Wake Forest University.

SENTENCING REFORM *in* OVERCROWDED TIMES

ONE

Sentencing Reform

Reform of sentencing has been on the American agenda for a quarter century. Because the United States is a large, noisy, and self-centered country, many Americans do not know that sentencing reform has also long been on the policy agendas of other countries. Finland in 1976 and Sweden in 1988 (Jareborg 1995), for example, substantially overhauled their sentencing laws. England did so in 1991 (Ashworth 1992). So, beginning with enactment of a "truth-in-sentencing" law in New South Wales in 1989 (Gorta 1992), did several Australian states (Freiberg 1995). Sentencing has been on the national agenda in Canada since the early 1980s, although, despite proposals from several blue-ribbon commissions and the introduction of proposed legislation by successive governments, by 1996 fundamental changes had not been made (Doob 1995). Many countries, including all those just mentioned and also Germany, the Netherlands, New Zealand, South Africa, and Switzerland, have made smaller but significant changes.

A story could be told that would stress continuities in sentencing policy changes in Western countries over a quarter century. Moved primarily by idealistic concerns about alleged arbitrariness and invidious bias in sentencing, American reformers in the 1970s called for creation of fairer, more principled policies. A variety of approaches were tried, and tested, and by the mid-1990s a consensus emerged that something called presumptive sentencing guidelines offered the greatest promise for making sentencing more consistent and predictable, reducing unwarranted and invidious disparities, and making judges more accountable. Reformers in other countries, beginning somewhat later but moved by similar concerns, are following a decade or more later in American footsteps to the same destination and are at various stages along the way. Just as it was once often said that California's present was America's future, America's today could be other countries' tomorrow.

The only problem with that story of steady progress in many countries toward achievement of fairer sentencing policies is that it is mostly untrue—both over time in the United States and over space in its depiction of parallels between countries.

The American sentencing reform movement of the 1970s and early 1980s bears little similarity to the movement of the 1990s. The early reform movement was centrally concerned with remedying unfairness to offenders—unwarranted disparities, invidious discrimination, and officials' arbitrariness. The contemporary movement is mostly concerned with making penalties harsher and with controlling government expenditures. No one objects if new, harsher laws are consistently applied, but that is seldom a primary aim.

Likewise, sentencing developments in other lands bear little resemblance to modern American developments motivated mostly by ideology and partisan politics. Instead, public safety in most countries, like public health, public education, and public transit, is among the unglamorous core functions of government; relying on the best available knowledge, officials attempt to devise fair, cost-effective ways to deal with chronic problems that face every developed Western country and always have and always will.

Crime control politics in the United States has long been disconnected from empirical knowledge. Every American state in the past fifteen years has enacted mandatory penalty laws; the best evidence everywhere shows such laws to be ineffective deterrents to crime (Tonry 1996, chap. 5), but they are enacted as proof of toughness. Other countries seldom enact such laws. Many American states over the past twenty-five years have enacted death penalty statutes, and increasing numbers of people are executed. Neither any European country nor any of the English-speaking countries with which the United States normally wants to be compared—Australia, Canada, England, Ireland, New Zealand—retains capital punishment (Hood 1992).

Sentencing changes in other countries have been neither as precipitate nor as radical as in the United States. Nowhere else, until recently in England and some Australian states, have sentencing policies been a regular feature of partisan, sound-bite politics. Partly this is because crime control, or "law and order," has nowhere else long been a central issue in electoral politics as it is in the United States. Crime control has been a partisan political issue since Barry Goldwater's 1964 presidential campaign against Lyndon Johnson. Since 1980, conservative politicians have continuously and disingenuously attacked their opponents as "soft on crime" (as if any mainstream political figure were in favor of crime or unconcerned about the suffering of victims), and the attacks have been successful (Edsall and Edsall 1991; Carter 1995).

Few elected public officials dare oppose any "toughness" proposal, whatever its unfairness, expense, or likely ineffectiveness. The result has often been a race to the bottom in which laws are enacted primarily because they are punitive. It is easy to seize the low ground in American political debates about crime.

Changes in crime rates have little to do with changes in sentencing laws, and most penalty changes accordingly have little demonstrable effect on crime; this has been the unanimous conclusion over two decades of National Academy of Sciences panels, each funded by a conservative Republican administration, on deterrence and incapacitation (Blumstein, Cohen, and Nagin 1978), criminal careers (Blumstein et al. 1986), and understanding and control of violence (Reiss and Roth 1993). The same conclusion was reached by the Australia Law Reform Commission (1980), the Canadian Sentencing Commission (1987), and the English Home Office while Margaret Thatcher was prime minister (Home Office 1990).

Nonetheless, the response of American politicians in recent decades has seldom been to wonder whether the reason why harsh laws fail to reduce crime rates is that they cannot, but instead it has been to enact still harsher laws. The result has been a sextupling of the number of people in American prisons (from 196,092 in 1972 to 1,104,074 on June 30, 1995) and crime rates in 1994 that were about the same as in 1980.

Sentencing changes in other countries tend to be incremental and to result from other than partisan political considerations. Many countries, including England and Wales, Germany, the Netherlands, Scotland, and Switzerland, have established new penalties with the aim of diverting people from imprisonment, and they have succeeded in doing so. Some countries, including England and Wales, Finland, and Sweden, have changed their laws to make it clearer that sentencing should be based on fairness concerns about proportionality, or "just deserts."

There are minor exceptions. Michael Howard, home secretary during the mid-1990s, has tried to bring American-style law-and-order politics, including boot camps, three-strikes laws, and claims that "prison works," to England. Newly elected conservative governments in the Australian states of New South Wales, Victoria, Western Australia, and South Australia have adopted truth-in-sentencing laws, which in that country means abolition of remission ("good time" in the United States), and in a variety of other ways have toughened sentencing laws, but the changes enacted to date are modest by American standards. Sentencing change in other countries bears more resemblance to the idealistic and principled U.S. initiatives of the 1970s than to the opportunistic and cynical changes of the 1990s.

The articles in this book attempt to fill a void in the academic and policy literatures about sentencing. Many are short sentencing policy histories for countries and states. Understanding how legal institutions and processes work often requires understanding how they relate to other institutions and processes, and how they came to be as they are. "History" and "process" are the key words, as any good manager knows, yet there is virtually no scholarly or professional literature on sentencing policy changes. There are few case histories; professional historians lack interest and incentive, and public officials lack time. Social scientists occasionally prepare evaluations of the effects of changes (though there are few because foundations seldom fund criminal justice research, and the federal research agencies have for many years had little interest in sentencing), but these tend to be quantitative and to test the effects of changes while paying little attention to how and why the changes were made.

When officials are given responsibility to oversee major sentencing changes, they first typically want to know what officials elsewhere have learned when attempting to effect similar new policies. All that can be done now is to get on the telephone, but this is seldom effective because key people have often either moved on to other jobs or don't remember, and a paper trail documenting the policy process is seldom available. As a result, the same mistakes are made again and again, often not from willfulness or arrogance but from lack of knowledge of how they could have been avoided.

Because most of the articles in this book are about sentencing policies in particular times and places, this introduction provides a general overview of the past quarter century's movements to reform sentencing. The first section concerns the United States, the second, other countries.

Sentencing Reform in the United States

Within the United States, every state has since 1975 considered fundamental sentencing changes, and most have enacted some. In 1975 every jurisdiction had an indeterminate sentencing system. Legislatures set maximum authorized sentences for offenses, judges could impose any lawful sentence and set minimum sentences in individual cases, parole boards decided when prisoners were released, and prison officials shortened sentences by as much as a third through award of time off for good behavior.

Sentencing was said to be "indeterminate" because the lengths of prison sentences could not be determined at sentencing. They became known in individual cases only when the last of the judicial, correctional, and parole board decisions was made. In recent years that system, which in broad outline continues in at least twenty American states, is commonly disparaged as "bark-and-bite sentencing." The judge's ten-year bark often resulted in a one- to three-year bite.

Indeterminate sentencing was premised on the desirability of "individualization." Judges and other officials were to tailor decisions to offenders' unique circumstances and characteristics, including the extent to which they could be or had been reformed. It was also premised on the notions that correctional programs could rehabilitate offenders and that corrections, parole, and treatment officials knew how to match offenders to treatments and how to recognize when treatment had worked (Blumstein et al. 1983, chap. 3). Indeterminate sentencing was also convenient for officials; they were granted wide, unreviewable discretions, which freed them from others' second-guessing and made it easier to reconcile treatment with management and budget concerns (e.g., Rothman 1980).

Indeterminate sentencing, which exists in various forms in most Western countries, has fallen into disrepute in the United States. This results partly from doubts about the rehabilitative effectiveness of correctional programs, partly from concern that officials should not be given wide, unreviewable discretion over others' lives, partly from apprehension that those discretions are often exercised randomly, arbitrarily, or invidiously, and partly from populist belief that bark-and-bite sentencing is dishonest and deceptive. That is a powerful indictment, and in the United States indeterminate sentencing crumbled before it (Tonry 1996, chap. 1).

Parole release went first, beginning in the early 1970s. Research revealed that few correctional programs appeared to reduce recidivism of participants, that parole boards had no special talent for identifying rehabilitated prisoners, and that the criteria parole boards used to make risk predictions were known when offenders entered prison and could as easily, and much earlier, be used by judges at sentencing (Morris 1974; von Hirsch and Hanrahan 1979). The solution in some jurisdictions—initially the federal system, Oregon, Minnesota, and Washington—was to establish guidelines to reduce release disparities. Eventually at least twenty states did likewise. The more radical solution, initially adopted in Maine and California but eventually adopted in at least ten jurisdictions including the federal system, was to abolish parole release altogether.

Prison officials' discretion also came under attack (Jacobs 1982). In most states, officials could give prisoners time off for good behavior, "good time," and often they could reward prisoners for work or educational participation with additional sentence reductions. Officials in addition generally had authority to grant various kinds of furloughs and to transfer prisoners to halfway houses, home detention, or other community placements. Although strong arguments, based on management, programming, and humanitarian considerations, can be made for these correctional powers, they also came under attack. Some jurisdictions cut back on correctional officials' release authority and some eliminated it altogether. The Federal Sentencing Reform Act of 1984 limited maximum good time to 15 percent of the sentence; a prisoner sentenced to twenty years' imprisonment must serve at least seventeen. Responding to federal incentives, many states in the 1990s reduced corrections officials' powers in order to assure that state prisoners also serve at least 85 percent of their announced sentences.

Judicial discretion was the next target. Research on disparities documented sometimes sizable differences in sentences received by comparable offenders convicted of similar or the same crimes (Blumstein et al. 1983, chap. 2). In addition, a general movement in the legal system toward greater procedural protections for individuals, liberal activists' belief that invidious race and class discrimination was commonplace, and conservative activists' belief that judges were exercising their discretion in unduly lenient ways combined to delegitimate judicial sentencing discretion.

The initial solution was to develop "voluntary" sentencing guidelines to reduce disparities (Gottfredson, Wilkins, and Hoffman 1978; Kress 1980). They were soon shown not to work and, in any case, in the eyes of both liberal and conservative activists offered insufficient restraints on judges. Although voluntary guidelines were adopted or seriously considered in every state in the late 1970s and early 1980s, usually at county or judicial circuit levels, the movement soon lost steam and most were abandoned (Blumstein et al. 1983, chap. 3). Beginning in the late 1980s, states again began to establish voluntary systems. New systems were adopted in Louisiana, Virginia, Arkansas, and Ohio. No evaluations concerning their effectiveness have been published.

Next in order were systems of "statutory determinate sentencing," usually coupled with abolition of parole release, in which legislatures amended their penal codes to specify penalties that should be imposed on offenders convicted of designated offenses. The first such law, California's Uniform Determinate Sentencing Law, enacted in 1976, for example, provided that robbers ordinarily be sentenced to three years in prison, or to two or four years in mitigating or aggravating circumstances. Illinois, Indiana, North Carolina, Colorado, and Arizona enacted similar laws. It soon became apparent that such laws provided only crude guidance to judges and were easily evaded (Tonry 1993). No new ones were enacted after 1983.

The third and last comprehensive approach to control of judicial discretion was to establish administrative agencies called sentencing commissions, which developed "presumptive sentencing guidelines." Proposed guidelines were submitted to legislatures and, depending on the enabling legislation, took effect if they were approved, or not disapproved. The guidelines were presumptive in the sense that they set sentencing standards for individual cases that were presumed to be appropriate and that judges were expected to follow unless they provided reasons for doing something else. The adequacy of those reasons was subject to appeal by the parties and to review by higher courts. Minnesota and Pennsylvania promulgated presumptive guidelines in the early 1980s and were soon followed by Washington, Oregon, Kansas, and other states.

Evaluations showed that some presumptive guidelines systems made sentencing more consistent, reduced disparities in general and in relation to race and gender, and made sentencing sufficiently predictable that states could make realistic projections of their needs for new prisons and other corrections programs (Tonry 1996, chap. 2). By 1996, nearly twenty-five jurisdictions had created sentencing commissions. Some failed to develop guidelines or to persuade legislatures to adopt them, but a number succeeded, and early in 1996 new commissions were at work in Massachusetts, Missouri, Montana, New Mexico, Oklahoma, and South Carolina.

Thus, by 1996, there was no standard American approach to sentencing as there had been twenty-five years earlier. Some states had parole boards that made ad hoc release decisions, some had parole guidelines, and some had abolished parole release. In some states, judges retained all the discretion they had possessed for nearly a century. Of the other states, some had voluntary sentencing guidelines, some had presumptive guidelines, and some had statutory determinate sentencing laws. Most states' sentencing systems, of whatever kind, were overlaid with mandatory sentencing laws, usually for drug and firearms crimes, and some in the mid-1990s had enacted newly popular three-strikes-and-you're-out laws.

A sizable American literature exists on the effects of changes in sentencing laws and practices, on the crime-preventive effects of changes in penalties, and on the operation and effects of community penalties. From them, we know how to create sentencing systems that are just and efficient and protective of public safety. Policy makers in many states are trying to use that knowledge to design better systems. Whether they succeed will depend on which of the pressures for change—partisan political calls for ever-increasing severity or nonpartisan calls for more just, efficient, and effective systems—proves more powerful and lasting.

Sentencing Reform outside the United States

Four themes stand out concerning recent sentencing changes outside the United States. First, idealistic and human rights concerns have shaped changes in many countries. In the 1976 and 1988 overhauls of Finnish and Swedish laws, achievement of greater proportionality in punishments—of closer ties between the relative seriousness of crimes and the relative severity of punishments—was a paramount objective. So, too, in England, the Home Office White Paper (Home Office 1990) that provided the rationale for the Criminal Justice Act 1991 explained that greater proportionality was a primary objective. The Finnish government in the mid-1970s decided that the country's incarceration rate, higher than those of other Scandinavian countries, was too high; through a variety of legal changes, the rate was reduced by a third and has since remained stable. The Federal Republic of Germany in the late 1960s and early 1970s decided that short prison sentences are a bad thing. On the "rationale that short-term imprisonment was incompatible with respect to rehabilitation due to the short period available for treatment and the corruptive effects of the prison environment" (Albrecht 1995, p. 7), Germany greatly reduced use of prison sentences under six months. Sentences to immediate imprisonment (that is, excluding suspended prison sentences) fell from 92,576 in 1967 to 40,270 in 1970 and varied between 30,000 and 40,000 for the following twenty years. These developments are discussed in the articles in this volume dealing with those countries.

Second, the principal focus of reform in most countries has been on development of noncustodial penalties rather than on comprehensive refashioning of entire systems. England, for example, has since the early 1970s implemented community service, combination orders, and day centers, conducted pilot projects on day fines, intensive supervision probation, and electronic monitoring, and implemented and within a year repealed a day-fine system. Germany successfully adopted day fines in the 1970s and successfully implemented a "conditional dismissal," or "prosecutorial fine," system in which accused persons agree to pay a fine in lieu of prosecution. The Dutch adopted both community service, patterned on the English system, and prosecutorial fines, patterned on the German. The Scots emulated the English approach to community service, and pilot projects are underway in a number of Swiss cantons. Australia, New Zealand, and South Africa adopted intensive supervision probation patterned on American examples. These developments are discussed briefly in the articles in this volume dealing with those countries and at greater length in articles on those countries in Tonry and Hamilton (1995) and in Tonry (1996, chaps. 4 and 7).

Third, with only minor exceptions, substantially increased severity of penalties and substantial increases in use of imprisonment have been major reform goals only in the United States, and only in the United States has sentencing policy become a major political issue. "Reforms" such as establishment and expansion in coverage of the death penalty, three-strikes laws, mandatory minimum sentence laws, and boot camps so far have not been widely emulated elsewhere, and they have been seriously proposed as government policy only in England. This does not mean that government policies do not change when conservative parties or coalitions replace more liberal ones, and vice versa, but that policy changes are typically marginal and incremental rather than fundamental and radical. Articles in this volume on many countries describe such changes.

Fourth, no other country has as yet shown interest in American-style sentencing guidelines in which presumptive sentences are set out in a numerical grid. Why this is we do not know, though reasonable guesses can be made. One is that indeterminate sentencing in the United States led to imposition of sentences that at least nominally are much longer than sentences elsewhere, and as a result the potential for extreme disparities is greater in the United States than elsewhere. In 1991, 43 percent of American state prisoners were serving terms of ten or more years (Beck et al. 1993); by contrast, in 1991 in Sweden, 95 percent of prison sentences ordered were for two years or less (Jareborg 1995), in 1991 in Germany, only

1 percent of sentences were to prison terms of two years or more (Albrecht 1995), and in 1991 in the Netherlands, only 14 percent of prison sentences exceeded one year, and less than 1 percent exceeded six years (Tak 1994).

Disparities can be and look much more extreme when sentences are commonly ten or twenty years long, as in the United States, rather than a few months or at most a few years as in most other countries. Long nominal sentences are largely a product of indeterminate sentencing, which was more extensively and enthusiastically adopted in the United States than anywhere else. Nominal sentence lengths became very long—ten, fifteen, thirty years—in part because the parole board set actual release dates and judges could impose essentially meaningless sentences for symbolic reasons, and in part because judges sometimes imposed long sentences to lengthen the minimum that must be served before release eligibility ripened (if prisoners became eligible for parole after serving a third of the nominal sentence, a judge who wanted a prisoner to serve at least six years could order a twenty-year sentence).

In most European countries, including England (which did not create a system of discretionary parole release until 1968), release eligibility ripens only after a third or a half of the term has been served, and prisoners must be released when two-thirds has been served because of good time (called "remission" in most countries). Thus, discretionary release cannot substantially reduce sentence lengths, and nominal sentences never became inflated.

Another reason why numerical guidelines have not caught on elsewhere may be that prosecutors and judges are not elected in most countries, and in most European countries they are career civil servants. Judges everywhere resent and resist creation of new constraints on their discretion. Judges in other countries, because they are nonpolitical, may be more effective in opposing guidelines by invoking ideas of judicial independence and the need to impose individually appropriate sentences in every case. Numerical sentencing grids look and are mechanical, and most judges and lawyers believe that questions of justice require individualized human, and not mechanical, solutions.

Whatever the reason, no other country has adopted or seriously considered any of the comprehensive American approaches to structuring sentencing discretion—voluntary guidelines, presumptive guidelines, or statutory determinate sentencing. Nor has the United States had much success at adopting European noncustodial penalties. Efforts have been made to create community service programs patterned on the English programs and day fines patterned on the German and Swedish experiences, but with little success (Tonry 1996, chap. 4).

Mass air travel and international telecommunications are making the earth a smaller and more homogeneous planet. The Internet and the emergence of English as the common tongue in Western countries will hasten the shrinkage. Ten years from now, English-language scholarship and informed reportage on criminal justice policy developments in different countries will be readily available, but they are not now. We hope this book, which pulls together knowledge from many countries, will accelerate the process.

Sentencing Reform in the United States

Twenty-five years ago, indeterminate sentencing with its broad discretions and overlapping powers of judges, parole boards, and corrections managers was *the* American sentencing system, and it had changed little since 1930 (Rothman 1980). In 1996 there is no single American system. Some states retain indeterminate systems. Some have determinate systems in which parole release has been abolished. Others have sentencing guidelines of various sorts with or without parole and, if parole, with or without parole guidelines. Every state has mandatory minimum sentence laws for some crimes, and some have "three-strikes-and-you're-out" laws.

Supreme Court Justice Louis Brandeis once observed that "it is one of the happy incidents of the federal system that a single, courageous state may, if its citizens choose, serve as a laboratory and try social and economic experiments without risk to the rest of the country" *New State Ice Co. v. Lieberman*, 285 U.S. 262, 311 (1932).

Many kinds of sentencing experiments are underway, although remarkably little systematic effort has been made to evaluate or document the results. A number of federally funded evaluations of sentencing policy changes were carried out in the late 1970s and early 1980s, which were summarized by a National Academy of Sciences panel on sentencing research (Blumstein et al. 1983). The broad pattern of findings was discussed in the introduction to this volume. No major evaluations of comprehensive state policy changes have since then been funded by government agencies or private foundations, although there have been a number of secondary analyses of agency data by outsiders (e.g., Boerner 1993; Frase 1993; all of the evaluations are discussed in Tonry 1996, chaps. 2 and 3). There has also been a major self-evaluation of the federal sentencing guidelines by the U.S. Sentencing Commission (1991), which was then expanded with major reanalyses of data by the U.S. General Accounting Office (1992).

Thus there is surprisingly little evidence available on the comparative success of different sentencing reforms. A conventional wisdom has emerged that presumptive sentencing guidelines, backed up by appellate sentence review, can reduce racial, gender, and other unwarranted disparities and, by making sentencing more consistent and predictable, enable states to tie their sentencing policies to their existing and planned corrections resources.

The evidence that presumptive guidelines have reduced disparities is strong in Minnesota, Washington, and Oregon, weak but plausible in Pennsylvania, and so ambiguous concerning the federal guidelines that no firm conclusions can be drawn. Despite a U.S. Sentencing Commission (1991) conclusion that the federal guidelines reduced disparities, the U.S. General Accounting Office concluded that data limitations "made it impossible to determine how effective the sentencing guidelines have been in reducing overall sentencing disparity" (U.S. General Accounting Office 1992, p. 10).

The evidence concerning the effectiveness of "capacity constraint" policies, under which sentencing standards are tied to corrections resources, is also strong. Minnesota, Washington, and Oregon adopted such policies with notable success and, while those policies were in effect, managed to hold prison population increases well below national averages and to hold prisoner numbers within institutional capacities. Pennsylvania has revised its guidelines to incorporate population constraint policies, as do newer guidelines in Kansas and North Carolina.

As the twenty-first century approaches, presumptive sentencing guidelines developed by sentencing commissions are the principal survivor of the comprehensive approaches to sentencing reform that have been attempted in the past quarter century. Many years have passed since the last statutory determinate sentencing system was enacted. A few states, including Arkansas and Ohio, have recently promulgated voluntary sentencing guidelines, despite considerable evidence that voluntary guidelines have few or no effects on sentencing patterns. In those states, judicial resistance to constraints on judges' discretion was powerful enough to forestall more ambitious (and promising) presumptive guidelines.

Three major issues confront newly appointed sentencing commissions. Should guidelines be presumptive or voluntary? Should capacity constraint policies be adopted? How can controls over judicial discretion that have successfully governed decisions about prison use be extended to include decisions about community penalties and intermediate sanctions?

The articles in this chapter discuss sentencing developments in thirteen states. For several states, notably Washington, Minnesota, Oregon, Pennsylvania, and North Carolina, a series of articles describes events as they unfolded, thereby both explaining the policies that were adopted and describing the processes by which they were adopted. The first section begins with articles by Richard Frase and Kevin and Curtis Reitz providing, respectively, an overview of sentencing guidelines developments in the states and a discussion of the American Bar Association's sentencing standards, which provide a blueprint that future guidelines states can consider.

National Developments

Sentencing Guidelines Are "Alive and Well" in the United States
Richard S. Frase

At least seventeen states have adopted sentencing guidelines, and the American Bar Association (ABA) recently renewed and strengthened its support for this approach.

State guidelines are diverse in their origins, purposes, and provisions, but they also have important features in common with each other and with the revised ABA Standards. Many of these features are lacking in the much-criticized federal guidelines. For sentencing reformers at both the state and federal levels, there is much to be learned from the experiences of the guidelines states.

State guidelines increasingly emphasize the importance of avoiding prison overcrowding and of making broader use of intermediate sanctions. State guidelines also tend to reflect a better balance than the federal guidelines on such key sentencing issues as the relative weight given to offense versus offender variables; the amount of remaining judicial discretion; the degree of sanction severity; and the allocation of sentencing power among the legislature, the sentencing commission, judges, prosecutors, defense attorneys, and correctional officers.

Where, When, and What Kinds of State Guidelines? The seventeen states with existing or proposed sentencing guidelines are listed in table 2.1 in the order in which their guidelines were (or will become) effective. The most important variations are discussed next.

Nature of the Sentencing Commission. All guidelines states except Alaska have established a permanent sentencing commission or similar body with authority to study sentencing practices and recommend guidelines. All of these commissions have some degree of legislative support, although Virginia's is located entirely within the judicial branch. Most sentencing commissions are broadly representative, including a mix of judges, prosecutors, defense attorneys, correctional officials, public members, and sometimes legislators.

Legislatively created commissions differ greatly in their roles relative to the legislature. Minnesota's enabling statute gave the commission relatively little guidance. In recent years, the legislature has taken back some of the authority it delegated, but the Minnesota commission still retains primary control over the formulation of statewide sentencing policy. In contrast, other state legislatures have played a much more active role, either by carefully structuring the commission's mandate (e.g., Arkansas) or by dominating the guidelines revision process (Washington).

Binding Force of the Guidelines. Most state guidelines recommend presumptively appropriate ranges of sentences that judges are bound to follow unless they depart, but several states (Delaware, Michigan, Utah, Virginia, and Wisconsin) have purely voluntary guidelines. Even within the group of states whose guidelines are not formally voluntary, standards for departure and for reversal on appeal vary widely. Pennsylvania guidelines departures are rarely reversed on appeal except for procedural reasons (e.g., failure to state any reasons), whereas reversal on substantive grounds (improper sentence) has often occurred in states such as Washington and Minnesota, each of which now has a large body of substantive appellate case law. But even in the latter states, trial courts retain substantial areas of discretion as to both the type and the severity of sanctions. In this respect, the federal guidelines appear to be uniquely and unnecessarily rigid.

Scope of Guidelines Coverage. Most state guidelines systems cover felony crimes only. All guidelines regulate aspects of both prison commitment and prison duration, and some also control the use of consecutive sentences. The states differ in

Table 2.1. Summary of State Sentencing Guidelines Systems

Jurisdiction	Effective Date	Scope and Distinctive Features
Utah	1979	voluntary; retains parole board; no permanent commission until 1983; linked to correctional resources since 1993
Alaska	1/1/80	no permanent sentencing commission; statutory guidelines' scope expanded by case law
Minnesota	5/1/80	designed not to exceed 95 percent of prison capacity; extensive data base and research
Pennsylvania	7/22/82	covers misdemeanors; retains parole board, encourages nonprison sanctions since 1994; substantially revised effective 1995
Florida	10/1/83	formerly voluntary; overhauled in 1994 and 1995
Michigan	1/17/84	voluntary; retains parole board; commission to submit guidelines to legislature in July 1996
Washington	7/1/84	includes upper limits on nonprison sanctions, some defined exchange rates, and vague, voluntary charging standards; resource-impact assessment required
Wisconsin	11/1/85	voluntary; descriptive (modeled on existing practices); retained parole board (abandoned 1995)
Delaware	10/10/87	voluntary; narrative (not grid) format; also covers misdemeanors and some nonprison sanctions; linked to resources
Oregon	11/1/89	grid includes upper limits on custodial nonprison sanctions, with some defined exchange rates; linked to resources; many new mandatory minimums added in 1994; commission abolished 1995 (but guidelines kept)
Tennessee	11/1/89	also covered misdemeanors; retains parole board; sentences linked to resources (abandoned 1995)
Virginia	11/1/91	voluntary; judicially controlled, and parole board retained, until 1995; resource-impact assessment required since 1995
Louisiana	1/1/92	includes intermediate sanction guidelines and exchange rates; linked to resources (abandoned 1995)
Kansas	7/1/93	sentences linked to resources
Arkansas	1/1/94	voluntary; detailed enabling statute; resource-impact assessments required
Massachusetts	1994	commission established 1994; interim reports submitted to legislature in 1995; draft guidelines include intermediate sanctions and linked to resources
North Carolina	1/1/95	covers misdemeanors; sentences linked to resources; includes intermediate sanctions
Missouri	7/1/95	voluntary guidelines took effect in July 1995
Montana	1996	commission to report legislature on advisability of guidelines in May 1996
Ohio	1996	presumptive guidelines take effect July 1996; include intermediate sanctions; linked to resources; parole abolished
Oklahoma and South Carolina	(in process)	enabling statutes encourage resource matching

Source: Adapted from Tonry (1996).

the extent to which statutorily based mandatory-minimum prison terms determine or override guidelines rules. Such statutes appear to play a much smaller role in some states (e.g., Kansas, Minnesota) than they do in the federal system.

Some guidelines states (Pennsylvania, Michigan, Tennessee, Utah, Virginia, and Wisconsin) have retained parole release. In these states, the guidelines determine either the minimum or the maximum prison term to be served, but not both. The new ABA Standards now recommend abolition of parole. This change from the prior (1979) edition may be related to the ABA's current focus on "front-end" resource matching, discussed below.

As for nonprison sentences, many guidelines states (as well as the new ABA Standards) give greater emphasis than do the federal guidelines to probation and other intermediate sanctions. This difference is probably due in part to differences between state and federal caseloads, but it may also reflect a generally less punitive approach by the states, as well as a greater emphasis on the goal of preventing prison overcrowding. The states differ greatly in the degree to which they regulate the conditions of nonprison sentences and decisions to revoke probation and post-prison release.

Nature and Priority of Sentencing Reform Goals. All state guidelines reforms reflect a desire to make sentencing more uniform and to eliminate unwarranted disparities. Beyond this, however, the declared or apparent goals and priorities of these reforms are diverse. Those states that have abolished parole release and sub-stituted limited good-time credits were often responding to the desire for truth in sentencing: the length of prison sentences imposed by courts should correspond closely to the amount of time inmates actually serve.

A few states have largely "descriptive" guidelines, designed to help judges follow existing sentencing norms more consistently. But even these states usually seek to make some "prescriptive" changes in prior norms (especially to eliminate existing racial disparities). In other states, the most common prescriptive changes have involved increased sentence severity for violent and drug crimes. Minnesota, Washington, and Kansas explicitly based their guidelines on retributive, or just-deserts, theories of punishment, placing greater emphasis on the severity of the current offense and less on offender characteristics. However, even these states (unlike the federal guidelines) leave substantial room for offender-based sentences and the pursuit of rehabilitative, incapacitative, special deterrent, and other practical (nonretributive) goals.

Controlling Prison Populations. Increasingly, states are turning to sentencing guidelines with a primary goal of using them to gain better control over rapidly escalating prison populations. Such control is made possible by the greater uniformity and predictability of guidelines sentences, in comparison with the prior indeterminate sentencing regimes. Minnesota pioneered this approach in 1980 and explicitly adopted a goal of never exceeding 95 percent of available prison capacity. That goal was achieved throughout the first decade of guidelines sentencing; prison populations did increase, but at rates much slower than in other states. Prison construction and expansion were thus able to accommodate inmate population increases without overcrowding or multiple-bunking of high-security inmates. In contrast, Pennsylvania did not recognize resource matching as a goal until the start of its second decade of guidelines sentencing; by that time, prisons and jails were operating at about 150 percent of capacity.

Starting in the mid-1980s, as prison overcrowding problems grew around the country, a number of states (Kansas, North Carolina, Oregon, Tennessee, and Washington) adopted guidelines linked to available resources. This trend received substantial recognition and support in the new ABA Standards. Resource matching is a central principle of these Standards but was not even mentioned in the previous (1979) version.

A related and very important principle found in both the prior and the current edition of the ABA Standards is the concept of "parsimony" in sentencing: sanctions should be the least severe necessary to achieve the purposes of the sentence. This principle was adopted by Minnesota in 1980, and it has been recognized in several other states (but not at the federal level, despite statutory authorization). The principle is grounded in reasons of economy and humane treatment of offenders, but it also has implications for punishment theory. If sentences may be mitigated for reasons of parsimony, this implies that just-deserts considerations do not, in many cases, narrowly limit the permissible range of sanction severity.

Principal Determinants of Guidelines Sentences. All guidelines states base their recommended sentences primarily on the conviction offense and the offender's prior conviction record. Although nonconviction offense details play some role (e.g., enhancements for weapon use, regardless of whether such use was an element of the charged offense), the guidelines states are unanimous in rejecting the broader "real offense" approach of the federal guidelines, which permit frequent and quite substantial enhancements based on uncharged "relevant conduct."

The federal approach was apparently designed to prevent prosecutors from granting undue or inconsistent leniency by means of selective charging and plea bargaining concessions. Except for Washington, the guidelines states place no limits on prosecutorial discretion; even Washington's limits are vague and are not judicially enforceable. Nevertheless, this seeming loophole does not appear to have caused any widespread dissatisfaction with state guidelines. Apparently, state prosecutors are not often willing, or able, to use their charging powers to dictate unreasonable sentences that courts are powerless to avoid (as often seems to happen in federal courts).

Most states have promulgated guidelines in the form of a two-dimensional grid, but a few employ narrative rules for each offense or offense group. State grids vary widely in their layouts and "cell" ranges. There are also major variations in severity ranking of offenses, formulas for computing prior record, good-time credit amounts, and the nature and extent of listed factors that permit (or do not permit) departure. Criminal history scoring is particularly diverse. These variations reflect differences in sentencing goals and state traditions, as well as the relatively primitive state of case-level sentencing jurisprudence (which, prior to determinate sentencing, was rarely addressed in opinions or scholarship).

Case Monitoring, Research, and Evaluation. One of the most important features of sentencing guidelines reforms is their empirical research component. Most permanent, legislatively created guidelines commissions have been given a mandate to collect and analyze sentencing data, not only as background for development of the initial guidelines but also as a means of monitoring implementation and proposing guidelines revisions. This empirical component has become more and more important, as states have begun to focus on the goal of predicting and preventing future prison overcrowding. Such predictions require detailed information on current sentencing practices and the development of sophisticated, computerized models combining data on expected caseloads, presumptive sentences, and other factors known to have an effect on the size of inmate populations.

Despite these important applications of guidelines data, and the research mandates of most state commissions, there is still relatively little published data, and there are even fewer published evaluations by researchers independent of the commissions. In some cases, this is because the guidelines are too new to have generated significant sentencing data; in older systems, data may not be collected due to inadequate commission staffing and budgets. When data are collected, they are not always known, or fully available, to outside researchers. Finally, such data, even when available, are usually not in a form that permits meaningful cross-jurisdictional comparisons.

Conclusion Presumptive guidelines developed by an independent sentencing commission have, since the late 1970s, represented the dominant approach to sentencing reform in the states. Such guidelines have earned greater acceptance than have the federal guidelines, and more and more states are moving to adopt them. The states have followed a wide variety of approaches, and much has been learned. At the same time, research and policy development are still too limited. Much more complete and more standardized data are needed. Outside researchers need to analyze and compare state data. Sentencing theorists need to focus their attention on the crucial, unresolved issues of sentencing policy that are highlighted by guidelines sentencing (e.g., offense severity ranking; criminal history scoring; the proper weight of various offender characteristics; intermediate sanction severity scales).

The experience of the states suggests that presumptive sentencing laws can reduce sentencing disparities without imposing excessive rigidity, promote truth in sentencing (matching time imposed to time served), encourage wider use of intermediate sanctions, and help states avoid prison overcrowding by linking sentencing policy to available resources. "Front-end" resource management has become one of the most important reasons for states to adopt guidelines. Parole and other "back-door" release mechanisms can deal with prison overcrowding, but they cannot achieve either truth in sentencing or the most efficient use of limited prison capacity.

American Bar Association Adopts New Sentencing Standards
   Kevin R. Reitz and Curtis R. Reitz

The American Bar Association late in 1994 published a third edition of its Criminal Justice Standards for Sentencing Law and Procedure. The second edition, currently in libraries, is now fifteen years old. Looking back, we can see that the period since 1979 has been one of intense activity and controversy in the sentencing field. All existing guidelines systems have come on line since 1979, including the much-vilified federal program. The politics of sentencing has become increasingly polarized. Across jurisdictions, there has been considerable experimentation toward structuring the use of intermediate sanctions but, so far, there have been few success stories. Against this backdrop, it is mildly amazing that the ABA was able to promulgate a new set of Sentencing Standards at all. More notable still, the Standards endorse a strong sentencing reform agenda that will challenge existing practice in most jurisdictions.

Four Highlights   The new Standards assert four "big-picture" positions that bear highlighting.

First, the Standards advocate creation of a permanent sentencing commission, or equivalent agency, in every jurisdiction, with responsibility to effectuate legislative policies. In the view of the Standards, the legislature should make broad policy decisions regarding the goals of sentencing, but it should not fix sanctions for particular cases, through mandatory minimum penalties or otherwise. The Standards maintain that an "intermediate" sentencing agency ("in between" the legislature and sentencing courts) is best positioned to develop a systemic approach to sentencing policy. The agency must have the resources to monitor and evaluate the operation of the system and to make credible predictions regarding its future operation. The agency should have permanent, not temporary, existence because the demands of criminal justice policy are constantly changing, as is the knowledge base affecting decisions.

Second, the sentencing agency is charged with drafting determinate sentencing provisions, including presumptive sentences for ordinary offenses and offenders, that judges should impose unless there are "substantial" reasons for departure. (In many jurisdictions such sentencing provisions are called "guidelines." For reasons explained below, the ABA chose more generic terminology.) Judges' decisions regarding presumptive sentences and departures become matters for appellate review. The Standards envision that the judiciary and the sentencing agency will play coequal roles in the evolution of a common law of sentencing. The agency is particularly competent at considering systemwide issues. The courts have unique expertise concerning the imposition of punishment in individual cases.

Third, the Standards call for the extension of presumptive sentencing provisions across the full range of criminal sanctions—not just imprisonment. This is part of the Standards' general emphasis on the increased use of nonprison sanctions. Separate Standards are included for compliance programs for individuals, compliance programs for organizations, restitution or reparation, fines, community service, acknowledgment sanctions, intermittent confinement, home detention, and total confinement. Recognizing that efforts to expand the creative use of nonprison sanctions are still at an early stage, the Standards do not require any specific mechanism for structuring their application. During the drafting stage, the "sanctions unit" approach was considered by the ABA, along with other proposed systems. None of the suggestions, however, has moved beyond the experimental stage. In writing standards for the nation, the ABA concluded that the project of devising a system of presumptive sentences for all sanctions should be encouraged, but that no clear solution had yet emerged that could be urged on all jurisdictions.

Fourth, the Standards assert that all sentencing provisions should be drafted so that the total of sentences imposed (incarcerative and nonincarcerative) does not exceed the funded resources available for their execution. For states with sentencing commissions, "The legislature should provide that the commission may not promulgate sentencing provisions that will result in prison populations beyond the capacity of existing facilities unless the legislature appropriates funds for timely construction of additional facilities sufficient to accommodate the projected populations" (ABA Standard 18-14.4[c][i]). In the Standards' view, such an approach is a fundament of fiscal responsibility—but one that is frequently ignored in matters of criminal justice. The policy of matching sentences to available resources does not foreclose the possibility of prison growth or other increases in sentence severity. Rather, it requires that policy initiatives be coupled with necessary funding allocations.

In all of these respects, the Standards are not a pioneering document. They endorse or build on the work of Marvin Frankel, Kay Knapp, Michael Tonry, and others, and on the experience of states such as Minnesota, Pennsylvania, Washington, and Oregon. What is significant is that the new Standards assert strong positions as to matters of live dispute in most jurisdictions, and they place the weight of ABA credibility behind those assertions. The major structural recommendations of the Standards have considerable import in many regions of the country. Only a minority of states (albeit a substantial and growing minority) have instituted the kind of commission-based presumptive sentencing scheme envisioned by the Standards. Most jurisdictions continue to operate under traditional regimes of indeterminate sentencing. Most states do not have permanently chartered sentencing commissions or a culture of active appellate review of sentences. No state has developed an adequate plan for structuring the use of intermediate punishments. Most states allow criminal justice policy to lurch forward without systemic oversight and without coordination with available resources. Thus, to the extent the Standards raise questions and provide information, they will challenge policy makers in many jurisdictions to reexamine current sentencing practices.

Federal Guidelines Distinguished   The Standards provide numerous points of contrast with the present federal system. Indeed, one principal theme in the drafting and adoption of the Standards was that they clearly distinguish themselves from present federal law. On a cosmetic level, the drafters avoided the word "guidelines," in favor of terms such as "sentencing provisions," to avoid any possible association with the federal guidelines. This excision was made even though, in the eyes of the drafters, some *state* guidelines systems provided highly attractive models for sentencing reform.

On a substantive level, the Standards differ in many important respects from the federal guidelines. They suggest a framework for sentencing decisions that is simpler and less mechanical than the federal guidelines and preserves greater discretion for sentencing judges. The Standards recommend a broader departure power than federal law, and they allow consideration of personal characteristics of offenders ordinarily inadmissible to federal sentencers. The Standards disapprove of the federal commission's decisions to promulgate guidelines without regard for anticipated prison capacities, and undirected to nonprison sanctions. Indeed, the Standards question the institutional makeup of the federal commission as a whole, and they posit that members of diverse criminal justice constituencies should be represented, including prosecutors, defense lawyers, and corrections officials. The Standards reject the "relevant conduct" provision of the federal guidelines in favor of the uniform approach, in other guidelines jurisdictions, of "offense-of-conviction" sentencing. The Standards provide that, unlike the federal system, sentencing provisions should be based on explicit choices concerning the purposes of sentencing and should not rely on a reproduction or averaging of past sentencing practices.

Conclusion   The Standards speak to all American jurisdictions. They provide a basic template for states that have not started down the road of sentencing reform or are at an early stage of development. They are perhaps most urgently addressed to jurisdictions that continue the outmoded practices of indeterminate sentencing, without the oversight and assistance of a sentencing agency, and without a mechanism for matching sentences to funded resources. In five years of preparation of the Standards, no individual or group argued in defense of indeterminacy. The debate was always over what kind of determinate system, not whether determinacy itself was a proven idea. The Standards do not neglect existing determinate systems, however, and they offer suggestions for adjustment or improvement applicable to many states. They also offer guidance for reworking the current federal guidelines.

Washington State

Washington State: A Decade of Sentencing Reform
  Roxanne Lieb

Washington in 1984 was one of the first states to promulgate presumptive sentencing guidelines and to link its sentencing policies to correctional resources. The guidelines achieved many of their goals. Sentencing disparities were reduced. Prison beds were—as intended—increasingly used for violent offenders and decreasingly used for property offenders. And incarceration rates declined, falling from 156 per 100,000 in 1984 to 124 per 100,000 in 1988.

Washington is looking again at sentencing policy, for a number of reasons. Legislation enacted in the late 1980s increased sentences for a number of crimes. Prison populations have grown rapidly since 1988, increasing by more than a third between December 31, 1988 (5,816), and December 31, 1990 (7,995). This article describes the Washington experience during the 1980s and the challenges it faces as new policies are considered for the 1990s.

Legislative Intent, 1976–1981   Washington's sentencing system dates to the passage of the Sentencing Reform Act of 1981. The legislature had been considering sentencing reform proposals for five years. Washington had operated with an indeterminate sentencing system for eighty years. In the late 1970s, sentencing was debated across the state, involving voices from both sides of the political aisle, and from conservative and liberal viewpoints.

Persons committed to individual rights and a "more equal society" argued that indeterminate sentencing was not fair; that persons who committed similar offenses received widely different sentences; and that minorities and the poor received harsher sentences than did white middle-class defendants.

Organizations and individuals who were primarily concerned with public safety made two arguments: that indeterminate sentencing allowed dangerous criminals to be released into the community, and that the system paid too much attention to the offender's rehabilitative potential and progress and not enough to the harm he or she caused.

By the early 1980s, state prisons were operating at 135 percent of capacity, and the federal courts intervened. Discussions about emergency release mechanisms and sentence reductions surfaced, and judges became very aware that every prisoner sentenced to the state prisons might mean another prisoner's release.

A coalition provided the political momentum for legislative change. The state's elected prosecutors took a leadership role. They asserted that the criminal justice system was "bankrupt." They linked sentencing reform with prison construction, arguing that a sound criminal justice system must have adequate prison space as its foundation, and that the system's capacity was seriously inadequate.

Prosecutors argued for a system that structured decision makers' discretion, rather than an approach such as mandatory sentencing that tried to eliminate such discretion. They pointed to earlier reforms of the juvenile justice system, enacted in 1977, that created a statewide system for setting punishments but still allowed deviations to address unusual cases. The experience of Minnesota's sentencing commission was also discussed and close attention was paid to the initial operation and consequences of Minnesota's guidelines.

The Sentencing Reform Act of 1981 articulated the major purposes of sentencing, provided a framework for a presumptive system of sentencing guidelines, and created the Washington Sentencing Guidelines Commission. Implementation of the law was delayed until 1984, which allowed the commission sufficient time to complete its recommendations in two stages, presenting the bulk of the recommendations in 1983 and the remainder in 1984.

Development and Implementation, 1981–1986   The Washington reform concentrated primarily on the punishment aspects of sentencing, with specific focus on the harm done by the offender. Other goals included protecting the public and offering rehabilitative opportunities to offenders. The legislation directed that scarce prison resources be reserved for violent offenders and that alternatives to confinement be used for nonviolent offenders.

The Sentencing Guidelines Commission was created as a fifteen-member independent body, with four legislators serving as nonvoting members. The commission was given responsibility to develop guidelines, to recommend sentencing policies, and to monitor the implementation of the reform. All sentencing guidelines and rules had to be affirmatively adopted by the legislature.

The sentencing guidelines were to encompass all felonies, including those that traditionally carried a sentence of less than twelve months that was served in a local jail facility or as a probation term. In the past, 80 percent of adult felons served time in a local facility or on probation. Including this majority within the structured sentencing system significantly increased the reform's scope, complicating the task and the political challenges facing the sentencing commission.

Judges were given authority to depart from the standard guidelines range in exceptional cases, but such departures were to be in writing and based on findings of fact. The intent was to create a "common law of sentencing" through written appellate opinions and thus further guide the courts' decisions.

The rehabilitative aspects of sentencing were primarily restricted to first-time, nonviolent offenders. For this group, the legislature wanted the sentencing system to emphasize alternatives to total confinement. Treatment-oriented sentences were also authorized for sex offenders.

Parole release was eliminated, as was parole supervision, with voluntary services substituted.

The legislature directed the commission, in developing the first set of guidelines, to construct a preferred set of guidelines, then to project their impact on existing and planned prison capacity. If the preferred set required additional capacity, the commission was also to submit a second set of guidelines that did not require new construction, and the choice could be made by the legislature. Prosecutors had successfully lobbied to add 1,800 new prison beds to the state's system by 1985; thus, the capacity base used by the commission in its forecasting was approximately 25 percent over existing capacity.

A projection of the effects of the commission's preferred guidelines concluded that the guidelines could be implemented within existing and planned resources. Thus, only one set of guidelines was submitted to the legislature. These guidelines were promoted as "tough but fair" and were adopted by the Washington legislature in 1983 by a near-unanimous vote.

Between 1983 and 1984, the commission concentrated attention on sentencing policy for drug and sex offenses and proposed modifications to the original guidelines for these offenses. These recommendations were also adopted by a near-unanimous vote. The Sentencing Reform Act went into effect for crimes committed after July 1, 1984. Because of the typical delay between the date a crime is committed and the sentencing date, it took until 1985 for the majority of felony offenders to be sentenced under the new system.

The Amendment Era, 1986–1990   In states debating the merits of structured sentencing systems, people sometimes raise the concern that these systems cannot respond to shifts in criminal activity and public opinion. Washington's experiences should allay this fear. Since the guidelines went into effect in 1984, eleven major bills have passed amending the state's sentencing policies. Each bill addressing sentence lengths has increased the length of confinement for offenders. The combined effect of these amendments has been to increase the number of prison inmates by approximately 1,500 through the end of fiscal year 1990. This effect will increase over time and will result in an additional 3,400 inmates in Washington's prison system by 1995. Figure 2.1 shows the projected effects of recent legislative amendments of sentencing laws on prison populations through 1997.

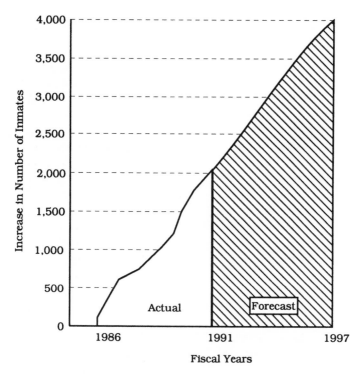

Figure 2.1. Legislative Amendments' Impact on Prison Population.
Source: Washington State Sentencing Guidelines Commission (1991).

During the first two years of experience with guidelines under the Sentencing Reform Act, the Washington legislature was very attentive to concerns for proportionality within the sentencing grid and to the goal of reserving prison space for the most violent offenders.

During 1985 and 1986, legislative discussions frequently mentioned the perils of "felony creep," where increasing the penalty for one offense leads to an argument for increasing penalties for a second and then a third offense in following years. In 1985, the Sentencing Guidelines Commission proposed amendments to increase sentence lengths for certain repeat criminals, presenting the bill as a correction to the commission's original scheme. This bill was not adopted in 1985, primarily because of the forecasted prison impact of 600 additional beds; it did, however, pass the following session.

In 1986, the commission recommended several adjustments for sentencing for sex offenses with a projected impact of over 250 additional prison beds. Again, the legislature did not pass the bill the first year but passed it with minor amendments in 1987.

The legislative atmosphere changed by 1987, when the state was in the luxurious position of having excess prison capacity. This resulted from two factors—

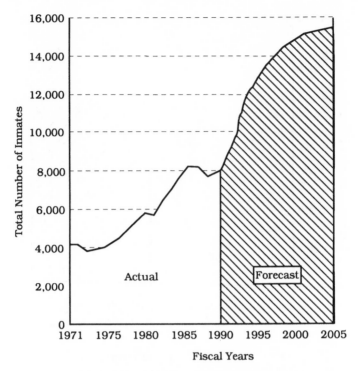

Figure 2.2. Inmate Population of Washington State. *Source:* Washington State Office of Financial Management, Forecasting Division (1991).

reduced admissions of nonviolent offenders to prison and the increased releases of inmates serving indeterminate sentences. The reform had successfully altered the type of offender sent to prison. In 1982, 48 percent of violent offenders were imprisoned; by 1987, the rate was 59 percent. Similarly, the imprisonment rate of nonviolent offenders declined from 13.3 percent in 1982 to 10.6 percent in 1987.

Prison populations were reduced by the effect of accelerated releases of inmates under the indeterminate system that applied to prisoners who had committed crimes before July 1, 1984. The Washington Supreme Court ruled that the parole board must consider the sentencing guidelines in setting release dates for these inmates, and this resulted in earlier release of many prisoners than might otherwise have occurred. Figure 2.2, showing past and projected prison populations from 1971 to 2005 in Washington, shows a dip in the mid-1980s caused in part by changed patterns of parole release.

The state rented its excess prison capacity to the federal government and other states from 1987 through 1989. The "Rent-a-Cell" program was born, involving approximately 1,000 beds.

In an atmosphere of excess prison capacity, even though it was generally understood to be a two-year phenomenon, the legislature began to take a different attitude toward sentencing guidelines. With extra "room in the inn," the legislative debate on crime and the need to toughen sentences was not tempered by concerns about prison population.

In addition, representatives of local government argued that the state had solved its crowding problem by simply shifting felons to local jails. That was not really true. A careful review of the statistics revealed that an increase in the number of felony convictions was the major cause of local jail population increases. In this atmosphere, bills to increase sentences often accomplished two political goals: getting tough on criminals and moving felons from local facilities to state beds.

In 1987, the legislature eliminated a rehabilitation option for first-time drug dealers and increased sentences for the crime of escape. In 1988, the penalties for sex offenses against children were increased, and postprison supervision was reestablished for the most serious offenders.

By 1989, concerns about drug offenses led to passage of higher penalties for cocaine and heroin dealers, with additional enhancements for drug crimes committed near schools. In addition, the legislature created a new category of residential burglary and increased its penalty.

Public attention in late 1989 shifted to sex offenses following a series of very brutal crimes against children and women. The governor appointed a task force on community protection; this group recommended several increases to the penalties for sex offenses, all of which were adopted in 1990.

The other major cause of prison population increases has been a significant increase in felony filings and convictions. Between 1980 and 1990, for example, drug convictions increased by 438 percent. Thus, the felony "base" increased at the same time that penalties were increased.

Challenges for the 1990s   The 1991 legislature was presented with an ambitious prison construction plan to accommodate the increases in prison inmates anticipated for this decade. The state is planning to add over 4,500 new prison beds by 1996.

Simultaneous with this construction and financing discussion has been attention on identifying the forces that are driving the prison population and on reviewing the policy options within the state's control. A legislative task force on sentencing of adult criminal offenders was established by the 1991 legislature, with the directive to review the original purposes of the 1981 sentencing reform and to compare them to incarceration patterns, to study the present use of alternatives to incarceration, and to determine if the courts' sentencing options should be expanded. In addition, the governor has asked the Sentencing Guidelines Commission to provide renewed emphasis on alternatives to jail and prison confinement for nonviolent offenders, particularly for substance abusers whose criminal activity is limited to or caused by that abuse.

Washington is also reviewing how discretion has been exercised under the sentencing guidelines and examining the population impacts caused by this discretion. Perhaps the most striking information concerns the state's increasing reliance on confinement, even when nonconfinement options are allowed by law.

Figure 2.3 shows the changing distribution of felony sentences in Washington from 1981 to 1990. In fiscal year 1982, 24 percent of sentences were to nonconfine-

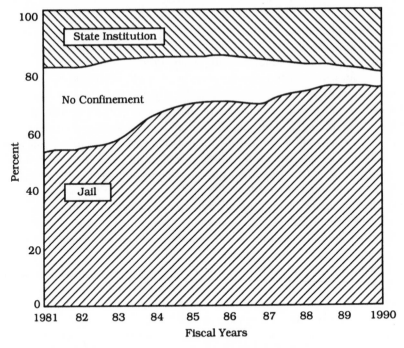

Figure 2.3. Distribution of Felony Sentences in Washington State. *Source*: Washington State Office of the Administration for the Courts (1991).

ment options, primarily probation. By fiscal year 1987, that percentage dropped to 10 percent, and in fiscal year 1991, it was 6 percent.

Some observers speculate that judges wanted to ensure punishment "up front," imposing jail terms instead of alternative sentences that require supervision but often result in jailing of offenders who do not satisfy their obligations.

Washington's sentencing guidelines system enabled the state during the mid-80s to avoid some of the problems of crowding and skyrocketing costs that bedeviled most American states. By 1991, faced with a rapidly increasing prison population, a major program of prison construction, and the need to create new and effective community-based sanctions, Washington is looking to adapt its innovative sentencing policies of the 1980s to the needs of the 1990s. A comprehensive introduction to the Washington guidelines system is contained in Boerner (1985).

Washington Prison Population Growth Out of Control

Roxanne Lieb

Washington's major prisons are operating at 167 percent of capacity, a massive construction project is underway, and by 1999 the operating budget for corrections will double to a billion dollars. In the ten years that Washington State has operated with

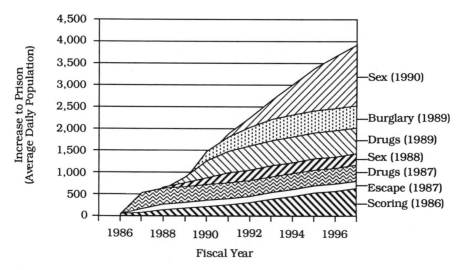

Figure 2.4. Cumulative Effects of Sentencing Reform Act Legislation, 1986–1990 Sessions. *Source:* Washington State Sentencing Guidelines (1991).

sentencing guidelines, the legislature has amended its adult felony sentences nearly every year. The new chair of the Senate Law and Justice Committee recently described the collection of bills that have significantly increased penalties as "crime-of-the-month bills." The legislature's fiscal committees are now holding hearings trying to understand how the state went from renting excess prison beds in the late 1980s to the current situation.

One rationale for sentencing commissions is that they may be able to insulate sentencing policy making from partisan politics and short-term pressures to toughen sentences in the emotional aftermath of notorious crimes.

This article describes the political forces that shaped Washington's sentencing guidelines over the last decade and the law changes that led to the current crowding. Figure 2.4 shows projected increases in the prison population due to changes enacted from 1986 to 1990.

The Early Years   Washington State's sentencing reform was passed in 1981 and went into effect in 1984. The sentencing guidelines were drafted by the Sentencing Guidelines Commission, but the legislature retained authority to adopt, amend, or veto the commission's proposals. In addition, the legislature was directly responsible for adopting any future changes to the guidelines. Thus, after the guidelines went into effect, the commission made recommendations to lawmakers regarding guidelines changes, but it was only one of several groups playing such a role. The original guidelines were linked to correctional resources, including newly available prison capacity that added approximately 25 percent to existing capacity.

In 1985, lawmakers met for the first time following guidelines implementation. The commission proposed several amendments, arguing that the sentencing rules for multiple convictions were flawed and threatened the reform's credibility. During hearings, legislators concentrated attention on the bill's projected prison impact of 600 additional beds over twelve years. Several speeches reminded members about several just-completed years of overcrowding, federal court intervention, and messy battles over siting of prisons and jails.

The commission's proposed bill did not pass in 1985. Reintroduced in 1986, the bill faced uphill battles at every turn, solely because of the prison population forecast. On almost the final hour of the last day of 1986, the bill passed.

The Rent-a-Cell Years, 1987–1989   By 1987, the legislative attitude toward correctional capacity had changed, as the state was in the luxurious position of having excess prison capacity.

Washington rented its excess prison capacity to other states and the federal government from 1987 to 1989. The "Rent-a-Cell" program involved approximately 1,000 beds. With extra "room in the inn," the legislative debate on crime and the need to toughen sentences was no longer tempered by concerns about prison population costs. In addition, representatives of local government argued that the state had solved its crowding problem by simply shifting felons to local jails. In this atmosphere, bills to increase sentences often accomplished two political goals: getting tough on criminals and moving felons from local facilities to state beds.

Escape Offenses   The state's prosecutors began meeting in 1986 to discuss complaints regarding the guidelines and to propose amendments in the 1987 session. The sentencing guidelines had balanced correctional resources primarily by emphasizing prison beds for violent offenders and county jails for nonviolent offenders. Several prosecutors disputed the wisdom of this policy. When the legislature learned that the prosecutors' bill required several hundred new prison beds, their receptiveness cooled. The final bill addressed penalties for offenders with a prior history who were convicted of escape, and rules for offenders with multiple current convictions. The bill passed with a prison impact forecast of 171 additional beds by 1997.

Drug Penalties   Drug penalties were revised by the legislature twice, first in 1987 and again in 1989. In early 1987, Washington State opinion makers began focusing on the dangers of drug crimes, particularly involving cocaine and heroin. Norm Maleng, the prosecutor from Washington's largest county, had been the primary political force behind the passage of the Sentencing Reform Act. He became concerned that the legislature could unravel the guidelines' proportionality with drug sentence increases. He offered another solution: to eliminate drug dealers from an alternative sentencing option allowed under the law called the First-time Offender Waiver. This waiver allowed judges to keep a narcotics dealer convicted for the first time in a local jail, typically with a requirement of drug treatment. The waiver had been used for half of the eligible drug dealers, and Maleng argued that by "closing this loophole," the drug sentences would be toughened adequately. In 1987, this bill passed.

By late 1988, concerns about drug crimes again captured the attention of lawmakers. Prosecutors recommended amendments that doubled the penalties for cocaine, heroin, and methamphetamine dealers, with significantly higher penalties for those with prior drug convictions.

A group of school administrators and citizen activists recommended that offenders who dealt drugs within 1,000 feet of a school or school bus stop be given enhanced sentences. Independently, state prison directors recommended a similar enhancement for certain drug crimes committed within a county jail or a correctional facility. The omnibus bill was worked out behind closed doors, and the legislators from the "Four Corners" (House, Senate, Democrat, and Republican) emerged committed to the entire package. The final bill was massive in scale, providing a mixture of treatment resources, community mobilization funds, and expanded police and prosecutor resources, in addition to increased penalties.

Sexual Offenses   Both in 1988 and 1990, sexual offenses were the subject of legislative attention. The chair of the Sentencing Guidelines Commission initiated a review of sex offenses two years after the guidelines took effect. The commission worked closely with prosecutors and victim advocates to conceptualize and draft criminal code amendments, in addition to penalty increases. When the 1987 legislature considered the bill, two concerns dominated: the costs of additional prison beds (forecast at 260 beds by 1997) and the way consensual sexual acts between teenagers would be handled. When the bill was stalled in committee, some prosecutors attempted to rescue it by contacting popular television news commentators to identify the specific legislator keeping the bill in committee and suggesting that he might be indifferent to child sexual abuse. The legislator introduced a slightly revised bill in 1988 and it passed easily.

In late 1989, a massive public outcry arose when a released sex offender with a twenty-four-year history of violent sex crimes sexually assaulted a seven-year-old boy and cut off his penis. This incident, and the murder of a young Seattle woman by an offender on work release, resulted in intense pressure on the governor to call a special session and immediately pass legislation to protect citizens better. One victims' rights group dumped 8,000 sneakers on the steps of the state capitol in Olympia, saying that the sneakers were symbols of children who deserved to walk without fear.

The governor quickly created a special Task Force on Community Protection, which held two sets of public hearings throughout the state. During the task force's deliberations, Westley Dodd was captured and charged with sexually assaulting and killing three young boys. The task force's recommendations became a 130-page bill covering penalty increases for sex offenses, treatment resources for juvenile offenders and victims, sex offender registration, and civil commitment authority to retain dangerous sex offenders indefinitely for treatment following expiration of their criminal term. Again, the bill was packaged as an omnibus bill and legislative leaders locked arms. A budget surplus made money available for the initial costs of the bill. The task force's recommendations were unanimously adopted. The resulting prison population increases are forecast to add 1,400 prisoners by 1997.

Burglary   The felony that affects the greatest number of citizens is burglary. The original proposed set of guidelines caused controversy because the second-degree burglary penalties were not viewed as sufficiently tough. The commission modified the guidelines so third-time burglars went to prison, with jail penalties for preceding offenses. In 1989, a legislator decided that these penalties were not high enough and sponsored a bill to separate second-degree burglary into two categories: residential burglary and other forms of burglary, and to raise the penalties for each. The bill was forecast to add inmates to the county jails and over 502 prison inmates by 1997. A "sleeper" during the first and middle parts of the session, the bill was not expected to pass. When it did pass, the governor promptly issued a veto, arguing that the prison and jail population consequences were significant and were not adequately considered. The legislature, still in session, immediately overrode the veto. This was the single veto override in the governor's eight-year term, and it sent a loud message.

Summary   The crimes that have preoccupied the Washington State legislature are ones that have concerned most of the country: sex offenses, drug offenses, and burglary. Washington's story with sentencing guidelines can be told from many points of view. Some would argue that Washington proves that state legislators can set sentencing policy and, thus, citizens in a democracy can determine how much punishment they want to afford. Others conclude that politicians are too politically vulnerable to control sentencing policy, that only punitive bills can be passed, and that full costs are never factored into political equations because the complete bill only shows up years after the legislative vote.

Sentencing Policy in Washington, 1992–1995
David Boerner

The reformer's zeal shows no signs of abating as Washington enters its second decade of sentencing reform. The Sentencing Reform Act, enacted in 1981 and implemented in 1984, has been amended every year since. The past two years have been no exception.

Asserting the inherent political power reserved to them in the state constitution, Washington's citizen's twice amended the Sentencing Reform Act by the initiative process, thus ushering in a third phase of sentencing reform in Washington.

During the early years—through 1987—sentencing reform was dominated by the Sentencing Guidelines Commission. Its initial proposal and the revisions it proposed in 1985 and 1986 were adopted without change by the legislature.

The second phase, beginning in 1987 and continuing through 1992, as described by Roxanne Lieb, was characterized by the legislature asserting its power directly. Rejecting a series of proposals by the commission, the legislature, with strong bipartisan majorities, enacted a series of amendments to the Sentencing Reform Act that significantly increased sentence severity for drug, sex, and burglary offenses.

In the third phase, citizens, using techniques enshrined in the Washington Constitution by their populist forebears, took it upon themselves to twice amend the Sentencing Reform Act. Following a history of national leadership in sentencing reform, the three-strikes-and-you're-out concept originated in Washington. Seventy-five percent of voters approved it in the November 1993 election. In 1994, signatures were gathered for an initiative to the legislature, "hard time for armed crime." The legislature adopted the initiative without change, again by an overwhelming majority, in April 1995.

The legislature, however, accompanied each of these initiatives with legislation which, for the first time in the Sentencing Reform Act's history, lessened sentence length.

This article describes this third phase of sentencing reform in Washington, assesses its impact, and reviews the cumulative effect of the past decade's "reforms" of the Sentencing Reform Act.

Three Strikes   The nation's first three-strikes-and-you're-out proposal grew out of frustration with the legislature's failure to enact a similar proposal during the 1992 session. The initial proposal called for mandatory life sentences on the third conviction of a "most serious offense," which included most crimes of violence. A counterproposal, supported by the Sentencing Guidelines Commission among others, narrowed the focus, but both proposals failed when the legislature could not resolve the differences.

The supporters of three strikes, a coalition of victim advocacy groups led by a local conservative think tank, turned to the initiative process. In Washington, any proposition may be placed on the ballot with the signatures of 8 percent of the voters at the last general election. As in many western states, initiatives are regular features of Washington's political landscape. The process has been used in recent years to enact a wide variety of proposals, including a limitation on state taxes and expenditures in 1993, and term limits and a freedom of reproductive choice measure in 1992.

The three-strikes initiative proved very popular. It easily qualified for the November 1992 ballot and passed with over 75 percent of the vote, carrying each of Washington's thirty-nine counties.

As adopted, three strikes imposes a mandatory life sentence, without reduction by good time or parole, on the third separate conviction for a designated "most serious offense." The list includes homicide, serious assaults, most sex offenses, robbery, any crime committed with a deadly weapon, and repeat drug offenses. The requirement that each conviction be of a most serious offense narrows the scope considerably when compared to Washington's former habitual criminal law, repealed in 1984 by the Sentencing Reform Act. That provision applied to the third conviction of any felony. Initial projections were that eighty offenders a year would be eligible to be sentenced under the new law. Because offenders sentenced under three strikes would have received significant prison sentences under prior law, the impacts on prison population occur far in the future. Estimates were that it would increase prison population by 134 in 2000, 407 in 2005, and 673 in 2010.

As of March 1995, sixteen offenders had been committed to prison under three strikes. The defendants ranged in age from twenty-five to fifty-three, with the median age of forty. Table 2.2 displays the additional sentences to be served by these offenders, assuming a seventy-year life expectancy and that each would have received a typical sentence under prior law for their crime and criminal history.

The 1993 legislature also, for the first time since enactment of the Sentencing

Table 2.2. Impact of Three Strikes on Sentences, First Sixteen Cases

Offense	Age	Sentence under Former Law (in months)	Three-Strikes Sentence (in months)	Increased Sentence Length (in months)
Murder 1°	28	312	503	191
Rape 1°	35	222	420	198
Rape 1°	41	102	348	246
Rape 2°	53	153	204	51
Kidnapping 1°	45	175	300	125
Child Molestation 2°	39	72	372	300
Robbery 1° (armed)	41	50	349	299
Robbery 2° (unarmed)	35	26	420	394
Robbery 2° (unarmed)	45	12	300	288
Robbery 2° (unarmed)	29	50	493	443
Robbery 2° (unarmed)	32	42	457	415
Robbery 2° (unarmed)	32	68	456	388
Robbery 2° (unarmed)	25	26	540	514
Robbery 2° (unarmed)	40	26	360	334
Robbery 2° (unarmed)	41	50	349	299
Burglary 1°	49	45	252	207

Source: Washington State Department of Corrections (unpublished data provided to author).

Reform Act in 1984, adopted a "reform" that was projected to reduce prison population. It created "work-ethic camps." Eligible defendants—those who received sentences from twenty-two to thirty-six months for nonviolent offenses—have their sentences cut to four to six months upon successful completion of a work-ethic camp. Projections were that this would reduce prison population by 22 in the first year, rising to 350 by the fifth year.

The 1994 legislature, responding to concerns over youth violence, provided for mandatory transfer from juvenile to adult court of sixteen- and seventeen-year-old youths charged with their second violent felony offense. This "reform" was projected to increase prison population by 51 in the first year, rising to 176 by the fifth year.

Hard Time for Armed Crime   Encouraged by the success of three strikes, its sponsors returned in 1994 with an initiative to the legislature styled "hard time for armed crime." This initiative called for amending the Sentencing Reform Act and the criminal code in a number of ways. First, it increased the length of the existing sentence enhancements for crimes committed while armed with firearms or other deadly weapons, broadened the sweep of the enhancements from designated violent felonies to all felonies, made the enhancements mandatory rather than presumptive, and eliminated any good-time credits on the enhancement portion of the sentence. Second, it increased the presumptive sentence ranges for reckless endangerment, theft of a firearm, and unlawful possession of a firearm, and it also broadened the definition of first-degree burglary to include those occurring in any building rather than the previous limitation to residences. Third, it required prosecutors to make public the reasons for plea bargains and required the Sentencing Guidelines Commission to compile records comparing the sentences imposed by individual judges with the applicable sentence ranges.

The projected impact of this initiative is far greater than three strikes. The Sentencing Guidelines Commission estimated that it would increase prison population by 209 in the first year, 810 by the fifth year, and 1,145 by the tenth year. It would require capital and operating expenditure increases of $64 million the first biennium, $57 million the second, $68 million the third, $50 million the fourth, and $55 million the fifth; a total increase of $294 million over the first decade.

The necessary signatures were gathered and the legislature was faced with the choice of adopting the initiative as proposed or adopting an alternative and placing both the initiative and the legislative alternative on the ballot. With the memory of the success of three strikes at the polls fresh in their minds, by strong bipartisan majorities the legislature chose to adopt the initiative.

Drug Offender Sentencing Alternative   The Sentencing Guidelines Commission, after sponsoring a variety of proposals to modify the increases in sentence severity for drug offenses enacted in 1989—they increased the percentage of inmates who are drug offenders from 16 percent in 1990 to 25 percent at the end of 1994—achieved success in 1995 with its Special Drug Offender Sentencing Alternative. The alternative gives judges discretion to reduce sentence lengths for first-offense drug offenders by up to 50 percent and requires that the offender "receive, within available resources, treatment services appropriate for the offender." Alternative sentences must include a year of postrelease supervision on condition that the offender receive "appropriate outpatient substance abuse treatment" and not use illegal drugs. Violation of any condition of supervision may result in return to prison for the balance of the presumptive sentence not originally imposed.

The alternative was championed by a coalition led by Norm Maleng, the prosecutor from Washington's largest county and a member of the Sentencing Guidelines Commission. The coalition included support from prosecutors and police, as well as the sponsors of the three-strikes and hard-times initiatives. Arguments in support of the proposal included saving expensive prison capacity for violent criminals.

The sentencing alternative is projected to reduce prison population by 196 its second year, 240 in its third, and 258 by the fourth year, before stabilizing at a reduction of 275 into the future. Its significance and that of the 1993 work-ethic camp option, however, lies not in the magnitude of the reductions but in the fact that they are the first "reforms" in the history of the Sentencing Reform Act to decrease sentence severity.

Cumulative Effect   The impact of the two initiatives and the other changes were, of course, added to the impact of the series of amendments from 1987 to 1992.

Figure 2.5 displays the consequences of the "reforms" of the Sentencing Reform Act. The cumulative effect of these statutory changes when added to the results of a steadily growing state population was enormous. These increases were predicted. The legislature knew, each step of the way, what the impacts would be. They

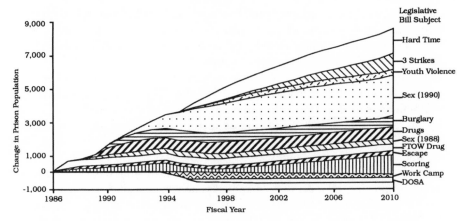

Figure 2.5. Cumulative Effects of Criminal Justice Legislation, 1986–1995 Sessions. *Note:* FTOW = First Time Offender Waiver; DOSA = Drug Offender Sentencing Alternative. *Source:* Washington State Office of Financial Management, Forecasting Division (unpublished data provided to author).

responded by embarking on the largest prison construction program in Washington's history. During the two years from May of 1992 through April 1994, prison capacity was increased by 4,032 at a capital cost of $263 million. Washington's next prison—a 2,000-bed facility—is scheduled for completion in 1997, as is expansion of minimum security space by approximately 1,000. Figure 2.6 displays capital expenditures. Figure 2.7 displays the past and projected growth in operating expenditures. The operating budget of the Department of Corrections increased from $561 million in the 1993–1995 biennium to a projected $740 million for the 1995–1997 biennium.

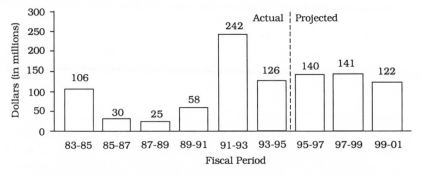

Figure 2.6. Department of Corrections: Capital Expenditures, Historical and Projected, Washington State. *Source:* Washington State Department of Corrections (unpublished data provided to author).

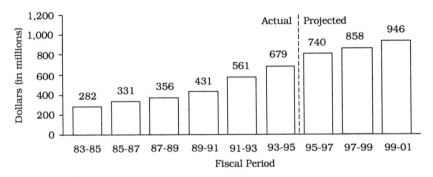

Figure 2.7. Department of Corrections, Operating Expenditures, Historical and Projected, Washington State. *Source:* Washington State Department of Corrections (unpublished data provided to author).

The Future?   For those who view Washington's recent history with something less than unmitigated approval, there may be some reason for hope. First, the actions of the legislature can be seen as returning to the policy judgments of the original Sentencing Reform Act that sentences "should emphasize confinement for the violent offender and alternatives to total confinement for the nonviolent offender."

Second, throughout the past decade the basic structure of the Sentencing Reform Act has remained intact. With the exception of the hard-time initiative's addition of mandatory minimums for crimes committed with firearms and deadly weapons, sentences remain presumptive, thus allowing variations, up or down, via the structured discretion of sentencing judges.

The Sentencing Reform Act has proven remarkably effective at transmitting legislative judgments, necessarily expressed in general terms, into specific sentences. In the early years those judgments were that prison population be reduced, and that happened. Over the past eight years that judgment was that sentences should be increased, and that happened as well. Both results were predicted and the costs of the increases in prison population were paid.

Should Washington's citizens decide the price of continued severity is too high, in either economic or human terms, the Sentencing Reform Act can readily be altered to implement that decision. Washington citizens are responsible for the massive increases in prison population and they have been responsible about paying the bill. They are also responsible for the future of sentencing reform in Washington.

Minnesota

Minnesota's Sentencing Guidelines—Past and Future

Debra Dailey

Minnesota's presumptive sentencing guidelines took effect on May 1, 1980. Since then, nearly twenty states have created sentencing commissions. Although many failed to develop politically salable guidelines, Pennsylvania, Washington, and Oregon succeeded, and commissions are today at work in another half dozen states.

By most measures, Minnesota's guidelines have been remarkably successful. Sentencing disparities declined. Racial differences in sentencing patterns lessened. The commission's policy choices to de-emphasize imprisonment for property offenders and to increase use of imprisonment for violent offenders were reflected in prison admissions. A decision to tie sentencing policy to prison resources enabled Minnesota to avoid major programs of prison construction or jerry-rigged early release programs. Minnesota's incarceration rate, like Minnesota's crime rate, remains among the lowest in the country.

However, Minnesota in 1980 and in 1992 are different places. The guidelines and the political climate have changed. The legislature has repeatedly enacted laws to make sentencing tougher, and prison populations are, as a result, likely to rise steadily in the 1990s. For some types of crimes and offenders, compliance with guidelines has decreased.

This article provides an overview of Minnesota's experience with sentencing guidelines, with emphasis on the late 1980s and the 1990s.

The Minnesota legislature created a sentencing commission and mandated that sentencing guidelines be developed. The aim was a system that would achieve more uniform and determinate sentencing. Although the guidelines set presumptive sentences for individual offenders, judges retained authority to "depart" from the guidelines and impose some other sentence.

Sentencing Patterns during the 1980s The early successes were clear. Consistency in sentencing increased by 50 percent. Proportionality in sentencing emphasized a just-deserts philosophy, and significantly more violent and fewer nonviolent offenders went to prison. Prison populations remained manageable.

After the first years, however, problems became apparent. Minnesota experienced a dramatic increase in the number of convicted felons sentenced in the 1980s. The number of cases grew from approximately 5,500 in 1981 to nearly 8,000 by 1989. Crime rates fluctuated but were no higher in 1990 than in 1980. Arrest rates increased, however, by 46 percent between 1983 and 1989.

Because of the increased volume of cases, the number of offenders sentenced to prison increased steadily, rising from 825 in 1982 to 1,752 in 1989. Imprisonment rates fluctuated only slightly, holding at approximately 20 percent of felony convictions.

Prison populations rose from approximately 2,000 in 1980 to nearly 3,400 in 1991. A mental health institution was recently converted to a prison and will eventually house 500 medium-security inmates.

Increased prison use for repetitive property offenders, violent offenders, sexual offenders, and drug offenders offset increased "departures" from guidelines that reduced penalties because of mitigating circumstances.

Repetitive Property Offenders During most of the 1980s, the proportion of property offenders subject to presumptive prison sentences increased. A trend toward a higher proportion of convicted felons with higher criminal history scores was reported in evaluations of the Minnesota experience. The percentage subject to presumptive prison sentences increased from 7 percent in 1981 to 12 percent in 1983 and 15.4 percent in 1985, stabilizing at 16.8 percent in 1988. The imprisonment rate for property offenders mirrored this trend, increasing from 9.2 percent in 1981 to 16.2 percent in 1983 and stabilizing at 18.7 percent in 1987.

Violent Offenders   The imprisonment rate for offenders convicted of more serious crimes is significantly higher than before the guidelines took effect. However, as the 1980s progressed, smaller percentages of these offenders were sentenced to prison than during the first two years of the guidelines. The imprisonment rate for the four most serious levels of offense fluctuated over time but dropped substantially from the 85.9 percent rate experienced in 1981 to 76.5 percent in 1983 and 72.5 percent in 1989.

Aggravated Departures   Aggravated dispositional departures occur when the guidelines recommend a stayed sentence and the judge imposes a prison term. Mitigated dispositional departures occur when the guidelines recommend prison and the judge stays the sentence and imposes some other sanction.

The overall dispositional departure rate was 6.2 percent of all sentenced felons in 1981, and it increased each year through 1984, when it reached 10.2 percent. The rate fluctuated between 10 and 11 percent through 1989. The aggravated dispositional departure rate has held steady at 3 to 3.5 percent since 1981, while the mitigated dispositional departure rate more than doubled from 3.1 percent in 1981 to 6.9 percent in 1989.

Aggravated dispositional departures are even rarer than the figures suggest: 80 percent occur when offenders insist on going to prison. This generally happens when offenders believe they are going to be revoked to prison on a current probationary sentence or that they will be sentenced to prison on another offense and may as well serve the sentences concurrently. Another reason to prefer prison is that some offenders believe probationary conditions will be more onerous than serving a prison term.

Mitigated Departures   In 1989, approximately one-third of the mitigated dispositional departures were for assault in the second degree (assault with a dangerous weapon). The mitigated dispositional departure rate for this offense alone was 67.8 percent. This crime would not carry a presumptive prison sentence were it not for a mandatory minimum law that requires that offenders convicted of weapons offenses serve time in prison. The law permits the judge to impose a different sentence if reasons are given. This offense is unique in merging sentencing guidelines policy and statutory minimums.

Child sexual abuse cases are another regular source of mitigated dispositional departures. These cases usually involve adults related to the minor. Substantial increases in the number of sentenced sex offenders began in 1984 and was due exclusively to increases in these types of sex offenses. Departure rates are high because judges often believe that a long period of probation with time in jail and treatment is more appropriate than prison, particularly for offenders with no criminal record, and is in the best interest of the family and the child victim.

Jail Sentences   The most dramatic change in Minnesota sentencing during the 1980s involved jail as a condition of probation. The sentencing guidelines set presumptions for who should go to prison and for how long. For those offenders not presumed to receive a prison sentence, the judge may order any intermediate sanction, require time in a local jail or workhouse, or both. There are no guidelines for judges regarding intermediate sanctions and jail.

The percentage of felons receiving jail terms climbed from 46.2 percent in 1981 to 58.6 percent in 1989. Before implementation of the guidelines, jail rates were 35.4 percent in 1978. This increased use of jail, coupled with the growth in the volume of cases, has resulted in an increase in the number of felons sentenced to jail each year, from 2,539 in 1981 to 4,669 in 1989.

Major Legislative and Commission Policy Changes in the 1980s   Sentencing guidelines cannot and should not be static. Minnesota's have changed continuously. For the most part the changes involved technical and implementation problems. Two changes in the 1980s, however, were extensive and substantive.

Drug Crimes   Since the mid-1980s, the commission has changed the guidelines with regard to drug offenses each year. Before 1985, the commission treated cocaine offenses as comparable to marijuana offenses, but cocaine offenses were then raised in severity to match heroin offenses. The legislature fine-tuned the drug laws each year until 1989 when the controlled-substance laws were entirely rewritten. Minnesota now has five degrees of drug crimes elaborately described on the basis of sale or possession, type of drug, and amount of drug.

These changes have elevated certain drug crimes to very high severity-level rankings. The greatest controversy has concerned street-level dealers. Despite pressure to do otherwise, the commission has generally ranked drug offenses so that probation is the presumptive sentence for most first-time street-level dealers; street-level dealers with a prior conviction are subject to a presumptive prison sentence.

Judges also appear to be ambivalent about drug cases. In 1989, the mitigated dispositional departure rate for drug offenses with a presumption of prison was 31.1 percent. However, while most drug offenders are not subject to a presumptive prison sentence, nearly 90 percent of all drug offenders sentenced in 1989 served time in jail or prison.

Sex Crimes   Violent offenders, particularly sex offenders, have been the second focus of major concern and change. Following a series of horrible crimes against women and children in 1988, numerous groups and individuals recommended major changes in sentencing guidelines and state laws.

The commission adopted changes that went into effect in August of 1989. These were the most substantive and far-reaching changes since the guidelines were first implemented. These changes not only addressed the demand for harsher penalties for violent offenders but also dealt with the increasing prison populations and increasing incarceration of property offenders.

Presumptive prison sentences increased substantially for offenders convicted of serious crimes such as armed robbery, first-degree assault, many sex crimes, and high-level drug crimes. In addition, the legislature directed the commission to increase presumptive prison sentences for second- and third-degree murder significantly and mandated lengthy sentences for certain murderers and sex offenders.

The commission also adopted a weighting system for criminal history that increased the impact of prior high-severity crimes but decreased the weight of prior low-severity crimes. These changes de-emphasizing nonviolent criminal history somewhat tempered the substantial effects of increases in the severity of sanctions discussed above. Prison populations will grow extensively due to the overall impact of the changes made in 1989, but the increases would have been more immediate and less manageable were it not for the moderating changes.

Current and Future Issues   Minnesota experienced several notorious murders in the summer of 1991 that once again aroused public concern. The legislature will likely consider more increases in presumptive sentences.

Increased Sentence Lengths   The commission is proceeding with caution and will not recommend increases in presumptive sentences at this time. Sentencing data for 1990 will be examined to learn more about the effects of the changes made in 1989. Preliminary data suggest that lengths of prison sentences for some offenses have increased substantially but that departure rates have also increased.

Guidelines for Intermediate Sanctions   Minnesota's guidelines create presumptions only for prison sentences. The absence of standards for nonprison sentences has contributed to numerous problems, including lack of statewide consistency and proportionality in nonprison sentences, increasing use of jails and workhouses, a widening of the net in the use of intermediate sanctions, and a widespread tendency for prison to be viewed as the only legitimate means of punishment and control.

Despite these problems, there has been widespread resistance to development of statewide guidelines for nonincarcerative sanctions. The commission has not favored statewide guidelines for intermediate sanctions but has expressed support for local guidelines. The primary reason for inaction has been a belief that statewide policy could not effectively govern the use of locally funded correctional resources and that additional guidelines would unduly complicate the system. But as increasing demands have been placed on community corrections, and as price tags have increased alarmingly, the need for standards for intermediate sanctions has become increasingly clear.

The 1991 legislature stopped short of mandating statewide intermediate sanctions guidelines, but it did direct each judicial district to develop a criminal justice resource management plan. The commission was told to develop principles for the judges to incorporate into their management plans. The plans are to take account of the role of individual judges' discretion in the use of resources within the district, the role and use of intermediate sanctions, and the feasibility of sharing correctional resources between counties within multicounty judicial districts.

Day Fines   The commission is developing a day-fine proposal to be presented to the legislature by 1993. A recent study conducted by the commission has indicated that fines are rarely used for convicted felons except in some rural areas. Key legislators are interested in seeing a greater use of fines, and they prefer a day-fine approach in which the fine is proportional to the offense of conviction and is relative to the person's ability to pay. It is unclear when the policy will be implemented.

As the 1990s unfold, the commission's agenda contains many complex issues. Many issues are broader than a sentencing commission alone can resolve. Legislators, commissioners, practitioners, and concerned citizens must work together toward common goals. Minnesota has been fortunate to have progressive leadership and strong commitment to ideals and principles. The challenges of the future will need no less.

Prison Population Growing under Minnesota Guidelines
   Richard Frase

A major goal of Minnesota's sentencing guidelines is to ensure that sentencing practices stay within the state's correctional resources, especially available prison capacity. This goal was achieved throughout the first decade of guidelines sentencing. However, prison population began to increase rapidly in the late 1980s. By the end of 1992, the number of prisoners was almost double the number in 1980, and the state's prisons were full. This article traces the evolution of the Minnesota guidelines' "prison-capacity constraint" and the factors that explain recent prison population increases.

The Prison Population Policy   The prevention of prison overcrowding is one of the most important advantages of sentencing guidelines. Using an independent commission to set presumptive sentences maximizes the potential to avoid prison overcrowding. Such a specialized agency can develop sophisticated measures to predict future prison populations and draft sentencing rules that stay within expected capacity. In addition, appointed commissioners are less subject to short-term political pressures, which often lead elected officials to propose unrealistically severe penalties.

The enabling statute directed the new commission to take existing prison capacity and other resources into "substantial consideration," but the commission chose to go further: it made prison bed space a controlling factor in drafting and implementing the guidelines, and it established a goal of never exceeding 95 percent of state prison capacity.

The commission developed a computerized sentencing and inmate population model to evaluate the population effects of each presumptive sentence being proposed. As the new guidelines took shape, these inmate population projections often required the commission to moderate its sentencing policies to stay within prison capacity. After the guidelines took effect in 1980, the commission continued to project future prison populations and predict the effects of proposed guidelines changes and major crime bills. For example, when potential overcrowding problems appeared in 1982, the commission scaled back some of the presumptive prison durations to head off the crisis.

Effects on Inmate Populations As a result, Minnesota's state prison population increased very slowly in the early 1980s and stayed well within capacity. At a time when prisons in most other states were overcrowded and subject to court intervention, Minnesota faced neither problem. Between 1980 and 1984, the prison population increased by only 8 percent, while the total U.S. prison population went up by 41 percent.

In the late 1980s, however, things began to change. Minnesota's prison population rose by 29 percent from 1984 to 1988; it increased by another 37 percent from 1988 to 1992. (The national increases were about 35 and 41 percent, respectively.) By the end of 1992, Minnesota still had the second-lowest per-resident imprisonment rate in the country—one-fourth the national average—and it was still managing to avoid serious prison overcrowding problems. But its prisons were full, and an additional 28 percent increase in inmate populations is expected by 1996.

Reasons for Recent Population Increases Three factors appear to explain the major increase in Minnesota's prison population since the mid-1980s: a substantial growth in felony convictions; increased rates of probation and parole revocation; and major increases in sentencing severity adopted by the commission and the legislature in 1989.

Between 1981 and 1986, case volume fluctuated between 5,500 to 6,200 cases per year. But starting in 1987, felony convictions rose substantially each year, reaching 9,161 cases in 1991 (the most recent year for which guidelines sentencing data are available). Much of the increase was due to the delayed but inevitable arrival in Minnesota of crack cocaine and other "big-city" crime problems.

As for revocation of probation and parole (technically, "supervised release," the duration of which is equal to good-time credits earned in prison), such prison admissions remained fairly constant through most of the 1980s and constituted a lower percentage of prison commitments in 1990 than in 1978. But in 1991 and 1992, revocation rates of both types increased substantially. The reasons are not clear, but it seems likely that more frequent use of intensive supervision, increased random drug-use testing, or both are revealing violations of conditions that went undetected in the past.

In the long run, however, the most important cause of increased prison populations may prove to be more severe guidelines and statutory penalties enacted in 1989. Prior to that year, changes occurred gradually and did not have a major impact on prison populations. But in 1988, political pressures for substantially increased penalties began to escalate. The commission had overcome similar crises in the past, but the "crime wave of 1988" proved too broad and too sustained to resist.

Pressure began to build in late spring, with a series of sexual attacks and rape-murders in Minneapolis parking ramps. The Minnesota attorney general appointed a task force on sexual violence against women that issued strongly worded reports, demanding substantially increased rape sentences. At the same time, the city of Minneapolis was experiencing what seemed like a general increase in violence and drug crime: the 1988 murder rate was 50 percent higher than in 1987, and 1988 drug arrests were up 60 percent.

In mid-November, the guidelines commission responded by proposing to increase prison sentences for many violent crimes. For example, the presumptive sentence for a first-degree rapist with zero criminal history would have increased from three and a half to four and a half years (not counting good-time reductions of up to one-third). However, these proposals did not satisfy the hue and cry and were met with calls to double sentences for all violent crimes. In December, the commission responded by increasing presumptive sentence durations at Severity Levels VII (e.g., armed robbery) and VIII (first-degree rape). Durations were doubled for defendants with zero criminal history and were increased substantially for those with higher criminal history scores.

The guidelines changes (along with other, mitigating changes designed to reduce the need for additional prison beds) were scheduled to take effect the following summer. However, pressure for increased severity did not let up. In the spring of 1989, the legislature considered a number of "get tough" crime bills, including two that would have brought back the death penalty, not used in Minnesota since 1911.

The Omnibus Crime Bill finally adopted included a number of severe measures: life without parole for certain first-degree murderers; mandatory maximum terms for other recidivist murderers and sex offenders; minimum prison terms for certain drug crimes; and increased statutory maximums for other violent and sex crimes. The bill included extended terms and mandatory minimum prison terms for certain sex-motivated crimes if the court finds that the defendant is a danger to public safety.

A number of provisions in the 1989 crime bill suggested that the legislature no longer trusted the commission to set sufficiently severe presumptive sentences and had decided to take back some of the delegated power to set specific guidelines and overall sentencing policy. Thus, the bill amended the guidelines enabling act to specify that the commission's "primary" goal in writing guidelines should be *public safety*; correctional resources and current practices remain as factors, but they are no longer to be taken into "substantial" consideration. In addition, the commission was directed to increase penalties at Severity Levels IX and X by specified amounts and to add a specific provision to the guidelines list of aggravating circumstances. Finally, judges were given authority in certain cases to impose the statutory maximum prison term, apparently without regard to ordinary guidelines rules governing departure and degree of departure.

The commission made the changes ordered by the legislature, and these changes (along with the amendments previously adopted by the commission) became effective for all crimes committed on or after August 1, 1989. The changes in presumptive prison durations are shown in figure 2.8. All cases falling below the heavy black line are presumptive prison terms (cases above the line are presump-

Severity Levels of Conviction Offense		Criminal History Score						
		0	1	2	3	4	5	6 or more
Sale of a Simulated Controlled Substance	I	12	12	12	13	15	17	19 *18-20*
Theft Related Crimes ($2,500 or less) Check Forgery ($200 - $2,500)	II	12	12	13	15	17	19	21 *21-22*
Theft Crimes ($2,500 or less)	III	12	13	15	17	19 *18-20*	21 *21-23*	25 *24-26*
Nonresidential Burglary Theft Crimes (over $2,500)	IV	12	15	18	21	25 *24-26*	32 *30-34*	41 *37-45*
Residential Burglary Simple Robbery	V	18	23	27	30 *29-31*	38 *36-40*	46 *43-49*	54 *50-58*
Criminal Sexual Conduct 2nd Degree (a) & (b)	VI	21	26	30	34 *33-35*	44 *42-46*	54 *50-58*	65 *60-70*
Aggravated Robbery	VII	~~24~~ *~~23-25~~* 48 *44-52*	~~32~~ *~~30-34~~* 58 *54-62*	~~41~~ *~~38-44~~* 68 *64-72*	~~49~~ *~~45-53~~* 78 *74-82*	~~65~~ *~~60-70~~* 88 *84-92*	~~81~~ *~~75-87~~* 98 *94-102*	~~97~~ *~~90-104~~* 108 *104-112*
Criminal Sexual Conduct, 1st Degree Assault, 1st Degree	VIII	~~43~~ *~~41-45~~* 86 *81-91*	~~54~~ *~~50-58~~* 98 *93-103*	~~65~~ *~~60-70~~* 110 *105-115*	~~76~~ *~~71-81~~* 122 *117-127*	~~95~~ *~~89-101~~* 134 *129-139*	~~113~~ *~~106-120~~* 146 *141-151*	~~132~~ *~~124-140~~* 158 *153-163*
Murder, 3rd Degree Murder, 2nd Degree (felony murder)	IX	~~105~~ *~~102-108~~* 150 *144-156*	~~119~~ *~~116-122~~* 165 *159-171*	~~127~~ *~~124-130~~* 180 *174-186*	~~149~~ *~~143-155~~* 195 *189-201*	~~176~~ *~~168-184~~* 210 *204-216*	~~205~~ *~~195-215~~* 225 *219-231*	~~230~~ *~~218-242~~* 240 *234-246*
Murder, 2nd Degree (with intent)	X	~~216~~ *~~212-220~~* 306 *299-313*	~~236~~ *~~231-241~~* 326 *319-333*	~~256~~ *~~250-262~~* 346 *339-353*	~~276~~ *~~269-283~~* 366 *359-373*	~~296~~ *~~288-304~~* 386 *379-393*	~~316~~ *~~307-325~~* 406 *399-413*	~~336~~ *~~326-346~~* 426 *419-433*

Figure 2.8. Minnesota Sentencing Guidelines Grid (showing durational changes as of August 1, 1989). Presumptive sentence lengths are in months. *Source:* Minnesota Sentencing Guidelines Commission (1990).

tive stayed or suspended terms). Cells below the line contain both a single presumptive prison sentence duration and a range within which the sentence is not considered a departure. The numbers crossed out in figure 2.8 are the presumptive durations that applied prior to the 1989 changes.

Only about 5 percent of cases sentenced in 1989 were governed by the new guidelines and statutes enacted in that year. But in 1990, the initial effects of these changes began to appear: sentence durations at Severity Levels VII through X increased substantially for "new law" cases. At the same time, the legislature apparently began to realize the serious prison population consequences of the increases enacted in 1989. The principal 1990 crime bill created a new program of "intensive community supervision" (ICS). It was limited to offenders already in prison or facing probation revocation, which suggests that the legislature was primarily interested in diverting offenders from prison.

In 1991, the legislature continued its rediscovered policy of penal and fiscal restraint, but events after adjournment that year began the cycle all over again. Strong law-and-order pressures were precipitated by a wave of high-profile violent crimes and media reports. In the summer of 1991, two college women were kidnapped, raped, and murdered, both in seemingly "safe" small-town areas. In late fall of 1991, the state's principal newspaper published a sensational series of articles entitled "Free to Rape," arguing that Minnesota's sex offender penalties were too lenient. Finally, in December, the Minnesota Supreme Court held that statutory and guidelines rules imposing much heavier penalties on crack cocaine offenses than on powdered cocaine violated the state constitution (largely because of the disparate impact on blacks, who constitute the vast majority of crack offenders).

The latter decision provoked an immediate, and predictably severe, response. In January 1992, without waiting for the commission's statutorily mandated report on drug sentencing (due in February), the legislature raised powdered cocaine penalties to equal those applicable to crack. Later in the spring, a general crime bill was enacted. Although the bill expanded treatment, education, and social service programs, it also contained numerous "get tough" provisions for sex offenders: mandatory doubling of some presumptive sentences; lengthier supervised release terms; increased statutory maximums; and mandatory life and thirty-year prison terms for recidivists. Discretion to grant early release to certain "amenable" sex offenders was eliminated. However, the commissioner of corrections was given discretion to select other offenders for early release to a new boot camp ("challenge incarceration") program. The latter option, like the ICS program adopted in 1990, suggests that the legislature was aware of the potentially dramatic budgetary and prison-overcrowding consequences of further severity increases.

Conclusion Minnesota's experience shows that sentencing guidelines incorporating an explicit prison-capacity constraint can help to contain growth in inmate populations and prevent prison overcrowding. However, they cannot guarantee zero population growth, nor is commission-based sentencing policy immune from the kind of law-and-order politics that, if unchecked, can cause inmate populations to increase rapidly.

Nevertheless, as of 1992, Minnesota still imposed prison sentences far less often than other states, and it had managed to slow the growth of its inmate population enough to stay within expanding capacity—without resorting to double-bunking in any of the state's maximum-security prisons. Although the severe penalties adopted in 1989 and 1992 have greatly increased the risk of future overcrowding, there are signs that key legislators recognize the importance of avoiding overcrowding and limiting further growth in prison construction and operating costs. Recent public opinion polls in Minnesota show a preference for restitution and delinquency prevention programs over more imprisonment, which suggests that harsh "lock 'em up" sentiments reflect short-term hysteria. It is precisely such temporary political pressures that commission-controlled sentencing guidelines are designed to buffer.

The Minnesota commission's task in the years ahead is to make sure that legislators understand the value of commission-controlled sentencing policy. The commission must also make effective use of its inmate population prediction model and other data, so that legislators, the media, and the public appreciate the full costs of proposed prison sentences and the availability of effective nonimprisonment sanctions. In the absence of such data, politicians will continue to engage in irresponsible "credit card" sentencing policy—enactment of "get tough" crime bills without regard to whether the state will ever have the funds to pay for them.

Minnesota's Sentencing Guidelines—1995 Update
Debra Dailey

Minnesota's experience with sentencing guidelines now exceeds fifteen years, and the experience has not been dull. Policies dealing with the sentencing of criminals tend to be among the most controversial in government, and they set the stage for highly charged politics and a watchful media. While controversy has surrounded the sentencing guidelines over time, Minnesota is fortunate to have a solid framework for rational decision making. The sentencing guidelines have helped Minnesota avoid "striking out" with simplistic and extreme sentencing laws; mandatory minimum sentences enacted by the legislature have been kept to a minimum. This article summarizes developments in the last few years and highlights challenges for the sentencing guidelines commission in the latter half of the 1990s. The story of Minnesota's guidelines is a story of success. Consistency, proportionality, and neutrality in sentencing increased under the guidelines, and prison resources have been coordinated with sentencing policies. Crime rates fluctuated but remained relatively stable and always have been low when compared with most other states'.

In the late 1980s, however, Minnesota began to show increasing intolerance for criminals, particularly for drug and violent offenders. Minnesotans were becoming more and more troubled by deteriorating neighborhoods dominated by drug dealing. In 1988, extensive media coverage focused on two heinous back-to-back murders. Policy makers rushed to respond.

Increased Severity after 1989   The 1989 legislative session represents the first significant change in sentencing policies since the enactment of the guidelines in 1980. Dramatic proposals were introduced to toughen sentences. One proposed bill would have abolished the sentencing guidelines and created a mandatory minimum sentencing scheme requiring every convicted felon to serve time in prison. Impact analyses showed that this bill would increase the prison population fivefold. The bill did not pass, but there was a tie vote in the senate, demonstrating how eager legislators were to be seen as tough on crime.

One of the sentencing guidelines commission's functions is to conduct impact analyses on pending bills. Thus, the 1989 legislature was aware of the present and projected future costs of its bills with respect to prison resources. What passed the legislature that year was significant but far more selective and moderate than the mandatory minimum bill. The changes mainly focused on heinous murderers and repeat and violent sex offenders. The controlled-substance laws were overhauled, creating five degrees of drug crimes differentiated on the basis of the type and amount of drug involved. Crack cocaine offenders were treated significantly more harshly than powder cocaine offenders.

In addition to legislative sentencing changes, the sentencing guidelines commission also reexamined sentences for violent offenders. It determined that particularly with regard to sex offenders, the guidelines did not carry a degree of punishment proportionate to the seriousness of the crime. Presumptive sentences were increased substantially for offense severity levels that included many sex crimes, serious assaults, and armed robbery. The legislature directed the commission also to increase presumptive sentences for second- and third-degree murder.

The commission moderated its changes to toughen the guidelines for violent offenders by adopting a weighting system for determining criminal history scores. Originally, all prior felonies were counted as 1 point. Under the weighting scheme, prior serious violent felonies are weighted at 1 1/2 or 2 points, and prior lower severity-level nonviolent felonies are de-emphasized.

In subsequent years, the legislature continued to pass major crime bills, but each year the changes were specific to certain sex offenders and murderers. Sweeping and simplistic laws such as three strikes were averted. However, in November 1991, the Minnesota Supreme Court ruled that the drug laws were unconstitutional in that they treated crack cocaine offenders, who were predominantly black, more harshly than powder cocaine offenders, who were predominantly white. The 1992 legislature responded by equating the powder cocaine penalties to those of crack, thus substantially toughening the penalties for powder cocaine. With all of these changes, Minnesota offenders now serve some of the toughest sentences in the country, particularly certain murderers and sex offenders, and by most state standards, Minnesota's cocaine laws are comparatively harsh. The policy changes to sentence lengths resulted in a 42 percent increase in the average prison sentence over the last decade.

Independently of increased severity of sentences for some crimes, increased numbers of arrests also contributed to prison population growth. The crime rate in Minnesota has fluctuated but remained relatively stable over time, but arrests for serious crimes rose by nearly 30 percent from 1983 to 1993, with arrests for drug crimes increasing by over 100 percent. Consequently, the number of sentenced felons increased by 73 percent over this same period.

As a result of the combined effects of tougher penalties and increased arrests, the prison population grew from approximately 2,000 in the early 1980s to a current population of over 4,700. The population is expected to grow by another 1,000 by the end of the century.

With prisons consuming an increasing proportion of criminal justice spending, the appetite for tougher sentences began to diminish. By the 1994 legislative session, some legislators asked what could be done to slow or stop this trend in increased spending. However, they did not want to undo the tough laws they had recently passed and, because they valued the goal of truth in sentencing, they did not want to enact early release mechanisms. A bill passed that session directing the sentencing commission to "evaluate whether the current sentencing guidelines and related statutes are effective in furthering the goals of protecting the public's safety and coordinating correctional resources with sentencing policy." Based on this evaluation, they asked the commission to "develop and recommend options for modifying the sentencing guidelines so as to ensure that state correctional resources are reserved for violent offenders."

1994 Proposals for Change A subcommittee of the commission put together a package of modifications to the guidelines that focused on preserving a greater proportion of prison space for violent offenders. The subcommittee determined that some violent crimes needed to be increased in severity and some property and drug crimes decreased. These proposed changes would also affect offenders' criminal history scores. Prior (mostly property) crimes would be reduced in the weight assigned to them when computing criminal history scores and result in fewer property offenders who would be recommended to go to prison under the proposal. The proposal also limited the inclusion of misdemeanors to violent crimes when calculating criminal history scores. In addition, the subcommittee recommended that durations be reduced somewhat at lower severity levels.

The subcommittee's proposal, while increasing sentences for some violent offenders, overall would reduce the future need for prison space by more than 500 beds, or one whole prison. The full commission moved the proposal forward to a public hearing and received strong but mixed reactions.

Reactions Support came from judges, some community corrections representatives, the present and a former governor, academics, and some citizens. They viewed the proposal as a way to balance Minnesota's response to correctional issues instead of putting all of the state's money into prisons.

Opposition came from county attorneys, sheriffs, county government representatives, the attorney general's office, some community corrections representatives, a key legislator in the house (chair of the Judiciary Committee), and some citizens. They viewed the proposal as shifting the burden of paying for repeat offenders to local taxpayers, and they believed the proposal would erode public confidence in the system at a time when the public is demanding more accountability for criminal activity.

The mixed reaction left the commission divided, but a majority wanted to move the proposal forward. However, the commission faced strategic problems. The law grants the commission the authority to adopt modifications to the guidelines and requires most modifications to be reviewed by the legislature before they can go into effect. This is a passive review process, and if the legislature does nothing, modifications will take effect August 1 of the same year. The legislator who testified against the recommendations at the public hearing made it clear that he opposed the proposal and that he believed the commission would be stepping outside the bounds of the 1994 legislative directive if it adopted the modifications. He pointed out that the directive had asked the commission to recommend options and not to adopt modifications.

Local Funding Concerns The commission also discussed concerns about local funding. It realized its proposal shifted responsibility for a greater number of offenders to local communities and wanted to see state funding follow the offender. It was concerned about adopting its proposal without the certainty of funding. The commission chose not to adopt the proposal but sent it to the legislature asking for its consideration.

1995 Proposals for Change Bills were introduced in the 1995 legislative session to enact the commission's proposal. The bills were discussed in committee in both the senate and the house, but the committees concluded that the proposal needed more study, particularly to address the concern that too great a financial burden would be placed on local jurisdictions. It was also clear that the underlying politics of crime would have made it very difficult for legislators to vote yes on a bill that would reduce sentences.

The commission took seriously the call for further study. It determined that while the original proposal was developed almost in isolation from other interested organizations, a successful revision of the proposal would require input from those groups who had originally raised concerns. The commission offered opportunities to a wide range of organizations to hear and ask questions about the original proposal, and it invited representatives from these groups to attend subcommittee meetings and join in the revision process.

The Basic Revised Proposal This process generated a revised proposal that appears to have support from a wide range of groups, including many that previously opposed the original proposal. However, the proposal is quite different from the original. The commission decided to take an incremental approach to modifying the guidelines, and there are no proposed changes to presumptive sentences for drug offenders. All of the proposed increases in severity levels for certain violent crimes have been dropped, as have many of the decreases in severity levels for certain property crimes. Decreases in the severity level of certain theft crimes remain, as do modest decreases in prison sentence durations for lower severity-level crimes. The projected impact of the revised proposal is not as significant as before, but it will still reduce the need for future prison space by about 240 beds and will create only a small impact on local jurisdictions.

The Intensive Supervision Proposal An important addition to the commission's revised proposal could result in prison bed savings as great as the original proposal (over 500 beds). This part of the proposal would require legislative action and involves a program now operating completely within the discretion of the Department of Corrections (DOC). Back in 1991, the legislature created the Intensive Community Supervision (ICS) program, which gives the DOC the authority to place certain inmates in the community after they have been committed to prison by the judge. The statute restricts possible candidates to those with sentences of thirty months or less and excludes those who have been convicted of certain serious violent crimes. The statute also specifies levels of supervision and requires offenders to spend the day at work or in an education or treatment program.

The Intensive Community Supervision program was established by the legislature to ease prison population pressures but, ironically, the DOC placed relatively few inmates into the program. In addition to statutory limits on eligibility, the DOC developed its own criteria that made it even more difficult for offenders to qualify. Under the commission's revised proposal, the discretion for placing offenders into the Intensive Community Supervision program would be shifted to judges. The sentencing guidelines grid would have a designated zone in which prison would be presumed but the judge could sentence an offender to ICS without departing. It is believed that giving judges discretion to place offenders in ICS will increase the program usage and preserve prison beds for violent offenders. Because DOC will operate and fund the ICS program, a financial burden will not be shifted to the local jurisdictions. Placing discretion on the front end of the system has the additional benefit of maintaining the integrity of the sentencing system and supports the goal of truth in sentencing.

The proposal is scheduled for a public hearing in December 1995, when the commission will provide an opportunity for organizations to go on record with their positions. If the commission decides to move ahead with this revised proposal, it will likely vote to adopt those changes as an exercise of its authority to propose changes, rather than only "report" them to the legislature. If the 1996 legislature does not pass a bill to stop the proposal, it will go into effect August 1, 1996. For the ICS component of the proposal to be implemented, the legislature must pass a bill to shift discretion over entry into ICS from DOC to the judiciary.

Minnesota, like most American states, has not avoided the increasing use of incarceration. However, the sentencing guidelines have helped policy makers remain informed about the impact of their decisions, and the guidelines provide policy options to help balance the use of correctional resources. If the commission's proposal passes all obstacles and is implemented next summer, it will help Minnesota move toward a greater balance. The next step will be to begin work on finding policy options for drug offenses. This will be an even greater challenge.

Oregon

Sentencing Reform in Oregon
Kathleen Bogan

In 1988, an American Correctional Association study reported forty-seven states with crowded prison systems. Corrections administrators were asked about the causes of prison crowding in their states. The number one cause, they said, was judicial sentencing practices. Building more prisons was recommended as their primary solution.

Oregon in 1989 chose to build prisons but also to control judicial sentencing by adopting sentencing guidelines for felonies. Oregon adopted "capacity-based" guidelines, in which sentencing ranges are based on the number of beds available within the state prison system. Capacity was defined to include existing and expected future capacity.

Oregon was the first state since Washington, which implemented its guidelines in 1984, to adopt presumptive capacity-linked guidelines for felonies. Oregon is one of several states, including Louisiana and Tennessee, in which guidelines based on capacity are being developed for already crowded systems. Oregon's guidelines apply to crimes committed on or after November 1, 1989.

Guidelines legislation was proposed in 1987 by the Oregon Criminal Justice Council, the statewide criminal justice policy planning agency, following a twelve-year period in which the prison population doubled from 2,000 inmates to 4,000 inmates. Through participation in a series of national conferences under the auspices of the National Prison Overcrowding Project, funded by the National Institute of Corrections and the Edna McConnell Clark Foundation, the council was exposed to the sentencing guidelines of Minnesota and Washington. The idea of sentencing guidelines as a control for prison intake was presented to the Oregon legislature as a partial solution to prison crowding. The legislature authorized guidelines development, and in 1989 it approved their implementation.

In adopting felony sentencing guidelines, Oregon was taking a logical next step. Oregon had been, in 1977, one of the first states to enact a state community corrections act. Also in 1977, Oregon was one of the first states to adopt parole guidelines based on the offender's criminal history and the severity of the offense. Since adoption of those guidelines, which increased the length of stay for the most serious felons, the average length of prison stay for these offenders increased by nearly 50 percent. The parole board nonetheless became the scapegoat of the criminal justice system, blamed for releasing convicted offenders early. Prior to the adoption of felony sentencing guidelines, offenders were generally released after serving an average of 24 percent of the imposed sentence.

Also contributing support to the council's recommendation for the guidelines was a 750-bed reduction at the state's prisons ordered in 1980 by the U.S. District Court for Oregon. This decision was vacated by the U.S. Court of Appeals for the Ninth Circuit, but this close call, and the fact that eighteen of Oregon's thirty-three county jails were under federal court orders, increased pressure to do something about rising populations.

The primary goal of Oregon's guidelines was not to end disparity in sentencing, for the parole guidelines had already achieved a great deal in equalizing prison terms for similar offenders. The goals of the felony guidelines were to treat like offenders alike in sentencing, but also to provide increased protection for the public by using the prison space for the most serious and dangerous offenders. The enabling legislation required consideration of "the effective capacity of state and local corrections facilities and other sentencing sanctions available." The council took seriously its responsibility to consider the impact of the guidelines on state prisons, as well as on county jails and field supervision caseloads.

The Oregon guidelines both resemble and differ from the capacity-based guidelines in Minnesota and Washington. Like those states', Oregon's are expressed in a two-dimensional grid, one axis of which is a crime seriousness scale and one a criminal history scale. Oregon, however, made refinements in both of these scales.

In ranking crime seriousness, Oregon subcategorized broadly defined criminal code offenses that encompassed a wide array of conduct. For instance, Burglary I can include behaviors ranging from entries into unoccupied dwellings with the intent to commit a crime to confrontations with occupants involving dangerous weapons. Oregon's guidelines distinguish among different kinds of behavior by sub-categorizing the offense at various levels on the crime seriousness scale. Accompanying legislation requires that facts that differentiate a subcategory of a crime be pleaded and proven at trial.

Oregon made a significant contribution to the evolution of sentencing guide-lines in design of its crime seriousness scale. The council established principles for ranking crime seriousness, an approach taken in neither Washington nor Minnesota, but now being considered in these and other states. Use of ranking princi-ples allowed for objective decisions about offense ranking and provided a way to estimate the ranking, and thus the potential system impact, of any new crime under legislative consideration. Sentencing commissions in Louisiana and Kansas have since adopted the "principled" approach to crime seriousness rankings.

Advances in the construction of a criminal history scale may be Oregon's most significant contribution to the development of sentencing guidelines. Most such scales are based on the number of prior convictions, without consideration of their seriousness or type. Oregon's goal was to develop a scale that is simpler to operate and reduces disputes at sentencing over the precise number of prior convictions. Oregon also wanted a scale that would be sensitive to the seriousness of prior offenses and to "patterning" in an offender's criminal history.

The Oregon scale distinguishes between prior person and nonperson felonies, and between misdemeanor and felony convictions. Each level of the criminal his-tory scale has a threshold number; once an offender has the threshold number of convictions for a given category, additional convictions do not increase the punish-ment until the threshold of a new level is reached. The thresholds in Oregon are quite low: an offender is at the most serious criminal history category with three or more prior felony convictions for crimes against persons. (Only 4 percent of Ore-gon felons fall into this category.)

Oregon's felony guidelines allow judges to impose sentences other than the guidelines sentence by means of a "departure," which requires substantial and compelling reasons in justification. Oregon's guidelines also provide a probation option for many offenders convicted of fairly serious crimes but who have modest criminal records. For these offenders, a probationary sentence may be imposed in lieu of prison if the judge finds that a community program will be more effective than prison at preventing recidivism by the offender and that space in the program is immediately available. The option may not be used if the offender was on super-vision status or armed with a firearm when the current offense was committed. Ini-tial data indicate that 63 percent of the eligible offenders are receiving probation.

Oregon's new felony sentencing guidelines resemble Minnesota's and Washing-ton's in the operation of presumptive sentences for offenders sent to prison. In Ore-gon, the development of a structured system for nonincarcerative sentences has been both the most promising and the most challenging aspect of the guidelines. Oregon's guidelines include a "sanction-unit" system of structuring the length of probationary terms and the use of jail and other custodial sanctions for the 80 per-cent of felony offenders who receive probation. The sanction units provide a medium for imposing different sanctions of comparable severity.

This aspect of the guidelines was greeted with enthusiasm by local county commissioners and sheriffs, who anticipated a reduction in the use of local jails. Community corrections officials were generally enthusiastic about the use of alternative sanctions as equivalencies for jail. Corrections officials were initially concerned, and continue to be concerned, about the record keeping required to track the actual sanction units served. Judges, to whom the idea of imposing sanction units at the time of sentencing is completely new, have had the most difficulty with the new system.

For each offender for whom the presumptive sentence is probationary, the sanction-unit system allows imposition of probation ranging from eighteen to thirty-six months, jail ranging from thirty to ninety days, or the use of sanction units (which can include jail) of from 90 to 180 units, based on the seriousness of the crime and the offender's criminal history. The guidelines set out equivalencies for sanctioning units. A day in a restitution center, a day in residential drug or alcohol treatment, or twenty-four hours of supervised work all satisfy one sanction unit. Local community corrections officers administer these units. Judges can sentence the offender to specific programs or impose a number of units that can be administered by the probation officer.

As part of the guidelines legislation, the Oregon legislature required the Department of Corrections (DOC) to prepare a directory of community programs for offenders. This directory is now available for use by judges, corrections officials, and others in structuring sanctioning options.

The effort devoted to structuring alternative punishments for felony offenders and to attempting to fit prison sentences to available space was complicated by the passage in November 1990 of a statewide property tax limitation. This vastly reduces the amount of money available to DOC and will cause significant reductions in funding for prison construction, for community corrections programs, and for field supervision staff. It remains to be seen how much of the existing system can be preserved and what shifts of offenders will occur if Oregon backs away simultaneously from prison building and community supervision.

Following the advice of other sentencing guidelines states, Oregon decided to develop misdemeanor guidelines as a next step in structuring proportionate sentences and managing correction resources effectively. The council was authorized in 1989 to develop guidelines for misdemeanors; legislation to implement these guidelines is being presented to the 1991 legislature.

Under Oregon's felony guidelines, many Class C felonies have a six-month prison term as a sanction, yet existing state law allows up to a year in local jail for misdemeanors. The 1989 legislature was receptive to the proportionality argument; during the two years while misdemeanor guidelines were being developed, a statutory six-month jail cap for most misdemeanor offenses was in effect. The council has asked that the cap be continued, and it is proposing to structure the use of jail, fines, community service and other sanctions for the 400 misdemeanors in the criminal code.

The council plans to work over the next two years to develop a statewide day-fine system of means-based fines, while at the same time working to improve the state's fine collection system.

Oregon Guidelines, 1989–1994
Kathleen Bogan and David Factor

Oregon in 1989 was the first state to develop felony sentencing guidelines in the context of an already overcrowded state prison system. Other states—Kansas, North Carolina, and Arkansas are recent examples—have now embarked on the same journey. The fifth anniversary of guidelines adoption has passed, and a look at past goals and present realities shows considerable success in achieving the initial policy goals, operating the state corrections system within capacity limits, and increasing the efficiency and effectiveness of the justice system.

The Oregon guidelines have four principal policy goals:

> *to provide proportional punishments* based on sanctions that become more severe as the seriousness of crimes increases and as defendants' criminal histories become more extensive;
>
> *to provide truth in sentencing* so that offenders serve the imposed prison terms rather than, as in the preguidelines system, a fraction of the sentence imposed;
>
> *to provide sentence uniformity and reduced disparity in sentencing* by using narrow sentencing ranges, so that like offenders receive similar sentences; and
>
> *to maintain a sentencing policy consistent with correctional capacity.*

Oregon's guidelines set presumptive sentences for all felonies based on crime seriousness and offenders' criminal histories, but in important ways they differed from guidelines in other states. Oregon used a simpler criminal history scale and followed carefully crafted principles for crime seriousness ranking. In addition, Oregon developed a "custody-unit" system for structuring use of jail and other local correctional sanctions as part of probation sentences.

Judges may depart from the presumptive sentences for "substantial and compelling" reasons stated on the record, and they are free to impose up to twice the presumptive sentence if aggravating factors are found. In mitigation, the judge may reduce the sentence to discharge. Some offenders subject to presumptive prison sentences may be eligible for probation if certain conditions are met. The guidelines also set presumptive terms of probation and postprison supervision.

Achieving Original Goals   During the guidelines' first five years, substantial progress was made toward achieving the four principal goals.

Truth in Sentencing. Before adoption of sentencing guidelines, imprisoned felons in Oregon served an average of 24 percent of the judicial sentence imposed. By contrast, guidelines offenders serve an average of 83 percent of their judicially imposed sentences, with the maximum potential of earning up to a 20 percent credit for participation in prison work or education programs.

Proportionality. Judges follow the guidelines and impose presumptive sentences in 87.5 percent of the cases, and they depart in 12.5 percent. The 12.5 percent departure figure for 1993 is higher than the 10 percent departure rate estimated in 1989. The rate has crept up slowly, from 6 percent in 1990 to 9 percent in 1991, as judges become more familiar with the guidelines and more serious cases reach sentencing.

Uniformity. One of the most important goals of the guidelines was to limit disparity in sentencing based on race or gender. This subject has been tracked closely by the research staff of the Oregon Criminal Justice Council. The staff calculated that dispositional variability for offenders with identical crime seriousness and criminal history scores was reduced by 45 percent in 1990, the first full year of guidelines operation.

Data for subsequent years show that disparity remains most pronounced in areas where judges retain most discretion, such as in decisions about departures and use of sentencing options. When significant disparity is found, Hispanic offenders tend to receive more severe sentences than African-American offenders, who receive more severe sentences than Caucasians. When there is a significant gender difference, male offenders tend to receive more serious sentences than similarly situated females. These findings are consistent with other states' experiences.

Percentages of women and minorities among sentenced felons have risen since the implementation of the guidelines. Women, who made up 14 percent of the sentenced felon population in 1986, were 16 percent of total sentenced felons in 1993. Minorities, 8 percent of Oregon's population, made up 13 percent of sentenced felons in 1986, and in 1993 made up 27 percent.

Effect on Correctional Capacity. One aim in adopting felony sentencing guidelines was to operate an expanded prison system in which the population remained within 97 percent of capacity. At the time guidelines were adopted by the legislature in 1989, prison capacity was expected to increase from 5,534 beds to 7,894 beds by 1999. Instead, because of a 1990 property tax limitation measure adopted by state voters, prison construction stopped at 6,874 beds. The legislature has funded operation of 6,625 of these beds, though the prison population was 6,844 on December 1, 1994.

Prison population forecasts and policy simulations rely on use of a Criminal Justice Council data base, which contains sentencing data received from the courts on approximately 13,500 felony cases per year. Forecasts based on these data have been accurate within 1 percent over the last forty months due largely to the predictability brought about by sentencing guidelines.

Designed for a prison construction plan that was never completed, the guidelines data base has provided the mechanism to model various policy options to account for unbuilt prison beds. Rather than disturb the structure of the guidelines, legislators have modified probation revocation procedures and authorized use of a boot camp program for certain prison-bound offenders.

Passage in 1994 of a mandatory minimum sentencing measure by voter initiative will destabilize the system. Until charging and plea practices change and a new equilibrium is reached, the accuracy of population projections can be expected to suffer.

Accuracy of Initial Projections   Researchers for the Oregon Criminal Justice Council, using a Structured Sentencing Simulation Model developed with a grant from the U.S. Bureau of Justice Assistance, developed impact projections for the original guidelines. In 1989, council staff made a number of predictions concerning sentencing and imprisonment patterns under the guidelines by 1994.

Table 2.3. Imprisonment Sentencing Patterns in Oregon, 1993

Offense	Preguidelines Rate (%)	Projected Rate (%)	Actual (%)
All Person	34	73	45
Drug Distribution	9	48	24
Rape/Sodomy	40	82	56
Assault	29	60	37
Homicide	62	81	80
Robbery	50	59	50

Source: Oregon Criminal Justice Council (unpublished data provided to authors).

The overall imprisonment rate would rise, from 18 percent of felons to 22 percent, as more serious felons received prison sentences. The actual imprisonment rate in 1993 rose to 20 percent.

Imprisonment rates for specific offenses would change, with the rate of imprisonment for person offenses more than doubling, from 34 percent to 73 percent, and the rate for distribution of controlled substances rising, from 9 percent to 48 percent. Anticipated and actual 1993 sentencing patterns are shown in table 2.3.

Prison terms for specific offenses would increase. Changes in length of stay anticipated and the actual 1993 sentencing patterns are shown in table 2.4.

Imprisonment rates are lower than expected, in part because of judges' use of departure sentences. Judges used dispositional departures to impose probation instead of presumptive prison sentences in 15.2 percent of felony cases in 1993. The highest prison-to-probation rate was for sodomy, where 26 percent of offenders received a departure sentence to probation.

Prison terms for many offenders sent to prison, however, are longer than anticipated. This may be accounted for by several factors, among them the use of consecutive sentencing and upward durational departures. Departures to longer sentences occurred in 10 percent of cases involving person offenses in 1993.

Table 2.4. Sentencing Patterns in Oregon, 1993

Offense	Preguidelines Average (in months)	Projected Guidelines (in months)	Actual (in months)
Homicide	34	68	84
Rape/Sodomy	40	49	73
Sex Abuse	22	22	28
Drug Possession	9	19	6
Drug Distribution	10	22	14

Source: Oregon Criminal Justice Council (unpublished data provided to authors).

Beneficial Effect on Courts   Two beneficial aspects of the guidelines have been reductions in the number of trials and in time spent in jail awaiting sentencing. Oregon's trial rate prior to sentencing guidelines was 11 percent. Rather than rising under guidelines, as opponents feared, the trial rate fell to 8 percent of cases in 1993. Before guidelines, the average time from verdict to sentencing was forty-eight days; by 1993 it had dropped to twenty-three days. More than twice as many offenders are now sentenced on the same day as conviction—57 percent compared with 24 percent before guidelines. This has reduced demand on local jails holding felons awaiting sentencing. Both results were anticipated by Criminal Justice Council researchers.

Reduced Pressure on Local Sanctions   The impact of guidelines on local jails was a major issue in the adoption of the guidelines. Sheriffs were concerned that guidelines would increase the number of felons serving local jail sentences. The council anticipated a decrease because of increased rates of imprisonment and limitations on use of jail as part of probation sentences. A jail monitoring system implemented by the council shows the result expected: there was a decline of 17 percent on any given day in the number of felons in local jails serving a sentence.

Before guidelines, judges could impose a term of probation up to five years for felony offenses; the average duration of probation sentences was forty-three months. In 1993 the average probation sentence was twenty-six months. Probation sentence lengths have been shortened across all classes of offenders, most notably for non-person offenders.

Attitudes of Bench and Bar   In a survey by the legislature's Senate Judiciary Committee one year after implementation of sentencing guidelines, 53 percent of judges, prosecutors, and defense attorneys indicated that guidelines were an improvement over the prior indeterminate sentencing system; 90 percent agreed with the goals of fairness, proportionality, and truth in sentencing. Seventy percent rated the guidelines as successful at achieving the stated goals. Only 37 percent preferred the prior system.

Appellate Review   When adoption of felony sentencing guidelines was being debated in the 1989 legislative session, one concern was that guidelines appeals would overburden an already overworked court of appeals. This has not happened.

During calendar year 1990, 4,593 notices of appeal were filed. Approximately 50 percent were criminal law cases. The first notice of appeal involving a guidelines issue was filed in June 1990, some eight months after implementation. Only 10 to 15 percent of criminal appeals involved guidelines issues; most of these also included other substantive issues and would have been in the court of appeals in any event. A tiny percentage exclusively involved guidelines issues.

The enabling legislation placed limits on the ability to appeal a guidelines sentence. Sentences may not be appealed if they are consistent with the presumptive guidelines range, were imposed pursuant to an accepted plea agreement, or were imposed under rules governing optional probation. In addition, sentencing courts retain jurisdiction for sixty days to correct arithmetic and clerical errors, further reducing appeals.

The only statutory grounds for appeal of a guidelines sentence, available to both the state and defense, are failure to comply with the guidelines rules, an error in crime seriousness ranking or criminal history classification, or imposition of a departure sentence not supported by "substantial and compelling" reasons.

Today the number of appeals concerning guidelines issues is negligible. Of these, most concern application of the rules in individual cases, rather than legal issues concerning the validity or meaning of the rules. Most substantive guidelines issues have been resolved.

Pennsylvania

The Evolution of Pennsylvania's Sentencing Guidelines
John Kramer

Pennsylvania's sentencing guidelines took effect in 1982. Developed at a time when Pennsylvania had excess prison capacity, they were intended to reduce sentencing disparities and perceived "leniency." Those goals were met. Disparities were reduced, and in the early 1980s sentencing became more severe. More recently, other developments, including preeminently the war on drugs, have made sentencing more severe still and have produced overcrowding in both prisons and jails. In its second decade, Pennsylvania's sentencing commission is working to reconcile sentencing guidelines with correctional resources and to incorporate intermediate sanctions into sentencing policy.

Pennsylvania's guidelines differ in important respects from other well-known guidelines systems. First, the guidelines create a presumptive sentencing system but *not* a determinate sentencing system like those in Minnesota, Oregon, Washington, and the U.S. federal system. Parole release was not eliminated, and the guidelines set presumptions for minimum sentences only. The parole board need not release prisoners when the minimum is completed, and judges may impose any lawful maximum sentence.

Second, Pennsylvania's guidelines provide three separate ranges of presumptive sentences for any individual case—a standard range, a mitigated range, and an aggravated range (most guidelines systems have only a single range). Figure 2.9 shows the ranges in the current grid for nondrug crimes. The judge has unreviewable discretion to decide which range to apply.

Third, judicial compliance with the guidelines is less aggressively monitored by appellate courts than in some jurisdictions. Although Pennsylvania statutes create a right of sentence appeal, appellate review has tended to focus on compliance with procedural rules rather than on the adequacy of reasons for departing from guidelines. Pennsylvania courts have applied a "manifest abuse of discretion" standard to substantive review of departures.

Origins The Pennsylvania General Assembly created the Commission on Sentencing in 1978, culminating several years of debate. Legislation establishing sentencing councils, which required a judge to obtain the sentencing recommendations of two other judges whenever he or she considered imposing a maximum

Figure 2.9. Current Pennsylvania Guidelines Sentence Ranges.

Each cell lists (top to bottom): Aggravated range / Standard range / Mitigated range.

Offense Gravity Score #	Offense	0	1	2	3	4	5	6
10	Third-degree murder	120 48 - 120 36 - 48	120 54 - 120 40 - 54	120 60 - 120 45 - 60	120 72 - 120 54 - 72	120 84 - 120 63 - 84	120 96 - 120 72 - 96	120 102 - 120 76 - 102
9	For example: Rape; Robbery inflicting serious bodily injury	60 - 75 36 - 60 27 - 36	66 - 82 42 - 66 31 - 42	72 - 90 48 - 72 36 - 48	78 - 97 54 - 78 40 - 54	84 - 105 66 - 84 49 - 66	90 - 112 72 - 90 54 - 72	102 - 120 78 - 102 58 - 78
8	For example: Kidnapping; Arson (Felony I); Voluntary manslaughter	48 - 60 24 - 48* 18 - 24*	54 - 68 30 - 54 22 - 30*	60 - 75 36 - 60 27 - 36	66 - 82 42 - 66 32 - 42	72 - 90 54 - 72 40 - 54	78 - 98 60 - 78 45 - 60	90 - 112 66 - 90 50 - 66
7	For example: Robbery threatening serious bodily injury	12 - 18* 8 - 12* 4 - 8	29 - 36 12 - 29* 9 - 12*	34 - 42 17 - 34* 12 - 17*	39 - 49 22 - 39* 16 - 22*	49 - 61 33 - 49 25 - 33	54 - 68 38 - 54 28 - 38	64 - 80 43 - 64 32 - 43
6	For example: Robbery inflicting bodily injury; Theft by extortion (Felony III)	12 - 18* 4 - 12* 2 - 4	12 - 18* 6 - 12* 3 - 6	12 - 18* 8 - 12* 4 - 8	29 - 36 12 - 29* 9 - 12*	34 - 42 23 - 34* 17 - 23*	44 - 55 28 - 44 21 - 28*	49 - 61 33 - 49 25 - 33
5	For example: Criminal mischief (Felony III); Theft by unlawful taking (Felony III); Theft by receiving stolen property (Felony III); Bribery	11 1/2 - 18* 0 - 11 1/2 Nonconfinement	11 1/2 - 18* 3 - 11 1/2 IP - 3	11 1/2 - 18* 5 - 11 1/2 IP - 5	11 1/2 - 18* 8 - 11 1/2 4 - 8	27 - 34 18 - 27* 14 - 18*	30 - 38 21 - 30* 16 - 21*	36 - 45 24 - 36* 18 - 24*
4	For example: Theft by receiving stolen property, less than $2,000, by force or threat of force, or in breach of fiduciary obligation	11 1/2 - 18* 0 - 11 1/2 Nonconfinement	11 1/2 - 18* 0 - 11 1/2 Nonconfinement	11 1/2 - 18* 0 - 11 1/2 Nonconfinement	11 1/2 - 18* 5 - 11 1/2 IP - 5	11 1/2 - 18* 8 - 11 1/2 4 - 8	27 - 34 18 - 27* 14 - 18*	30 - 38 21 - 30* 16 - 21*
3	Most misdemeanor I's	6 - 12* 0 - 6 Nonconfinement	11 1/2 - 18* 0 - 11 1/2 Nonconfinement	11 1/2 - 18* 0 - 11 1/2 Nonconfinement	11 1/2 - 18* 0 - 11 1/2 Nonconfinement	11 1/2 - 18* 3 - 11 1/2 IP - 3	11 1/2 - 18* 5 - 11 1/2 IP - 5	11 1/2 - 18* 8 - 11 1/2 4 - 8
2	Most misdemeanor II's	IP - 6 0 - IP Nonconfinement	3 - 6 0 - 3 Nonconfinement	11 1/2 - 12* 0 - 11 1/2 Nonconfinement	11 1/2 - 12* 0 - 11 1/2 Nonconfinement	11 1/2 - 12* 0 - 11 1/2 Nonconfinement	11 1/2 - 12* 2 - 11 1/2 IP - 2	11 1/2 - 12* 5 - 11 1/2 IP - 5
1	Most misdemeanor III's	IP - 3 0 - IP Nonconfinement	3 - 6 0 - 3 Nonconfinement	6 0 - 6 Nonconfinement	6 0 - 6 Nonconfinement	6 0 - 6 Nonconfinement	6 0 - 6 Nonconfinement	6 0 - 6 Nonconfinement

Notes: * Indicates eligibility for boot camp programs. # Offenses are listed for illustrative purposes only. IP = Intermediate punishments. There is a weapon enhancement of at least twelve months and up to twenty-four months of confinement to be added to sentence lengths when the offense involves a deadly weapon. All of the guideline's sentence ranges are months of minimum confinement as defined in 42 Pa. C.S.§9756(b) (relating to partial and total confinement). *Source:* Pennsylvania Commission on Sentencing (1991).

sentence of seven years or longer, was enacted in 1975; it was soon ruled unconstitutional by the Pennsylvania Supreme Court. Within a year, the Pennsylvania senate passed mandatory sentencing legislation and the house came within four votes of passing it. The obstacle to passage appeared to be an impact assessment that projected a need for additional prison space costing between $28 and $93 million.

A task force recommended creation of a sentencing commission and adoption of sentencing guidelines; legislation to create America's second such commission soon passed. Prison overcrowding and correctional resources were not among the issues the commission was directed to address. Pennsylvania then had a prison population of 8,000 inmates and a capacity of 10,000. The overriding concerns were disparity and leniency, with the primary issue being leniency.

Guidelines Development Pennsylvania did not establish determinate sentencing. The legislature kept the indeterminate sentencing system in which judges set minimum and maximum sentences, but a parole board decides when prisoners are released. The parole board generally released about 80 percent of prisoners when they first became eligible, and the average prisoner served 110 percent of the minimum sentence, making the minimum the actual sentence for most prisoners. The commission accordingly decided to develop guidelines for minimum sentences.

The commission, composed of four judges, four legislators, and three gubernatorial appointees, began work in 1979. To inform its work and to be able to assess the potential impact of guidelines, it undertook an extensive study of past sentencing practices. Guidelines development, however, began before data were available. The commission decided not to develop descriptive guidelines based on past practices but prescriptive guidelines that reflected the commission's policy choices and its legislative mandate. Legislative members of the commission stressed that the guidelines were expected to make sentencing more severe.

Rejection The commission submitted its proposed guidelines to the legislature in January 1981. They specified relatively long incarcerative sentences for violent offenders, particularly those with an extensive prior record. They also called for relatively short but more certain incarcerative sentences for offenders convicted of moderately severe offenses and for nonconfinement for relatively minor misdemeanors.

Seventy-five people testified at public hearings. Most were critical. The major complaints were that the guidelines constrained judicial discretion too much, were too lenient, and were too complicated. The legislative reaction was generally negative. By concurrent resolution both houses rejected the guidelines and directed the commission to revise them to increase both judicial discretion and sentence severity.

Revisions The legislative guidance was clear; what was not clear was how much discretion to provide and how much to increase severity. A few examples show how the commission responded. To increase discretion, the commission widened the guidelines ranges (for more serious crimes the ranges were widened from twelve to twenty-four months), deleted a list of proposed aggravating and mitigating considerations and left that subject entirely in sentencing judges' hands, and recommended

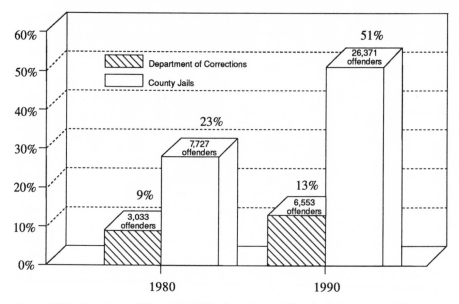

Figure 2.10. Percentage of Convicted Offenders Sentenced to the Department of Corrections and County Jails, 1980 and 1990. *Source:* Pennsylvania Commission on Crime and Delinquency (unpublished data provided to author).

significant jail sentences for offenders for whom the initial guidelines specified nonconfinement. The revised guidelines also increased severity by adding twelve to twenty-four months to any guidelines recommendation when the offense involved possession of a deadly weapon. Prior convictions also played a weightier role in determining the presumptive sentence. The tougher guidelines were readily accepted by the legislature in January 1982.

Overcrowding The guidelines were expected to raise prison populations by approximately 3,000 inmates. Prison populations did rise by 3,000 in the early 1980s; they then continued to rise to a 1991 level of 23,405 inmates. As figure 2.10 indicates, incarceration became much more common. In 1980, only 32 percent of offenders received incarceration; by 1990, that had risen to 64 percent. Other contributors to population growth were increases in convictions, decreases in parole releases, and increases in parole revocations.

The number of convictions grew rapidly between 1985 and 1990. As table 2.5 indicates, the growth was particularly acute for drug offenses (up 274 percent) and property offenses (up 149 percent). Table 2.5 also shows that convictions increased much more than did the number of cases processed and that incarcerative sentences increased much more than did convictions, particularly for drug offenders.

The parole board was the second source of growth. The percentage of the minimum sentence served increased from 111.5 percent in 1980 to 119.2 percent in 1990. The percentage of admissions resulting from parole revocations grew from 20

Table 2.5. Offender Processing by Offense Type, 1980 versus 1990

Offense Type	Processed			Convicted			Incarcerated		
	1980	1990	Change (%)	1980	1990	Change (%)	1980	1990	Change (%)
Violent	10,200	8,203	−20	3,685	3,750	2	2,096	3,135	50
Property	11,126	19,111	72	4,716	11,732	149	2,016	6,471	221
Drugs (total)	6,616	14,030	112	2,705	10,127	274	584	6,169	956
Sales	5,214	5,611	8	2,120	4,495	112	405	3,527	771
Possession	1,402	8,419	500	585	5,632	863	179	2,642	1,376
Other	60,902	49,687	−18	22,089	26,561	20	6,064	17,149	183
Totals	88,844	91,031	2	33,195	52,170	57	10,760	32,924	206

Source: Pennsylvania Commission on Crime and Delinquency (unpublished data provided by authors).

to 26 percent. In addition, the number of parole violators in state prisons rose 191.8 percent over the same period. (Caution must be exercised in interpreting these data as they may reflect changes in the standards of the board, or in the types of offenders and misbehavior of parolees, or both.)

Prison Riots at Camp Hill   In October 1989, the State Correctional Institution at Camp Hill experienced two riots, heightening the salience of prisons as a public policy issue. Two Republican members of the House Judiciary Committee wrote the commission, asking that the commission make recommendations regarding how the state could deal with the crisis in the state correctional system. The commission decided that overcrowding was a major contributor to the crisis and that the guidelines should be modified to help resolve overcrowding.

The commission considered changes that would reduce prison admissions. That, however, would shift overcrowding to the county jails and probation unless other simultaneous changes were made at the county level. The commission proposed two changes. The first was that guidelines ranges specifying state confinement for nonviolent offenders be lowered to call for county incarceration. Commission staff estimated that this would divert 750 offenders from state confinement.

The second change would have established presumptive nonincarcerative sentences for many offenders who formerly received short incarcerative sentences. The commission initially identified six cells in the guidelines grid to establish guidelines ranges that called for nonconfinement. This was estimated to reduce annual county jail admissions by approximately 3,600 and average daily population by 1,000 inmates.

Intermediate Punishment Act   Legislation was introduced to stimulate creation of community corrections programs across the state. In November 1990, the General Assembly passed the Intermediate Punishment Act, which included a provision that the commission identify offenders eligible for intermediate punishments.

The community corrections legislation set the stage for submission of the commission's proposals. Two arguments, however, convinced the commission to retreat from its decision to call for nonconfinement in six cells. First, the County Commissioner's Association argued that the counties needed time to develop intermediate punishment programs, a process hindered by the legislature's failure to provide financial incentives to the counties. Second, legislative staff argued that so drastic a change would hurt the commission's credibility. As a result, the commission reduced from six to two the cells limited to nonconfinement sentencing options. The revised proposals met no resistance in the legislature and took effect on August 9, 1991.

Current Activities   The commission's concern with correctional resources constitutes a major policy shift from its stance in the early 1980s that correctional resources should not shape sentencing policy. The commission is reviewing the guidelines and sentencing standards in light of correctional resources. The General Assembly has authorized the expansion of the state prison system to 23,800 beds, and the commission is considering how those beds should be used, with particular focus on reserving state prison space for violent offenders. More important, the commission is considering ways the guidelines can encourage the use of intermediate punishments for nonviolent offenders. During 1993 the commission expects to submit revised guidelines to the legislature that will fulfill these goals.

For many reasons, only one of which was the guidelines, prison populations have continued to rise and the policy agenda has begun to focus on optimal use of scarce prison capacity. Establishing public policies that are rational, fair, and just requires consideration of victims' injuries, offenders' culpability, and fair treatment of offenders. Determination of what is feasible is constrained by the availability of correctional resources.

There is still strong sentiment in Pennsylvania for "getting tough on crime" even though that goal was met in the 1980s. The Pennsylvania General Assembly funded the growth in correctional capacity to carry out those policies. Allocation of resources to support those policies has limited the legislature's ability to commit resources to improved education, enhanced health care, and reasonable taxes. The state must, and will, reexamine its sentencing and correctional policies during the 1990s if it is to reconcile the diverse needs of its citizens.

Pennsylvania's Sentencing Guidelines—The Process
of Assessment and Revision
   John Kramer and Cynthia Kempinen

Substantially revised sentencing guidelines took effect in Pennsylvania in August 1994. The new guidelines differ in major respects from their predecessors. They are premised on the "capacity constraint" notion that projected operation of the guidelines should be reconcilable with current and projected corrections resources. As a result, while prison use for violent offenders will grow, prison use for property and drug offenders will fall. Vastly greater use of intermediate and community sanctions has been embodied in guidelines recommendations, and steps have been taken to increase state support for the county programs that will serve those offenders. The guidelines grid has expanded, and the widths of presumptively appropriate sentencing ranges have been narrowed.

The Pre-1994 Guidelines In 1978 the Pennsylvania legislature created the sentencing commission to write sentencing guidelines to correct perceived problems of disparity and leniency. The first set of guidelines was rejected by the legislature in 1981 because they were viewed as too lenient and too restrictive of judicial discretion. The commission widened guidelines ranges, thus allowing for harsher sentences and greater discretion, and the legislature adopted the revised guidelines to take effect July 22, 1982. Since then the guidelines have been amended often. Most notable have been increases in drug offenders' sentences (1988) in response to legislative requests and identification of offenders appropriate for nonincarcerative sentencing options and a boot camp program (1990). These piecemeal amendments undermined the initial integrity of the guidelines, setting the stage for a comprehensive reassessment.

Reasons for the Reassessment In June 1989, commission staff proposed that the commission reevaluate the guidelines for four primary reasons. First, monitoring data showed that there were "problem areas" in the guidelines. One was a high mitigated departure rate for some serious offenses. The question was whether the departures represented inappropriately lenient sentences or inappropriately harsh guidelines recommendations. If the sentences were unjustified, then the departures represented a failure to achieve conformity with the guidelines. If the guidelines were inappropriate, then the commission needed to revise them.

Second, there was concern that changes in guidelines for drug offenses had resulted in undesirable disproportionality between sentences for drug offenses and for other offenses. Third, many other states had adopted guidelines since 1982, and the commission wanted to examine different approaches taken elsewhere and consider whether some of them should be adopted in Pennsylvania. Fourth, growth in prison populations during the 1980s had created a crowding problem, and the relevance of cost considerations in sentencing policy making became more pronounced.

Conducting the Reassessment The commission undertook four projects. These were a statewide survey of people who used the guidelines, a sentencing study in ten counties that involved interviewing key actors and observing the court process, visits to three states (Oregon, Washington, and Louisiana) that had taken innovative approaches to sentencing, and an analysis of current sentencing practices using the commission's monitoring system.

Statewide Survey. The statewide survey was intended to evaluate the current guidelines and to seek suggestions for change. Respondents were asked, for example, to rank sixty-seven of the most serious or most frequently committed offenses on a scale of seriousness, to assess whether the sentence recommendations for various types of offenses were appropriate, and to recommend whether various factors currently not considered by the guidelines, such as probation and parole status, should be incorporated.

The questionnaire was sent to all 366 trial court judges, all sixty-seven district attorneys, chief public defenders, and chief probation officers. It was also sent to a sample of 400 people representing assistant district attorneys, assistant public defenders, line probation officers, and private defense counsel.

The responses generally supported the guidelines and proved helpful in identifying specific problems. A majority of respondents thought that the guidelines were helpful in determining appropriate sentences and had increased fairness. Practitioners responded that, for the most part, factors determining the guidelines recommendations (e.g., calculation of the prior record score) were appropriate.

One issue before the commission was whether the guidelines' ten levels of offense seriousness adequately captured the diversity of offenses. The survey asked respondents to rank sixty-seven offenses on a fifteen-point scale. The results provided a significant benchmark for the commission for ranking offenses and comparing survey responses with current sentencing practices. The revised guidelines grid increased the number of offense severity levels to thirteen (see figure 2.11).

Interviews and Observations. Commission staff visited ten of Pennsylvania's sixty-seven counties. The visits included one-and-a-half to two-hour structured interviews with judges and others and observations of court proceedings to see how the guidelines were being used. The aims were to identify the process of guidelines application and variations in different counties, to obtain feedback from court practitioners concerning how the guidelines are operating, noting inefficiencies and practices that undermine fair sentencing, to receive suggestions for changes and improvements, and to investigate plea bargaining practices and how they affect guidelines application.

The ten counties represented various geographical regions and population sizes. Some were selected because they had high departure rates, thus enabling the commission to examine more closely the reasons for departures. The visits proved invaluable in obtaining a better understanding of how the guidelines were applied and how they could be improved.

The interviews substantiated questionnaire findings concerning the usefulness of the guidelines. It was particularly encouraging when newly selected judges indicated that the guidelines provided a good benchmark; one judge called them his "Sentencing Bible." Rural judges, who often do not have judicial colleagues with whom to discuss cases, said they took comfort in knowing that a standard existed for sentencing.

General Themes Emerged. A common suggestion was that guidelines recommendations of prison for certain drug and retail theft offenders be changed. This recommendation became a major part of revisions oriented toward reserving prison space for violent offenders.

Sentencing Data. Rich data involving current sentencing were a major source of information that was not available when the initial guidelines were developed. For the initial guidelines, commission staff collected information on a sample of cases sentenced during the late 1970s. In the 1990s, information on every felony and misdemeanor conviction in the state was available. Thus, the commission could examine current sentencing practices for specific offenses. The data also enabled staff to estimate the effects of proposed changes on the correctional system.

Information from Other States. After Pennsylvania's initial guidelines went into

KEY:

AGG = aggravated sentence addition
INCAR = incarceration
MIT = mitigated sentence subtraction
RFEL = repeat felony I and felony II offender category
RIP = restrictive intermediate punishments
RS = restorative sanctions
11-1/2 = denotes county sentence of less than 12 months

NOTES:

1. When the offender meets the statutory criteria for boot camp participation, the court should consider authorizing the offender as eligible.

2. Levels 1, 2, and 3 of the matrix indicate restrictive intermediate punishments may be imposed as a substitute for incarceration.

3. When restrictive intermediate punishments are appropriate, the duration of the restrictive intermediate punishment program shall not exceed the guideline ranges.

4. When the range is RS through a number of months (e.g., RS-6), RIP may be appropriate.

5. When RIP is the upper limit of the sentence recommendation (e.g., RS-RIP), the length of the restrictive intermediate punishment programs shall not exceed 30 days.

Level	Offense Gravity Score	Prior Record Score							AGG/ MIT
		0	1	2	3	4	5	RFEL	
Level 4 Incar	13	60-120	66-120	72-120	78-120	84-120	90-120	96-120	– 12
	12	54-72	57-75	60-78	66-84	72-90	78-96	84-102	– 12
	11	42-60	45-63	48-66	54-72	60-78	66-84	72-96	– 12
	10	30-48	33-51	36-54	42-60	48-66	54-72	60-84	– 12
Level 3 Incar Cnty Jail/ RIP	9	8-20	12-27	15-30	21-36	27-42	33-48	39-60	– 6
	8	6-18	9-21	12-24	18-30	24-36	30-42	36-48	– 6
	7	4-12	7-15	10-18	16-24	22-30	28-36	34-42	– 6
	6	3-9	6-11½	9-15	12-18	15-21	18-24	21-27	– 3
Level 2 Incar RIP RS	5	RS-6	1-6	3-9	6-11½	9-15	12-18	15-21	– 3
	4	RS-3	RS-6	RS-9	3-9	6-11½	9-15	12-18	– 3
	3	RS-RIP	RS-3	RS-6	RS-9	3-9	6-11½	9-15	– 3
Level 1 RS	2	RS	RS	RS-RIP	RS-3	RS-6	1-6	3-9	– 3
	1	RS	RS	RS-RIP	RS-RIP	RS-3	RS-6	RS-6	– 3

Figure 2.11. Pennsylvania Guidelines, Standard Ranges (August 12, 1994). *Source:* Pennsylvania Commission on Sentencing (1995).

effect, many states developed guidelines, providing a rich source of ideas and showing that innovations can be developed and implemented. Thus in the late 1980s, new ideas and experiences elsewhere could be drawn upon.

In 1992 a grant was awarded by the Bureau of Justice Assistance (BJA) to the National Council on Crime and Delinquency, the Pennsylvania Commission on

Crime and Delinquency, and the Pennsylvania Commission on Sentencing to conduct a nationwide study of structured sentencing. This project offered commission staff the opportunity to take an in-depth look at how other states developed guidelines. Commission staff visited many states to conduct interviews. These interviews gave the Pennsylvania commission detailed information on how other states dealt with guidelines issues.

Prior to the BJA study, commission members and staff visited Oregon, Washington, and Louisiana to study their approaches to the development of guidelines for community-based sanctions. Each state had developed an "exchange-unit" concept for nonincarcerative sentences, and Oregon had implemented such a system. Although the Pennsylvania commission eventually rejected the exchange-unit concept and developed its own approach, it benefited greatly from the experiences of those states.

Reassessment Subcommittee   The eleven-member sentencing commission consists of four judges, two state senators, two state representatives, a district attorney, a defense attorney, and a criminologist or law professor. Four commissioners served on the reassessment subcommittee that oversaw reconsideration of the guidelines. They were the commission chair, Judge Theodore A. McKee; the district attorney representative, Theresa Ferris-Dukovich; the defense attorney representative, John P. Moses; and a state representative, Michael E. Bortner.

The commission lacked representation from some criminal justice sectors, particularly corrections, and lacked broad geographical representation. To broaden representation, the commission invited the secretary of corrections, the chair of the board of probation and parole, and the president of the county chief probation officers' association to serve in an advisory capacity as members of the reassessment subcommittee. To address the lack of geographical representation, the commission invited judges from rural areas to serve as advisors during the evaluation project. A six-member "rural judges advisory committee" was established to represent five different sizes of counties. They took an active interest in the guidelines and offered many invaluable insights.

Outcome of the Reassessment   Pennsylvania's guidelines are similar to those in most states in that each sentence recommendation is based primarily on the current offense and prior record of the offender. The revised guidelines incorporated the following changes: first, careful reranking of all offenses on a thirteen-category scale and the subdividing of several offenses including arson and theft offenses to provide better assessment of their seriousness; second, giving more weight to prior violent offenses in the calculation of the prior record score and creating two new prior record score categories that targeted repeat serious offenders; third, providing an enhanced sentence that is proportional to the seriousness of the offense when the offender possesses a weapon during the commission of the offense; fourth, providing four levels of sentence recommendations based on the seriousness of the current offense and prior record; fifth, providing primary and secondary purposes behind the sentence recommendations at the four sentencing levels (e.g., retribution, treatment); and sixth, adopting a new guidelines grid that embodied the preceding changes and narrowed sentencing ranges for most offenders.

Harsher Recommendations for Violent Offenders. In a 1989 study comparing sentence recommendations among Minnesota, Washington, and Pennsylvania sentencing guidelines, staff demonstrated that the Minnesota and Washington guidelines were harsher on violent offenders while Pennsylvania was harsher on nonviolent offenders. In an era of "three-strikes legislation" and close attention to violent offenders and repeat violent offenders, Pennsylvania's guidelines seemed to have a misdirected focus and to make a poor use of prison capacity. Consequently, one goal, and subsequent achievement, of the reassessment was to increase the severity of sentence recommendations for violent offenders and decrease them for nonviolent offenders.

Expansion of Intermediate Punishments. A second major change was expansion of nonincarceration options for a wide range of offenders. Although the commission had incorporated intermediate punishment recommendations into the guidelines in 1991 (as a result of a legislative mandate), recommendations for nonincarceration options expanded dramatically in 1994. Though the commission had been reluctant to establish presumptive nonincarceration sentences, the 1994 revisions recommended presumptive nonincarceration in six of the guidelines cells. (The 1991 guidelines had originally established six cells that specified presumptive nonincarceration sentence recommendations, but this was eventually reduced to two cells to gain legislative approval.)

The new guidelines also divided intermediate punishments into two categories based on restrictiveness: "restrictive intermediate punishments" and "restorative sanctions" (204 Pa. Code Ch. 303.8 [a]). Restrictive intermediate punishments were authorized as day-for-day substitutes for incarceration. The commission considered the Oregon model of exchange units but decided that approach was too mathematical and opted for a broader exchange system. Restrictive intermediate punishments "provide for strict supervision of the offender such that they house the offender full or part time; or significantly restrict the offender's movement and monitor the offender's compliance with the program; or involve a combination of programs that meet the standards of the first two criteria" (Ch. 303.8 [c] [1]). Examples are work camps, inpatient treatment, day-reporting centers, and halfway houses. Restorative sanctions are defined as "the least restrictive in terms of constraint of offender's liberties, do not house the offender, and focus on restoring the victim to preoffense status" (Ch. 303.8 [c] [2]). Examples are outpatient treatment programs, community service programs, drug testing programs, and restitution programs.

Sentencing Levels. The guidelines revisions resulted in a guidelines matrix, shown in figure 2.11. It has four sentencing levels, each of which identifies the purposes the commission wished to achieve in that level and the appropriate sentencing options to be used to achieve those purposes. While retribution is the primary focus throughout the guidelines (based on seriousness of current and past offenses), the sentencing levels provide guidance throughout as to whether the predominant secondary purpose should be rehabilitation, incapacitation, or restoration. For example, Level 4 targets the most serious offenders, who have either a current violent offense or a serious history of violent offenses. Level 4 offenders are recommended for state imprisonment. The predominant purpose, other than retribution, is to protect the public through incapacitation for significant periods of time.

Level 3 targets serious offenders with potential for violence and moderately serious offenders with serious prior records. Because this level contains a variety of offenders, the commission established a relatively wide range of sentencing options, from state prison or county jail with the option to use restrictive intermediate punishments under certain conditions. Retribution is the defining purpose, but the guidelines also encourage the court to consider treatment considerations, particularly drug or alcohol treatment for drug-dependent offenders.

At Level 2 the guidelines emphasize the offender's rehabilitative needs. For the least serious offenders, at Level 1, the guidelines prescribe a restorative sanction that emphasizes sufficient supervision of the offender to assure fulfillment of restitution, community service, or treatment obligations.

In determining the offenses that fall within each level, and keeping in mind crowding problems, the commission concentrated on reserving state prisons for the most violent offenders and removing the less serious theft and drug offenders from that system. The projected consequences include reduction in admissions to state prison and in growth of the prison population. Prison populations are projected to grow but primarily because of an increase in convictions, increased sentence lengths as a result of new mandatory sentence statutes such as the three-strikes legislation, and a reduction in inmate releases by the parole board.

Factors Affecting Guidelines Recommendations. In evaluating the guidelines, the commission was influenced by several factors. First, the state prison population had grown excessively as a result of enactment of mandatory penalties, toughening of parole release and revocation standards, severity of the sentencing guidelines, and growth in the conviction rate. Second, many offenders received state and county incarceration who could more fairly and effectively be sentenced to intermediate punishments. Third, guidelines recommendations before the 1994 changes took effect focused very narrowly on retribution as a punishment goal. Yet statutes and the appellate courts prescribed that judges at sentencing consider the full range of purposes, including incapacitation, rehabilitation, and deterrence.

The commission chose to focus on providing harsher penalties for violent offenders, expanding sentencing options, expressing the basic sentencing purposes inherent in the guidelines, and encouraging judges to consider the priorities that the court wants to achieve in the imposition of various sentencing options. These focuses are connected. By focusing on incarceration for violent offenders, the commission was able to provide more sentencing options for nonviolent offenders. In expanding sentencing options for the court, the commission was concerned that the selective use of these options would result in disparate sentences. However, the commission viewed the expanded use of options as an important step toward achievement of a more comprehensive and rational sentencing system.

Adoption of the New Grid As required by law, guidelines amendments must first be subject to public hearings. The commission held several public hearings across the state, during which an extensive number of people testified, both for and against the revisions. In response to public testimony, the commission did make some changes to the proposed guidelines before submitting them for legislative approval during the summer of 1994. In Pennsylvania, the guidelines automatically become effective within ninety days of submission unless there is a legislative veto. The guidelines became effective in August 1994.

Postadoption Issues At the time the guidelines were adopted, the commission still faced two unresolved issues. First, the district attorneys' association argued that the guidelines did not always live up to their billing that they were more severe on the violent offender. The district attorneys recognized that the commission did provide for harsher recommendations for the vast majority of violent offenders but argued that the recommendations should be harsher in all violent cases. To address this concern, the commission met with district attorney representatives and evaluated the recommendations and agreed to "correct" areas where harsher penalties for violent offenders were needed.

The commission anticipates holding public hearings on the proposed changes in the spring of 1996. (Pennsylvania's new governor had called for the legislature to hold a Special Session on Crime, which officially ended in Fall 1995. The commission wanted to wait until this special session was over before moving forward with the public hearings so that any new legislation could also be incorporated into new guidelines proposals.)

Second, the commission has always been concerned that adequate funding be provided to counties to pay for programs required by the expected diversion of offenders from state prisons to counties. Pennsylvania has county-financed corrections. Thus, probation and county incarceration are locally funded. The commission expected a large number of people to receive county intermediate punishments under the new guidelines and has been working with the legislature, the county commissioners' association, and the district attorneys to secure adequate funding to ensure successful implementation of the programs. The commission has endorsed state subsidies to allow counties to develop the full range of intermediate punishment programs required to implement the revised guidelines.

Preliminary Impact of Revised Guidelines The ultimate question is whether the revised guidelines will achieve their goals. It is too early to tell, but preliminary data indicate that there has been a reduction in state prison admissions for the targeted drug and theft offenders. This, along with an increase in the use of intermediate punishments, suggests that the commission's goals are being realized. These are, however, preliminary findings, and the full effects will not be known until the revised guidelines have been in effect longer and more data are available.

North Carolina

North Carolina Prisons Are Growing
  Stevens H. Clarke

North Carolina controlled prison population numbers during the 1980s better than most states. In 1980, its incarceration rate of 261 per 100,000 was second in the nation (only the District of Columbia's was higher) and nearly twice the national average rate of 130. By 1990, the rate had increased to 278, somewhat below the average rate of 287 and lower than those of twenty states.

In other ways, however, North Carolina's corrections experience has paralleled most states'. Prison construction accelerated in the 1980s. Prison admissions increased, especially for minority offenders, and racial imbalances in admissions and populations worsened. What was distinctive about North Carolina was that state leaders, determined to avoid federal court oversight of prisons, used parole laws and population caps to limit prison populations.

Prison population in North Carolina in the 1980s was, accordingly, driven more by changes in capacity and in official policy than by crime. As the 1990s unfold, new prisons have opened and building continues, some officials complain that prison space is insufficient and causes premature release of prisoners, and North Carolina like most states is confronted by hard fiscal choices.

Prison Capacity   North Carolina's prison population has been increasing for some years. Litigation has threatened federal oversight and, to avoid this, the state agreed to consent decrees requiring it to build more prison space and to limit the number of prisoners. Between 1974 and 1984, appropriations for prison construction totaled $102 million and provided 3,604 new beds. Capital appropriations between 1985 and 1991 totaled $146 million, and an additional $75 million in prison construction bonds was approved; these funds will provide 5,000 beds. In 1990, voters authorized another $200 million in prison bonds. Operation of the new facilities over their useful lifetimes will cost at least seven times more than the construction costs. Prison construction poses a serious fiscal problem at a time of declining revenue growth and growing demand for other public services. As recently as 1988, North Carolina spent a larger percentage of its state budget on the justice system than any other state, largely because of corrections costs.

Prison Trends   The prison population has more than doubled since 1970, growing from an average 9,768 in 1970 to 20,400 by mid-1992. The increase occurred in four phases shown in figure 2.12, which shows annual arrests, prison admissions, and average prison population. Arrests are plotted on a different scale from admissions and population to show relative growth trends.

Phase 1 (1970–1978) might be headlined "Longer Prison Stays Increase Population; Sentencing More Selective." Between 1974 and 1978, arrests were fairly steady, then declined. Admissions generally declined between 1970 and 1978. But the average population rose substantially between 1972 and 1978 (from 12,063 to 13,799). The average stay must have increased. Why? One reason is that the number of felons admitted was growing and the number of misdemeanants admitted was declining; the average length of stay went up.

Phase 2 (1978–1985) might be headlined "Increasing Arrests Lift Prison Population; Sentencing Laws and Parole Practices Resist." Between 1978 and 1985, arrests and prison admissions generally increased. The population increased, but not quite as rapidly as admissions. The average stay must have declined. One reason is that misdemeanant admissions, which had been dropping until 1978, began an increase that lasted until 1982. Another reason is the Fair Sentencing Act (FSA), effective in

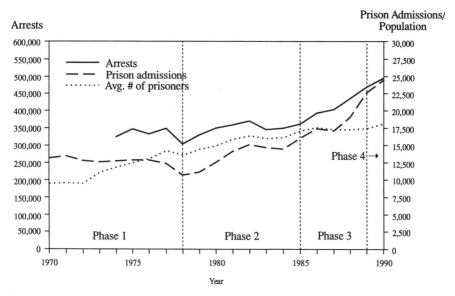

Figure 2.12. Arrests, Prison Admissions, and Average Prison Population in North Carolina, 1970–1990. *Sources:* North Carolina Department of Corrections and North Carolina State Bureau of Investigation (unpublished data provided to author).

1981, which—at least for several years—sharply reduced average active sentences for felonies. A third factor was the liberalized granting of good time, gain time, and parole. The average percentage of prison terms served had been declining for several years before the FSA took effect, and under FSA it continued to decline.

Phase 3 (1985–1989) could be headlined "Prison Admissions Increase; Time Served Decreases." The prison population held at approximately 17,500 during this period, despite more rapid rises in arrests and admissions than in the previous phase. This is the result of prisoner litigation. The state had entered into consent decrees calling, among other things, for increased space per inmate.

The General Assembly implemented these decrees in several ways. It authorized prison construction. It shortened prison stays and set a cap on population. These measures stabilized the population between 1985 and 1989, despite increasing admissions.

Phase 4, from 1989 to the present and probably continuing, might be headlined "Is There Any End in Sight?" It promises to be a period of rapid growth because new construction has allowed the General Assembly to raise the cap. The inmate population increased from 17,663 (average) in 1989 to 18,418 in 1990, and by mid-1992 exceeded 20,000.

Population Growth It should not be surprising that the number of prisoners increases as a state's population goes up—the more residents, the more crime, arrests, convictions, and prisoners. North Carolina's population increased from 5.1 million in

1970 to 6.6 million in 1990—about 29 percent. But prisoners increased by 89 percent, from 9,768 to 18,418. The incarceration rate went from 192 per 100,000 to 278, a 45 percent increase. Annual prison admissions per 100,000 declined until 1978, but by 1990 they had nearly doubled, increasing from 190 to 372.

Racial Disparity   Imprisonment and arrest rates behaved differently for whites and nonwhites. The incarceration rate for whites grew from 109 per 100,000 in 1970 to 136 in 1990—about 25 percent. The rate for nonwhites, already three times the white rate, increased more than twice as fast—by 62 percent (from 454 per 100,000 to 736).

The racial differences in incarceration rates result largely from dramatic differences in prison admissions (see figure 2.13). Admissions per 100,000 whites declined slightly between 1970 and 1978, then increased to 183 in 1990, just slightly above the 1970 level. For whites, the increase between 1978 and 1990 was 46 percent. For nonwhites, admissions declined between 1970 and 1978, but then increased 121 percent (from 395 per 100,000 to 960) between 1978 and 1990, with most of the growth after 1987.

Admissions are strongly affected by the number of arrests. Arrests per 100 white residents increased about 23 percent between 1970 and 1989 (from 4.0 to 4.9). For nonwhites, arrests per 100 residents increased 68 percent (from 8.4 to 14.1).

Why have nonwhites' rates of arrests, admissions, and incarceration increased much faster than whites' rates? One possibility is that law enforcement agencies have become more responsive to crime in nonwhite communities. Another is that the "war on drugs" has increasingly targeted blacks. In 1984 about twice as many whites (10,269) as blacks (5,021) were arrested for drug offenses. Thereafter, drug

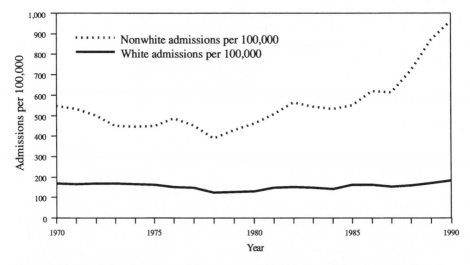

Figure 2.13. North Carolina Prison Admissions per 100,000 General Population, by Race, 1970–1990. *Sources:* North Carolina Department of Corrections and North Carolina Office of Budget and Management (unpublished data provided to author).

arrests of blacks increased much faster. By 1989 annual drug arrests of blacks had grown by 183 percent, reaching 14,192; drug arrests of whites increased only by 36 percent (to 14,007).

Changing Prisoners   Violent felons and drug law offenders make up an increasing proportion of prisoners. North Carolina's prison population consists mostly of felons, with increasing numbers serving long sentences for violent felonies. The percentage of admitted prisoners convicted of drug offenses climbed from about 2 percent in 1970 to 6 percent in 1980 and 14 percent in 1989. Most of that growth occurred in the last few years.

Changing admissions have affected the makeup of the year-end population. Violent felons went from 28 percent of the population in 1970 to nearly 46 percent in 1989. Drug offenders increased from about 5 percent to 13 percent; most of the increase occurred after 1985. Thus, in recent years, the war on drugs seems to have made increasing demands on prison space.

More Arrests: More Crime or More Police?   Arrests per 100 residents grew 49 percent between 1978 and 1990, from 5.04 to 7.51, and were the dominant factor in the growth of prison admissions. Why have arrests increased? Is it because crime has increased?

It is difficult to be certain whether crime per capita has increased because the leading sources of data disagree. The Uniform Crime Reporting (UCR) system based on police reports and maintained by the Federal Bureau of Investigation suggests that crime per capita has increased in the last twenty years. But the National Crime Survey (NCS) conducted by the United States Census Bureau indicates that crime victimization, though higher than UCR reports (because many crimes are not reported to police), has not increased since 1973. The NCS is done for the nation as a whole, but there is no reason to doubt that its findings apply to North Carolina.

If per capita crime really has not increased, why do the UCR statistics indicate that it has? The answer may be increased police reporting of actual crime.

The growth in the strength of police agencies has been documented. In the United States, recent publications of the Bureau of Justice Statistics show that real per capita expenditure (adjusted for inflation) on law enforcement began to increase in the 1950s and accelerated in the 1960s and 1970s, dropping somewhat after 1977. Sworn officers in eighty-eight large cities increased 41 percent, from 142 per 100,000 residents in 1970 to 200 in 1980, and real per capita spending on law enforcement increased at the same rate.

Law enforcement personnel in North Carolina increased 55 percent between 1975 and 1990—from 185 per 100,000 residents in 1975 to 272 in 1990. The same period saw improvements in training, pay, and equipment. Increased staff may make police agencies better able to respond vigorously to crime reports and to investigate and record them.

The available data on trends in index crime and law enforcement personnel suggest that increases in reported crime may have been a result, rather than a cause, of increased law enforcement personnel. Increases and decreases in personnel from 1970 to 1990 have been followed, not preceded, by corresponding changes in index crimes.

Trends in index crime arrests since 1975 have paralleled police personnel and UCR index crime increases, and they have been the major cause of increased prison admissions and population.

Tougher Prosecution and Sentencing? Increases in the ratio of admissions to arrests have increased prison population. Admissions per 100 arrests declined between 1974 and 1977 (from 4.0 to 3.6) and remained constant until 1980; thereafter they increased by 33 percent—to 4.8 in 1989. This suggests that sentencing or prosecution, or both, got tougher in the 1980–1989 period. However, the increase in the admissions-to-arrests ratio probably did not increase the prison population as much as the growth in arrests.

Conclusion North Carolina's incarceration rate has increased substantially since 1970. Arrests have increased, pushing up prison admissions. While other factors have resisted this pressure, it has persisted and increased, and the state has responded with a population cap and a surge in prison construction.

One interpretation of the data presented here is that increased incarceration rates result largely from increased crime, which has driven up arrests and prison admissions. Another explanation is that the state's response to crime has changed, and that the prison population growth has resulted from beefed-up law enforcement and tougher prosecution and sentencing rather than from a crime wave.

Has North Carolina any choice? Is massive prison construction now inevitable, as the price of a law-enforcement buildup begun years ago? If the answer is yes—if construction is seen as inevitable—then the state must be prepared to raise taxes or divert funds from other programs to build more prisons.

Our ideas of appropriate punishment for crime are not immutable; they change over time and vary among individuals. People differ in their beliefs. In a democratic society, the state's answer must be arrived at by political consensus. Most citizens would agree that there is a limit to the services the state can afford to provide, including the service of sanctioning criminal offenders. Most people would agree that there should be some clear, democratic system of punishment of criminals. These principles—scarcity of resources and the need for a clear system of punishment—suggest another approach to dealing with the correctional consequences of the law enforcement buildup, without continuing the massive expansion of prisons begun in the 1980s. This would be to allocate limited correctional resources in a principled way, according to clear policies and criteria established in a democratic fashion. Without a principled approach, the state's citizens may feel that the criminal justice system has let them down.

North Carolina Legislature Considers Sentencing Change
Robin Lubitz

The North Carolina General Assembly is considering a comprehensive package of proposals to revamp North Carolina sentencing and corrections. The proposals are the culmination of two years of study and deliberation by the twenty-seven-member

North Carolina Sentencing and Policy Advisory Commission. If enacted, the proposals will establish statewide structured sentencing policies linked to a comprehensive community corrections plan.

The commission presented lawmakers with two structured sentencing proposals. The first one (endorsed by the commission) would require additional funds for prison construction and expansion of intermediate programs. The second one (not endorsed by the commission) would put fewer offenders in prison and cost much less. Both would eliminate parole release and good time; both would create standards for sentences based on conviction offenses and offenders' prior criminal histories; both would tie sentencing policies to corrections resources; and both would allocate proportionately more prison use to violent offenders and less to nonviolent offenders.

The criminal justice system in North Carolina over the past seven years has become increasingly dysfunctional and inefficient. Problems accelerated in 1986 when, through a series of lawsuits and consent decrees, the state capped its prison population but took no steps to modify sentencing policies. The prison population held steady, but admissions skyrocketed. To comply with the cap, the Parole Commission released offenders earlier and earlier. This undermined the integrity of the system, subverted alternatives to prison, and eroded public confidence.

Prison sentences imposed by North Carolina judges today bear little relationship to times served. Felons typically serve less than 20 percent of the sentence imposed and misdemeanants less than 10 percent. To compensate, many judges have increased sentence lengths, further aggravating the gap between sentences imposed and punishments experienced.

The reduction in time served has undermined the viability of programs designed to divert offenders from prison. Many offenders find it less onerous to serve a short prison term than to comply with restrictive conditions. Increasingly, offenders reject probation and choose prison (a right they have under North Carolina law), or once placed in intermediate programs they refuse to abide by conditions. Admissions to prison for probation revocations have grown five times faster in the past year than admissions for new crimes.

The Proposals   The commission's proposals are intended to shift control of the system from the back end (parole release) to the front end (sentencing practices). They set priorities for the use of correctional resources and provide a means to balance sentencing policies with correctional capacity. Furthermore, they promise to help restore consistency, certainty, and truth in sentencing.

The system the commission proposed would resemble guidelines systems in other states.

Offense Classes. Crimes are divided into nine categories based on assessments of harm and risk of harm typically associated with the offense. Compared to current law, this process generally elevates the seriousness of violent and assaultive crimes relative to property and other nonviolent crimes.

Prior Record Levels. Offenders are divided into six prior record levels based on the extent and gravity of their prior records. Prior record is viewed both as a measure of culpability and as a predictor of future criminality.

Sentence Dispositions. For each combination of offense class and prior record level, sentence dispositions are prescribed. Dispositions include imprisonment, intermediate sanctions, or community sanctions. Intermediate sanctions are more restrictive and controlling than standard probation but less expensive than prison. They may include boot camps, intensive supervision, electronic monitoring, day reporting, residential treatment, and other similar programs.

Sentence Durations. For each combination of offense class and prior record, the judge would select a minimum sentence from one of three sentence ranges: a presumptive range, an aggravated range, and a mitigated range. Once the minimum is selected, the automatic maximum would be 120 percent of the minimum.

The commission's proposals are intended to restore truth in sentencing. The judge now imposes a single sentence that can be reduced by good time, gain time, and early parole. Under the commission's plan, the minimum sentence must be fully served. Good time and parole are eliminated. The maximum sentence, but never the minimum, may be reduced if the offender works or makes efforts toward rehabilitation. Following release from prison, the offender is eligible for voluntary aftercare services to help with reintegration into society.

The commission's proposals are intended to change the number and type of offenders sentenced to prison. Imprisonment will increase for violent and career felons but decrease for nonviolent and first-time felons. Fewer will be sent to prison and more will be sentenced to intermediate sanctions. Those imprisoned, however, will serve significantly more time than in the past. Under the commission's plan, expensive prison resources will be primarily allocated for violent and career offenders who pose the greatest threat to society. Less expensive intermediate sanctions will be allocated for nonviolent offenders who pose less of a threat.

The commission also recommended a separate and simplified sentencing structure for misdemeanants. Under this proposal, only repeat and career offenders can be incarcerated. Overall, fewer misdemeanants will be sent to prison, but those will serve more time. Because misdemeanants are generally less dangerous than felons, the commission recommends that they be housed in separate low-security and low-cost facilities. The goals of these facilities would be to encourage work and the payment of restitution and, when possible and financially feasible, to provide offenders opportunities for self-improvement through educational, vocational, or other treatment programs.

The commission used a computerized correctional population simulation model to project the correctional resources needed to implement the commission's proposals. Establishment of this model required the creation of a comprehensive statewide sentencing and prior record data base. This data base provided policy makers with an understanding of current sentencing practices and allowed the commission to simulate the impact of sentencing changes on the correctional system. This projection capacity is essential to the goal of balancing sentencing policies with correctional resources.

The commission also prepared a community corrections strategy. It describes existing community corrections programs in North Carolina and identifies potential new types of programs needed to create an "effective continuum of community corrections programs." The strategy broadly targets specific types of offenders for community and intermediate punishments. Furthermore, the plan includes legislation to create a state-county criminal justice partnership. This legislation, modeled after community corrections acts in other states, provides financial incentives to counties to encourage the implementation of new and innovative programs.

The North Carolina legislature will have to decide what sentencing policy it wants and how much it is willing to pay. The commission's proposals make these choices clear, explicit, and unavoidable. No matter what policy is finally enacted, structured sentencing offers the state an opportunity to retake control of its criminal justice system and to restore rationality, truth, and certainty to sentencing.

North Carolina Legislature Adopts Guidelines
   Stan C. Proband

The North Carolina legislature in July 1993 overhauled the state's sentencing and corrections policies, adopting many of the proposals of the North Carolina Sentencing and Policy Advisory Commission. The changes, which will apply to crimes committed on or after January 1, 1995, include creation of guidelines for felonies and misdemeanors, establishment of a new community corrections strategy, abolition of parole release, and recasting of good-time laws. The new system is intended to hold prison population within existing and currently planned capacity. This article summarizes highlights.

Presumptive Sentencing Guidelines   The new legislation establishes sentencing guidelines that build on those in other states. As in Minnesota, Washington, Oregon, and Kansas, the guidelines, which group offenders by reference to their conviction offenses and criminal histories, establish presumptions concerning whether offenders should be imprisoned and for how long. As in Pennsylvania, the guidelines (see figure 2.14) create three applicable ranges of presumptive sentence for each offender—a standard range, an aggravated range, and a mitigated range. The standard range normally applies, but the court, on finding that aggravating or mitigating circumstances exist, may sentence from within those alternative ranges.

The guidelines recognize three levels of sentences—custodial, intermediate, and community—and specify the applicable kind of sentence within each guidelines cell. As figure 2.14 shows, for example, an offender with a modest criminal record (category II), convicted of an E Class offense, could be sentenced to either a custodial or an intermediate punishment. For more serious offenses, only custodial sentences are authorized unless the judge finds "extraordinary mitigation," a finding the state may appeal.

Parole   Parole release is abolished for felons and misdemeanants. Felons in Classes B–E are subject to six months postrelease supervision and can be returned to prison for up to nine months. There is no postrelease supervision for misdemeanants.

PRIOR RECORD LEVEL

OFFENSE CLASS		I 0 Pts	II 1-4 Pts	III 5-8 Pts	IV 9-14 Pts	V 15-18 Pts	VI 19+ Pts	
	A	Mandatory Life or Death as Established by Statute						Disposition
	B	A 135-169 **108-135** 81-108	A 163-204 **130-163** 98-130	A 190-238 **152-190** 114-152	A 216-270 **173-216** 130-173	A 243-304 **194-243** 146-194	A 270-338 **216-270** 162-216	Aggravated Range PRESUMPTIVE RANGE Mitigated Range
	C	A 63-79 **50-63** 38-50	A 86-108 **69-86** 52-69	A 100-125 **80-100** 60-80	A 115-144 **92-115** 69-92	A 130-162 **104-130** 78-104	A 145-181 **116-145** 87-116	
	D	A 55-69 **44-55** 33-44	A 66-82 **53-66** 40-53	A 89-111 **71-89** 53-71	A 101-126 **81-101** 61-81	A 115-144 **92-115** 69-92	A 126-158 **101-126** 76-101	
	E	I/A 25-31 **20-25** 15-20	I/A 29-36 **23-29** 17-23	A 34-42 **27-34** 20-27	A 46-58 **37-46** 28-37	A 53-66 **42-53** 32-42	A 59-74 **47-59** 35-47	
	F	I/A 16-20 **13-16** 10-13	I/A 19-24 **15-19** 11-15	I/A 21-26 **17-21** 13-17	A 25-31 **20-25** 15-20	A 34-42 **27-34** 20-27	A 39-49 **31-39** 23-31	
	G	I/A 13-16 **10-13** 8-10	I/A 15-19 **12-15** 9-12	I/A 16-20 **13-16** 10-13	I/A 20-25 **16-20** 12-16	A 21-26 **17-21** 13-17	A 29-36 **23-29** 17-23	
	H	C/I 6-8 **5-6** 4-5	I 8-10 **6-8** 4-6	I/A 10-12 **8-10** 6-8	I/A 11-14 **9-11** 7-9	I/A 15-19 **12-15** 9-12	A 20-25 **16-20** 12-16	
	I	C 6-8 **4-6** 3-4	C/I 6-8 **4-6** 3-4	I 6-8 **5-6** 4-5	I/A 8-10 **6-8** 4-6	I/A 9-11 **7-9** 5-7	I/A 10-12 **8-10** 6-8	

Figure 2.14. North Carolina Felony Punishment Chart (numbers shown are in months).
Notes: A = Active Punishment; I = Intermediate Punishment; C = Community Punishment.
Source: North Carolina Sentencing and Policy Advisory Commission (1994).

Good Time; Determinacy   Judges will set the duration of the minimum custodial sentences. The maximum is an automatic multiple of the minimum. Existing good and gain time have been eliminated and replaced by an earned-time system that can reduce the maximum but not the minimum sentence. Earned time accrues for good behavior, work, and program participation.

Community Corrections   The State-County Criminal Justice Partnership Act encourages counties to develop community-based programs. The Department of Corrections will provide technical assistance, review local plans, and establish program standards. The department will provide funds to participating counties for new programs.

North Carolina Prepares for Guidelines Sentencing
Ronald F. Wright

In July 1993, the North Carolina legislature enacted new sentencing reform legisla-tion. This is the state's second recent effort to overhaul sentencing. The first, in 1979, worked roughly as intended for its first few years, but soon fell apart. Begin-ning in January 1995, North Carolina will start using presumptive sentencing guidelines similar to those in Oregon, Washington, Minnesota, Kansas, and Penn-sylvania. The legislation abolished parole release.

The politics of sentencing reform in North Carolina have been unusual. Because of concern for prison population and costs, key legislators, unlike legisla-tors in many states, have not called for ever-harsher penalties. Unlike Kansas and Pennsylvania, for example, where legislatures rejected initial guidelines proposals because they were not severe enough, North Carolina legislators rejected sentenc-ing commission proposals because they would place too great a burden on stretched corrections resources.

Criminal justice officials can expect the new system to bring relief to a severely damaged and barely credible system. As in Texas and Florida, North Carolina pris-ons have had revolving doors through which prisoners were released to accommo-date new admissions and comply with a population cap. Nonetheless, the sentenc-ing commission and the Department of Correction (DOC) must also brace themselves for difficult problems during the transition period before guidelines sen-tencing begins, and in the early years of guidelines operation.

During the transition period, DOC must respond to a potential loss of 3,000 prison beds (out of about 21,000 now in use). A 1986 consent decree is responsible for this loss of beds: in 1994, minimum square footage per prisoner goes up from 35 feet to 50 feet. The sentencing commission must also design and execute a training program for judges, lawyers, and court officials, and keep the General Assembly informed about guidelines changes that may be proposed during a special session on crime.

The 1979 Reforms   The 1979 Fair Sentencing Act specified presumptive sentences for each felony. Judges could sentence offenders to longer or shorter prison terms if they explained their reasons. The act abolished parole release, but the legislature restored "emergency" parole power a few years later.

The act promised to bring more uniformity and predictability to sentencing, but it delivered on these promises only for the first few years. As state population and convictions rose, the presumptive sentences produced overcrowded prisons. A fed-eral lawsuit over prison conditions ended in 1986 in a consent decree: the state agreed to a prison population cap (now embodied in statute).

The increase in convictions, the presumptive sentences, the population cap, and the failure to build new prisons combined to force the state to release prisoners early by use of a variety of "back-end" measures. Sentencing judges became frus-trated as they watched the gap grow between the sentences they imposed and the times served. They responded by imposing sentences longer than the presumptive levels, and the gap grew larger still.

As time served for active sentences declined, offenders sentenced to suspended terms with onerous conditions began to opt for active terms, or to violate the terms of their probation. This, of course, further worsened crowding.

The Commission and Its Proposals   In 1990, the General Assembly created a Sentencing and Policy Advisory Commission and charged it to propose changes in this deteriorating system. After two years of study, the commission proposed a sentencing structure now familiar in other guidelines states.

The proposed guidelines ranked felony offenses into nine levels of severity, giving a clear priority in prison space to violent offenders causing bodily harm and directing most property offenders into nonprison sanctions. Misdemeanors were ranked into three levels of severity, in order to prevent uncontrolled use of jails to bypass limits on the use of prisons.

The offense seriousness rankings created the vertical axis of a sentencing grid, a format now used in almost every guidelines state. The criminal record of the offender—the second major variable in the sentence—created the horizontal axis. The commission's proposal assigned points for every prior felony or misdemeanor, with more points assigned for more serious offenses (as in Washington State).

The sentencing grid sorted offenders into six different prior record levels. The proposed point system accelerated sentence lengths quickly as prior records lengthened. A simplified point system measured the prior criminal records of misdemeanants.

Most cells in the proposed grid specified a "community" punishment (such as outpatient drug treatment or unsupervised probation, providing minimal control of the offender), an "intermediate" punishment (nonprison sanctions requiring more intensive supervision), or an "active" punishment (a prison term). Community sanctions were available for the least serious crimes and the least extensive prior records. Some cells on the border between types of sanctions authorized judges to choose between them.

The commission's proposal gave judges no power to depart in unusual cases if the grid specified only one disposition. An active term could not be reduced to an intermediate sanction. The only departure power granted to the judge was to set the *duration* of the sentence outside the range specified in the grid box. The judge could choose from an aggravated or mitigated range (spanning 25 percent above and below the presumptive range) after explaining on the record why the case was unusual. Such decisions would be subject to appellate review.

After the commission provided an interim report to the General Assembly, key legislators sent a clear signal that new prison costs should be kept to a minimum. A new statute instructed the commission to submit at least one proposal that required no immediate prison construction beyond new facilities already financed by a recently (and narrowly) approved bond issue.

The commission complied grudgingly. It submitted and endorsed the plan already described, which called for over 10,000 new prison beds over five years: an increase in the felony population from 23,000 to 31,000, along with the space needed for the longest misdemeanor sentences. A minority report called for significantly greater reliance on prison.

The commission also submitted, without endorsement, a standard operating capacity (SOC) plan that did not require any new growth in prison capacity during the first few years. The SOC plan did, however, forecast a slower expansion of about 9,000 beds over ten years. The SOC plan reduced prison requirements by reducing the influence of prior record and by reducing durations in all cells.

The New Legislation A relatively small Democratic leadership group in both the house and the senate had drafted the statute creating the commission and had pointed it toward guidelines sentencing. The same group dominated the debate over the commission's final proposals. It soon became clear that the SOC plan would receive the closest attention in both houses. In a legislative session dominated by a search for health care and education dollars, there was no sentiment, within the leadership group or elsewhere, for funding much new prison construction.

There was serious question, however, whether the legislature would approve the concept of using sentencing guidelines to control prison population. Because the package eliminated parole release, the parole commission became its most influential critic.

The parole commission relied on anecdotal evidence to argue that sentencing guidelines in other systems have exacerbated prison crowding. More plausibly, the commissioners warned that predictions of future prison populations cannot be exact, and that some "safety valve" might be necessary. Finally, they pointed out that none of the proposals made any attempt to tie sentence lengths to the predicted dangerousness of individual offenders or to hold the most dangerous offenders for the longest times.

The legislature chose the public visibility and certainty of sentencing guidelines over the ability to hold potentially dangerous offenders for more indefinite terms. The parole commission's arguments, however, produced an important change. The final legislation provides for six months of mandatory postrelease supervision for all felons convicted of the most serious classes of offenses. Violators can be returned to prison for up to nine months.

Other deviations from the sentencing commission proposal tended to reduce the system's reliance on prison. Where the commission had treated attempted crimes and completed crimes as equivalent in seriousness, the legislation placed attempts one level below the completed crime.

Where the commission had given judges no power to change a disposition specified in the guidelines, the legislation allowed judges to depart from a prison sentence and impose an intermediate punishment where there is "extraordinary mitigation." Finally, where the commission had remained silent on the duration of nonprison sanctions, the legislation established ranges of possible lengths for probation terms. One effect will be to prevent judges from imposing such onerous probation terms that low-level offenders choose prison far more often than probation. Imposing guidelines on probation use acknowledges that probation, like prison, is a scarce resource and that funding for the system must match its anticipated use.

Hence, the legislature and the commission continued the roles they established earlier in the process. The commission called for faster growth in the prison system, and the General Assembly held out for cheaper alternatives to prison. In light of the greater insulation of the commission from political pressure, it is somewhat surprising that the General Assembly took the less politically popular position.

Although the legislature emphasized intermediate sanctions, it did not increase funding to match the anticipated needs. If caseloads increase for probation officers or other program officials, intermediate sanctions may become no more intrusive than community sanctions and may lose their already shaky credibility with the public.

Department of Correction Challenges   From DOC's point of view, the greatest disappointment is that the guidelines' effective date was pushed back from January 1994 to January 1995. This means one more year of operating a system that makes prison officials appear foolish or negligent. The accelerated early release practices of recent years will continue. There will be virtually no imprisonment for misdemeanors, and the gap between prison terms announced and actual times served will grow.

DOC must also deal with a public relations dilemma. The 1986 consent decree called for a further reduction in prison crowding by early 1994, with minimum allowable square footage per inmate up from thirty-five to fifty. This will mean a temporary loss of roughly 3,000 prison beds. The attorney general is seeking federal court approval to keep the square footage at thirty-five. The governor has explored the possibility of sending prisoners out of state.

Both the commission proposal and the legislation anticipated the loss of beds. It coincides, however, with a cluster of highly visible crimes, including the murder of Michael Jordan's father, committed by offenders with prior felonies, some released early from prison. These events may create pressure to rethink the new sentencing legislation before it takes effect, perhaps to reinstitute authority to hold dangerous inmates longer.

To complicate matters further, merchants in Durham have filed a lawsuit in state court seeking to invalidate the statutory prison cap under the North Carolina constitution. They claim that the "prisons" clause in the constitution ("Such . . . correctional institutions . . . as the needs of humanity and the public good may require shall be established and operated by the State") requires the state to respond to any prison bed shortage by building new prisons, not by shortening terms. If the court were to rule for the plaintiffs, it could call into question the constitutionality of the sentencing guidelines themselves. The guidelines, like the statutory prison cap, ration prison space rather than expand prison capacity.

Short-Term Challenges for the Commission   The governor recently called an emergency session of the legislature to address crime, and the odds of further changes in the sentencing guidelines have increased. The legislature could advance the effective date or change the classifications of particular crimes. It will almost certainly authorize new prisons.

The commission must anticipate various proposals for change and inform legislators about the systemic consequences of each. This is bound to distract them from another critical task—training judges, court personnel, and attorneys who will be using the guidelines. Two forces have made training unduly difficult.

First, the training will take place in an atmosphere of cynicism. The state has operated under a partially structured but failed sentencing regime for the past ten years. Judges will need to be convinced that this reform has greater prospects of success than its predecessor.

Second, the training will take place amid apprehension about sentencing guidelines, deriving mostly from controversies associated with the federal guidelines. Few court personnel and attorneys in North Carolina have firsthand experience with the federal guidelines, but they have heard plenty of complaints about them.

Long-Term Challenges   After the guidelines take effect, the commission's most important duties will be to monitor the system and anticipate any adjustments needed. Three areas will bear watching.

First, the commission must monitor the habitual felon provisions closely. Under the commission's proposal, any offender convicted of three prior felonies could be charged separately as a habitual felon, a Class D felony. The General Assembly limited this provision by insisting that at least two of the three prior felonies be more serious than a Class H or I felony (the two least serious classes). Still, offenders in more than one-quarter of the guidelines cells could have their sentences enhanced as habitual felons.

In the past, difficulties in proving prior felonies with authenticated records limited the use of the habitual felon statute. The new system, however, creates more incentive and, combined with the state's commitment to a cutting-edge information system, it may lead to an explosion in habitual felon charges. Moreover, prosecutors are hoping to convince the legislature to amend the statute to broaden the number of eligible felons.

The commission must also remain mindful of drug charges carrying mandatory minimums. The guidelines do not cover drug trafficking or drunk driving charges. A big change in the number of drug convictions could strain the prisons, as in the federal system. In the long run, it may be necessary to include drug trafficking and drunk driving within the guidelines system. Both were considered too controversial to include in the initial legislative package.

Finally, certain forms of judicial discretion must be monitored closely. The most important involves multiple offenses. Under the new statute, the court retains complete discretion to impose consecutive or concurrent sentences for multiple convictions. If judges start to impose consecutive sentences more often, prosecutors will respond by multiplying charges. The population estimates underlying the sentencing durations in the grid will become meaningless, and the system will spin out of control.

Judges also control the future of structured sentencing through their use of "border blocks," the areas in the grid giving them discretion to choose between active prison terms and intermediate punishments. If the judges choose active terms at a much higher rate than in the past, the commission's population projections will lose touch with reality.

When adjustments become necessary to keep the system afloat, the commission will not be able to make those adjustments alone. The enabling legislation gives the commission no independent power to amend its guidelines. It can only make recommendations to the legislature. Since the General Assembly meets in full session only every two years, it may be impossible to respond quickly to problems. Even if the commission can get a timely response from the legislators, it may not be able to obtain a satisfactory answer.

Because of lessons drawn from the failure of the 1979 act, it may be possible for North Carolina to keep a consistent approach to sentencing, to fund the strategy adequately, and to fine-tune the system as conditions change. The presence of the sentencing commission to frame and inform debate might make a difference this time.

North Carolina Avoids Early Trouble with Guidelines
Ronald F. Wright

The North Carolina legislature embraced structured sentencing in July 1993. During two sessions early in 1994, before the new sentencing guidelines took effect, the legislature had second thoughts. In the end, after some persuasion from the Sentencing and Policy Advisory Commission, the legislature kept its commitment to structured sentencing. Although it adopted a three-strikes-and-you're-in statute, it took a limited and realistic approach, and reaffirmed the need for sentencing guidelines. The guidelines began operation without incident in October 1994. For at least the next few years, they will operate amid large and persistent changes in crime policy. The 1995 legislature, dominated by Republicans for the first time this century, promises further scrutiny of structured sentencing and significant increases in the size of the prison system.

The 1994 Extra Session on Crime　After structured sentencing legislation passed in 1993, sentencing commission staff began to plan their agenda. They figured to spend much of their time in 1994 training prosecutors, defense lawyers, judges, and court personnel to use the new guidelines. But political events soon made it clear that training would not remain at the center of their attention.

North Carolina citizens, like those in many other states, were increasingly anxious about crime. After the murder of Michael Jordan's father and a spate of other notorious violent crimes, Governor Hunt called an extra session of the legislature to address crime. For seven weeks in February and March 1994, legislators introduced over 400 new crime bills. Many dealt with punishments for sexual assault and use of weapons. There were also many proposals for three-strikes-and-you're-out laws—sentencing repeat felons to life imprisonment without possibility of parole. A number of the bills proposed life imprisonment after two strikes or just one.

The commission worked long hours responding to the new sentencing proposals. A 1993 statute required the legislature to obtain a fiscal impact statement before making any changes to sentencing laws. Robin Lubitz, the executive director of the commission, delivered these impact statements in a way designed to make the cost of corrections a vital part of the lawmakers' deliberations, rather than just an afterthought.

The commission identified, whenever it could, the sentencing bills that would have the greatest impact on prison population. Lubitz or other commission staffers would approach the sponsoring legislator privately, as early as possible, to describe the likely impact of the bill. By delivering the impact statements early and in private, the commission hoped to make it easier for the sponsor to alter or withdraw the bill before taking a public stand.

By the end of the session, the legislature had passed a three-strikes law, a gun-enhancement law, and new punishments for rape. But in each case, the commission's analysis had a sobering effect on legislators, who amended the bills to reduce their impact on prison population. The bills introduced would have increased the population by well over 20,000 beds within ten years, doubling the current population. The bills that passed will now require about 2,000 new beds within ten years.

Some knowledge of North Carolina's system of offense classification is necessary to understand the new three-strikes and gun-enhancement laws. Felonies are divided into ten classes. Class A is reserved for first-degree murder. Classes B1 and B2 cover second-degree murder, rape, and serious sexual assaults. Crimes in Classes C, D, and E all result in "serious" personal injury (such as assault with a deadly weapon), or serious infringements on property that "implicate physical safety" (such as kidnapping, burglary of an occupied dwelling, or use of a deadly weapon during a robbery). Felonies in Classes F through I involve lesser physical harms, property injury, or "societal" injury.

The final version of the three-strikes bill mandated life imprisonment for felons convicted for the third time of a Class A–E felony (by and large, violent felonies). The commission estimates that the bill will have no short-term effect on prison population because the guidelines would have sent all such offenders to prison for lengthy terms anyway. Some unsuccessful three-strikes bills would have covered all third-time felons (not just those committing violent crimes) and could have increased the need for prison beds by as many as 16,000 over twenty years.

The gun-enhancement law requires the judge to add five years to any prison term imposed for a person who uses or displays a firearm while committing a Class A–E felony, unless the elements of the offense already include possession or use of a firearm. The enhancement does not apply to cases where the judge chooses a community or intermediate (nonprison) sanction. The law will require an addition of fewer than 150 prison beds. Earlier versions of the bill would have covered all felonies and would have added about 4,700 beds over ten years.

Finally, the legislature increased the punishments for first-degree rape and sexual assault. The 1994 law created a new class of felonies within the guidelines, B1 felonies, with higher penalty ranges than were available under Class B. The legislature had considered removing these two crimes from the sentencing guidelines entirely and simply imposing life terms without parole on all offenders convicted of them. By keeping the crimes within the guidelines, the legislature left judges the discretion to impose prison terms as low as 144 months in the most mitigated cases.

The commission spent most of its effort reacting to proposals that would increase the use of prison. But the appropriations decisions of the legislature were at least as important as the penalty decisions, and the commission again had a hand in the outcome. The General Assembly fully funded the prison construction necessary to operate within the sentencing guidelines until 2001. It also fixed a troubling omission from the 1993 statute: it provided over $20 million for additional probation personnel needed to implement new community and intermediate sanctions, and it added to the funding for development of community sanctions in the counties.

The commission's efforts were not always successful during the extra session. The legislature amended the existing habitual felon law in a way that poses a serious threat to structured sentencing. The 1993 version of this law allowed a prosecutor to charge a separate Class D felony on conviction of three prior felonies, but two of the three felonies had to be more serious than Class H. Two amendments in 1994 expanded prosecutorial power. The habitual felon crime itself was increased from Class D to Class C, and the new law now applies after any three prior felonies. This will dramatically increase the number of offenders eligible for the charge.

The commission did not publicly oppose the habitual felon amendment. Perhaps this was because prosecutors have seldom used the law in the past. However, the amended version gives prosecutors the power to remove offenders from a large number of low-level guidelines categories and push them up to a Class C felony (with presumptive terms of fifty to sixty-three months). In effect, prosecutors now have discretion to increase a presumptive sentence for almost all offenders in twenty of the fifty-four cells in the grid. The temptation to use the law more frequently may prove hard to resist.

Training   The need to train court personnel and lawyers to use the new guidelines became more urgent than ever at the close of the extra session because the legislature moved the effective date for the guidelines forward two months to October 1, 1994. The commission developed training materials, and during the summer and early fall it conducted sessions for judges, prosecutors, public defenders, clerks, probation officers, and corrections officials. The commission was prepared to train private defense attorneys, but several for-profit organizations specializing in "continuing legal education" for lawyers rushed in to fill the need. The use of well-chosen hypothetical cases in the commission's materials, combined with the relatively simple structure of the guidelines, made the training easier to carry out.

The most interesting aspect of the training, however, was not what the commission taught, but what it learned. Many prosecutors and judges expressed concern about misdemeanor assaults, which include most domestic abuse situations. The current guidelines contain a major discontinuity. Assault with a deadly weapon inflicting serious bodily injury is a Class E felony (with presumptive ranges starting at twenty to twenty-five months), but any simple assault on an ordinary citizen is a Class 1 misdemeanor. Such misdemeanors are punishable by nonprison community sanctions, unless the defendant has a prior record (and even then, only up to forty-five days' incarceration for those with one to four prior convictions). The frequency of complaints convinced the commission that change was necessary. The commission is proposing a new class of misdemeanors for assault crimes; one that will give judges the option of an active prison term for all offenders.

The Tranquil Start of Guidelines Operation   When the October 1 effective date for guidelines sentencing arrived, the new system faced tranquil and promising condi-

tions. The commission had reached agreement on controversial choices in design-
ing the guidelines and was working on much less divisive issues such as restitution.
The state was receiving favorable national attention in the wake of the 1993 decision to adopt guidelines. Glowing accounts of the system appeared in the *Washington Post* and the *Wall Street Journal*. Both Robin Lubitz and Judge Thomas Ross,
the chair of the commission, received national recognition for their work and made
speeches to national public policy groups.

The commission maintained its preguidelines level of staffing, after finding that
demands for data did not decrease as anticipated. A new computer system at the
Administrative Office of the Courts promised to make monitoring sentences easier
than originally planned. The Department of Corrections proceeded with plans to
increase the availability of intermediate and community punishments. And the
guidelines system made it appear likely that North Carolina would qualify for funds
under the 1994 federal crime bill.

The Prospects in the 1995 Legislature   The November 1994 elections introduced a
large element of uncertainty into this serene environment. For the first time since
Reconstruction, Republicans won a majority in the house of representatives. The
senate remained in Democratic hands, but by a slender margin. Although Republicans overwhelmingly supported the 1993 sentencing guidelines legislation, a few
Republican candidates vigorously attacked the guidelines. A campaign flyer produced by the Republican Party claimed that the new sentencing laws were too
lenient for crimes against children. It listed crimes such as child abuse that inflicts
serious injury and selling drugs to a minor under the heading "No Required Prison
Time." The flyer does not mention that active prison terms were not required for
these crimes under pre-1993 laws, either.

After the election, victorious Republican candidates softened their rhetoric and
suggested that wholesale repeal of sentencing guidelines was unlikely. They
planned to stress instead the construction of new prisons and also hoped to amend
the guidelines to allow prison terms for more categories of first-time felonies, especially breaking and entering.

Nonetheless, the new legislators probably will not be satisfied with changes at
the margins of sentencing policy. The House Republicans elected as majority
leader Representative Leo Daughtry, one of only two veteran Republican legislators
who voted against sentencing guidelines in 1993. Some are citing a recent survey
ranking North Carolina last in the nation in new prison construction.

Not all of the proposed short-term increases in prison usage will need to come
from new construction. The completion of a round of prison construction begun in
1990, together with modifications in the terms of a federal lawsuit, may increase
available beds from 24,000 to around 30,000. Some of this growth will be necessary
to keep pace with projections under the current guidelines. Some of it might also
be used to retain offenders sentenced under preguidelines law for longer periods.
Thus, if the legislators plan to change sentencing policy to rely more on prison in
the future, they will surely need to expand the system.

It remains to be seen how much influence the commission will have on the new
legislators as they make these choices. Surely political forces far beyond the commis-

sion's control will determine the chosen size of the state's prison system. The commission has existed only since 1991. Yet North Carolina has, whether by default or by choice, relied less on prison than its sister states over the last ten years. In 1972, North Carolina incarcerated 160 citizens per 100,000, the third highest rate in the nation (behind Georgia and the District of Columbia). While the rate climbed steadily in the state over the next twenty years, it grew far more quickly in other states. By 1987, North Carolina ranked fifteenth; in 1989, for the first time, the state's rate fell below the national average. By 1992, the state's rate of 290 placed it twenty-second among the states and fourteenth among seventeen states in the region. All these changes occurred before the commission even arrived on the scene.

The commission cannot and will not change public views on whether this moderating trend (relative to other states) should continue or reverse. It will, instead, only try to convince lawmakers to pay up front for the corrections resources it plans to use. Given this narrowly defined goal, the commission will likely succeed.

There are several positive signs. The same candidates who criticized the leniency of the sentencing guidelines were also most vocal about the need to reduce state spending. An effort to reconcile those two beliefs could lead new legislators to listen carefully to the impact projections that they must receive before they pass new sentencing legislation. New legislators have also spent more time talking about specific ways to create more prison space and surprisingly little time talking about specific crimes or criminals to punish more severely. They are not yet speaking about removing funds from nonprison sanctions to pay for prison expansion.

These legislators are by no means counting all the social costs of an enormous and expanding prison system, but they do show an appreciation for its direct cost to the state treasury. The sentencing commission is well placed to keep them focused on this subject.

Delaware

Sentencing Reform in Delaware
Richard Gebelein

Delaware's correctional system in the late 1970s was facing a crisis. Governor Pierre S. duPont IV decided that comprehensive changes had to be made. A sentencing reform commission was directed to develop a logical sentencing policy for the state. In 1983, the commission finished its work, adopting as its goals the incapacitation of violent offenders and the use of alternative sanctions for property and nuisance offenders. It suggested a ten-level continuum of sanctions, starting with fines and ending with incarceration. It proposed the creation of a permanent commission to formulate, monitor, and adjust sentencing laws and guidelines to achieve the goals enunciated. That commission, the Sentencing Accountability Commission, was established in 1984 by the General Assembly—SENTAC was born.

The commission was charged with establishing a system that emphasized accountability of the offender to the criminal justice system and accountability of the criminal justice system to the public. SENTAC was to develop sentencing guidelines consistent with the overall goal of ensuring certainty and consistency of punishment commensurate with the seriousness of the offense and with due regard for public safety and resource availability.

In May 1986, SENTAC published its Master Plan for Sentencing Reform. The plan, building on the earlier commission's work, called for a five-level continuum of sanctions ranging from unsupervised probation to incarceration, and for adoption of sentencing standards for the adult criminal courts. Sentencing standards suggested sentences up to 25 percent of the statutory maximum. Sentence lengths or levels are then enhanced or diminished based on aggravating or mitigating factors adopted by the commission.

The master plan called for sentencing offenders to the least restrictive (and, therefore, least costly) sanction consistent with public safety. The announced goals of SENTAC in priority order are incapacitation of the violence-prone offender, restoration of the victim as nearly as possible to his or her preoffense status, and rehabilitation of the offender.

SENTAC-endorsed legislation was enacted, and the system went into effect in October 1987. Judges may sentence people to one of the five accountability levels based on their criminal histories, the severity of the crimes committed, and enumerated aggravating and mitigating factors. The standards are voluntary, and judges retain full discretion to sentence within statutory limits.

The accountability levels are distinguished by the intensity of supervision an offender receives:

Level V, incarceration, involves twenty-four hours of daily supervision. Level V offenders are inmates; under limited circumstances, they may participate in transition programs such as work release or supervised custody.

Level IV, quasi-incarceration, involves nine to twenty-three hours of daily supervision. Programs include halfway houses, home confinement with electronic monitoring, and long-term residential drug treatment. Other programs—boot camp, work camp, or weekend incarceration—are being developed.

Level III, intensive supervision, involves one to eight hours of daily supervision. The officer-to-offender caseload ratio is 1:25.

Level II, field supervision, involves supervision from zero to one hour per week and is essentially regular probation.

Level I, unsupervised probation, includes an initial intake interview and subsequent monitoring to ensure fulfillment of conditions including participation in treatment programs, payment of fines, costs, and restitution, and avoidance of criminal activity.

At any level, the courts can order a variety of conditions. These may include substance abuse treatment, employment training or maintenance, payment of restitution, or community service. The crime and criminal history determine the level of sanction; the needs of the offender and the victim determine the details.

Evaluations, which include monitoring of judicial sentencing patterns and DOC records, have shown that many of SENTAC's goals are being met. The mid-level sanctions are being fully used by the judiciary, particularly for nonviolent offenders.

The incarcerated population has changed, as is shown in figure 2.15. Violent offenders, particularly felons, are being sentenced to incarceration more often and for longer; nonviolent offenders are more frequently sentenced to intermediate-

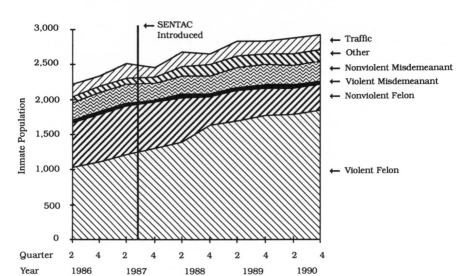

Figure 2.15. Composition of Delaware's Incarcerated Population before and after Effective Date of SENTAC Guidelines. *Source:* Delaware Statistical Analysis Center (unpublished data provided to author).

level sanctions. In the second quarter of 1986, before SENTAC took effect, 51 percent of Delaware prisoners had been sentenced for violent crimes. That percentage has increased. By the second quarter of 1990, 68 percent of Delaware inmates were convicted violent offenders.

There have been sizable changes in the numbers of people sentenced to each SENTAC level since the inception of the five-level system. The use of Level V, the most expensive correctional resource, has been well managed. Most growth has occurred in Levels I–IV.

Sentencing patterns have changed dramatically since the inception of SENTAC. Offender-specific sentences, often combining many levels of supervision with movement conditioned on compliance with supervision and treatment requirements, are now the rule. Sentences commonly are based on a stepwise movement to lower levels of supervision, offering the offender greater degrees of freedom as success in the community is evidenced.

Sentencing in Delaware is like a social contract. The onus is on the offender. Success can result in accelerated movement through and out of the criminal justice system; failure can result in additional sanctions, often at a higher level of supervision.

The sentencing standards have made the judge's role more activist. Although the standards suggest sentence lengths and levels, they have not made the judge's role more mechanical. Judges have become advocates for offender-specific nonincarcerative sanctions.

In part because the judiciary is structuring tightly controlled sentences requiring movement of the offender through the SENTAC system, the legislature, with the support of SENTAC, passed the Truth-in-Sentencing Act of 1989. SENTAC has always emphasized the importance of certainty of punishment, and the public, the legislature, and the system were tired of the uncertainty regarding when offenders would be released from incarceration under previous law.

In response to overcrowded conditions in the prisons, several early release programs and mechanisms were developed during the 1980s. In addition to parole, inmates earned a variety of early release credits and could be placed in transitional programs including work release, supervised custody, and extended furloughs. These programs provide some supervision of inmates in the community; however, neither the judge, the prosecutor, the defense attorney, nor the offender knew with any certainty how long an inmate would serve on any sentence imposed. Some inmates, even with lengthy sentences, were released almost immediately; others with shorter sentences served 80 percent or more of the sentence. Some long-term inmates earned over 600 days of early release credits per year served.

Truth in Sentencing was designed to eliminate uncertainty about sentence length while having a neutral effect on the prison population. While it abolished parole eligibility and restricted early release, at the same time statutory sentence lengths were reduced. The net effect should be that violent offenders serve more time, and nonviolent offenders serve shorter sentences or are diverted to nonincarcerative sanctions. Figure 2.16 shows how Truth in Sentencing will change lengths of sentence in time served for selected crimes. Truth in Sentencing took effect in June 1990. Early results show that judges are conforming to the new standards.

SENTAC is in its infancy and is evolving. The results have been promising, particularly in managing and reorganizing the population of incarcerated offenders. SENTAC has provided a forum through which the executive, the judiciary, and the legislature can meet, discuss sentencing policy, and work toward implementation and improvement of a sentencing system. Sustaining the progress made while expanding sentencing options is Delaware's challenge for the 1990s.

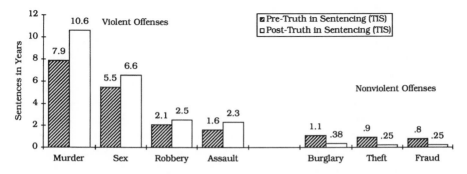

Figure 2.16. Projected Effects of Truth in Sentencing on Sentence Lengths for Selected Offenses. *Source:* Delaware Statistical Analysis Center (unpublished data provided to author).

Voluntary Guidelines Effective in Delaware
Thomas J. Quinn

Despite conventional wisdom to the contrary, Delaware has shown that voluntary sentencing guidelines can change sentencing policies and practices and still preserve judicial discretion. As a matter of conscious policy, Delaware has toughened sentencing of violent offenders by increasing their average prison sentences while reducing use of prison for nonviolent offenders and increasing use of various intermediate sanctions.

There has been a healthy debate for the last decade among criminal justice researchers and policy makers concerning the viability of voluntary sentencing guidelines as a means of changing sentencing patterns in a jurisdiction. Sentencing guidelines themselves emerged as a compromise approach to replace the unfettered discretion possessed by judges in indeterminate sentencing systems while avoiding the rigidity of mandatory sentencing schemes.

Guidelines usually classify an offender on two scales, one related to criminal history and the other to the current offense. The resulting calculation determines whether the offender goes "in" and for "how long." Improved systems in some jurisdictions provide an intermediate range that allows for a more graduated scale of punishment.

The voluntary aspect of voluntary guidelines refers to the latitude provided a judge in applying the applicable guidelines. If a judge may deviate without violating a statute or being subject to legal appeals, the guideline is voluntary. Many researchers question whether judges individually or collectively will change their sentencing philosophies or practices if there is no legal requirement to do so. Our experience in Delaware leads me to conclude that voluntary guidelines as applied here have changed sentencing practices. It should be noted, however, that the changes here were structural as well as procedural.

The Sentencing Accountability Commission (SENTAC) was charged by statute to revise both the structure and the process of sentencing in Delaware. In 1987, the "in/out" dichotomy previously available was replaced with a five-level continuum of punishment that ranges from Level I (unsupervised probation) through Level IV (house arrest or residential treatment programs) to Level V (prison). At the same time, the Delaware Supreme Court issued sentencing standards (narrative guidelines) drafted by SENTAC and required judges to give reasons on the record if they deviated from the standards. The goals of the standards were to "incapacitate violence-prone offenders" (at Level V—imprisonment) and to sanction other offenders for shorter periods at Level V or at intermediate levels.

Subsequently, at the request of SENTAC, the Truth-in-Sentencing (TIS) Act of 1989 was enacted and took effect in July 1990. TIS brought equity and proportionality to time served. TIS precludes release of an offender from Level V except during the last six months of a sentence. It also reduced the statutory maximum lengths of sentences and, for all practical purposes, abolished parole. It was intended to be resource-neutral.

As a result of these changes, violent offenders are being placed in prison at higher rates and for longer times. Over the period from June 1986 to September

Table 2.6. Prison Admissions, Time Sentenced and Time Served, for Violent and Nonviolent Crimes, Delaware 1986–1991

Types of Crimes	1986	1987	SENTAC Starts 1988	1989	TIS Phase-In 1990	TIS 1991
Violent Crimes:						
Murder admissions	30	14	15	8	17	20
Time sentenced	225.5	122.0	156.5	109.2	76.2	180.9
Time served	43.1	61.8	53.6	58.9	67.6	136.0
Sex admissions	55	80	69	71	84	58
Time sentenced	83.6	82.4	80.8	93.4	96.3	99.4
Time served	33.9	29.1	27.5	30.1	41.0	74.6
Robbery admissions	90	84	72	97	71	84
Time sentenced	103.0	98.7	86.1	96.3	71.6	81.4
Time served	38.0	51.3	50.0	42.5	47.8	61.1
Assault admissions	106	90	101	103	83	67
Time sentenced	61.8	51.5	54.2	51.9	47.0	42.2
Time served	25.6	27.1	27.1	23.8	29.6	31.7
Nonviolent Crimes:						
Burglary admissions	110	127	138	155	105	86
Time sentenced	75.9	71.7	64.7	58.8	54.5	38.3
Time served	30.3	31.7	35.2	29.7	30.1	28.7
Theft admissions	121	99	86	81	47	35
Time sentenced	42.1	37.6	33.6	34.4	37.9	31.6
Time served	18.7	22.2	25.0	18.9	23.0	23.7
Fraud admissions	28	38	41	35	28	12
Time sentenced	36.4	53.6	30.5	36.4	26.1	36.4
Time served	18.5	16.0	19.6	16.0	22.5	27.3

Notes: 1991 information is an estimate based on nine months data.
Time served 1986 to 1990 is based on releases within the year.
Time served 1991 is calculated as 75 percent of time sentenced.
Murder admissions do not include lifers.

Source: Delaware Statistical Analysis Center (unpublished data provided to author).

1991, the number of violent offenders in prison increased by 64 percent, from 1,284 to 2,009. This is partly due to TIS, which fixed the time to be served at 75 percent of the incarcerative sentence, which, as table 2.6 shows, has increased average sentences for violent offenses.

Nonviolent offenders are being supervised in effective community programs rather than occupying prison beds needed for violent offenders. According to Delaware's Statistical Analysis Center, the nonviolent felony population incarcerated (Level V) decreased from 460 in 1987 to 259 on June 30, 1991. TIS has reduced the average time of incarceration for many nonviolent offenses and reduced the number of admissions for others (table 2.6). As table 2.6 also shows, compared with 1987 before SENTAC took effect, in 1991, burglary admissions were down by one-third and theft and fraud admissions by two-thirds.

Table 2.7. Prison Population Growth (%), 1985–1990

Period	United States	Minnesota	Washington	Delaware
1985–86	8.6	5.1	–4.6	11.2
1986–87	7.4	3.4	–7.1	10.4
SENTAC starts				
1987–88	8.0	9.9	–5.1	1.9
1988–89	12.5	10.9	19.1	3.8
1989–90	8.6	2.4	15.4	–2.3
% change				
1985–87	16.7	8.7	–11.4	22.7
% change				
1987–90	31.9	24.7	30.4	3.3

Sources: Bureau of Justice Statistics, Prisoners in 1990, Prisoners in 1988, and Prisoners in 1986.

Prior to the SENTAC initiative, a shortage of sentencing options hindered the choices of judges. With the adoption of the five-level continuum of punishment, monitoring procedures showed judicial compliance with guidelines in 90 to 95 percent of cases. Judges value the purposes of alternative sanctions and are willing to use them. They are also willing to accept, and abide by, a system of sentencing standards that allows room for reasoned discretion in individual cases.

It is also instructive to review the growth trends in Delaware and elsewhere for prison population with sentences greater than one year. As evidenced in the Bureau of Justice Statistics Bulletin *Prisoners in 1990*, despite increasing the number of violent offenders behind bars, Delaware has managed its limited bed space effectively compared to the United States as a whole and two states with presumptive guidelines—Minnesota and Washington (table 2.7).

Delaware's Sentencing Accountability Commission has balanced concerns for punishment and treatment in a time of scarce resources while still addressing demands for public safety. The results demonstrate that a voluntary judicial guidelines system can indeed change sentencing patterns.

Other States

Sentencing Reform in Colorado—Many Changes, Little Progress
Marianne Wesson

Colorado has not gotten very far in its efforts to make sentencing fair and cost-effective, but it has taken a lot of time and effort to get there. Since the early 1970s, policy makers have proposed, passed, revised, repassed, and vetoed sentencing laws. As in many states, sentencing was on the legislative policy agenda throughout most of the 1980s. Judicial discretion has been narrowed, expanded, and narrowed again.

Parole release has been eliminated and reconstituted. Good time has been narrowed and broadened.

This article recounts Colorado's convoluted history of recent sentencing reform. Although reformers' goals have varied, some wanting heavy sentences, some wanting lighter ones, most over the years have sought greater certainty in sentencing, greater consistency, and more cost-effectiveness. Sentences have mostly become harsher, but Colorado appears as far away in 1993 from its goals of certainty, consistency, and economy as it was in 1972.

Colorado's 130 percent increase in its prison population between 1985 and 1991 led the nation. Local jails today hold 500 to 600 prisoners under sentence to state incarceration.

Colorado faces the same pressures on its prisons and jails that most states face. But since November 3, 1992, Colorado has faced an additional pressure not precisely matched anywhere else. Coloradans approved Amendment 1, a constitutional limitation on increases in taxing and spending by any governmental body, and Amendment 8, which requires the dedication of state lottery proceeds, formerly used in part for capital construction, to the purchase and maintenance of outdoor recreation spaces. The correctional system can no longer expect increased funding or substantial new construction to accommodate population growth.

Sentencing reform proposals that aroused little interest among legislators earlier in 1992 suddenly attracted attention. Without sentencing reform, Colorado faces a projected shortfall of 2,554 prison beds by January 1998. That assumes that three new facilities with 1,050 beds will be open.

The Colorado Criminal Justice Commission submitted several proposals to the legislature, and various other proposals also circulated. The commission's original proposal, to reclassify mostly property offenses, was projected to save 335 beds by 1998. A Department of Corrections proposal that would reduce some presumptive statutory sentencing ranges was predicted to bring the population within capacity by 1996, but no sooner. A Judicial Department proposal for moderate sentencing reductions coupled with mandatory parole supervision in every case would reduce the pressure but not eliminate the shortfall.

The most ambitious proposal, the Casey Bill, named after its sponsor Senator Lloyd Casey, would have brought population within capacity by 1994, but its provisions were breathtaking. Statutory sentencing ranges would be returned to their pre-1985 levels retroactively; all prisoners sentenced since 1985 would be entitled to resentencing under the new provisions. Mandatory minimum sentences for crimes of violence and habitual criminal sentences would be abolished retroactively. Parole release at first eligibility would be required for some classes of felons. Life prisoners would become eligible for parole after ten years (those sentenced to life after 1990 are not now eligible for parole).

Drastic measures seemed called for, but the impetus for serious sentencing reform petered out. The only legislation that passed in 1993 shaves a few years off the top of the presumptive sentencing ranges for felonies and enacts a period of mandatory parole supervision for each class. Law reform energy for sentencing was preempted by a successful drive by Governor Roy Romer to enact serious penalties for possession of firearms by juveniles, a measure that, whatever its virtues, will not alleviate Colorado's population crunch.

The 1972 Criminal Code The experience in 1993—seemingly irresistible pressure for sentencing changes producing anticlimactic legislation—has happened before. To make sense of 1993, and the problems Colorado faces, it is necessary to go back to 1972, when the first major modern sentencing reforms were enacted, and then to trace events forward.

In 1972, Colorado, like many states, adopted its first integrated penal code, codifying an indeterminate system of sentencing that was soon to go out of fashion. The code classifies felons into five classes. Class 1 felonies called for the imposition of either death or a life sentence, but for the other felonies, sentencing was indeterminate. The judge designated maximum and minimum sentences, subject to upper and lower statutory limits. Time served was determined by the parole board, taking into account time off for good conduct, or "good time," and time off for the diligent performance of certain duties, or "earned time."

The Gorsuch Bill A more nearly determinate sentencing law, the Gorsuch Bill, named after Ann McGill Gorsuch, then a Colorado representative and later director of the Environmental Protection Agency under Ronald Reagan, was enacted in 1979. As in many states, sentencing reform was supported by an unusual mix of liberals and conservatives, crime controllers and civil libertarians.

A local version of this strange-bedfellows scenario could be observed in Colorado, as the movement brought together such actors as Gorsuch, the District Attorney's Council, and the Colorado public defender. Gorsuch often said that her principal motivation was to show Colorado citizens, by importing truth in sentencing into the scheme, just how short criminal sentences really were. At other times her advocacy suggested that she was a true believer in the merits of determinate sentencing. The chief deputy public defender supported the proposed new law, explaining that his clients in prison suffered from the uncertainty created by indeterminate sentences, and that his office favored any reduction in the discretion of the parole board. Many prosecutors favored the bill because they believed that the indeterminate system encouraged inmate manipulation and deception of the parole board, and because they thought that new legislation might lead to longer sentences.

The Gorsuch Bill's Provisions The fundamental structural decision was to stay with the five-part classification of felonies adopted in the 1972 code. In the first version proposed, each classification was assigned a definite term of years, known as its presumptive sentence. The judge would have been required to sentence the defendant to the precise presumptive sentence designated. However, departures of up to 20 percent upward or downward would have been permitted if a finding on the record explained what aggravating or mitigating factors justified the departure.

The hard part was to fill in the numbers for the presumptive sentences, an exercise that created enormous confusion. By present standards, sentences at the time were not long. For a class 2 felony (for example, second-degree murder, or a violent sexual assault resulting in bodily injury), the presumptive sentence was seven years. This version of the bill did not seek to change the existing good-time practice that could shorten time served by more than a third; for a moderately tractable prisoner, the effective time served would have been between four and five years. For class 3 felonies, including first-degree arson and most violent sexual assaults, the presumptive sentence would have been three years, the effective time served less than two.

Problems with Projections. Gorsuch explained that she fixed these numbers by asking the Department of Corrections to compute the average time served for these various classifications of felony under the existing system.

The District Attorney's Council endorsed the idea of presumptive sentences but complained that the sentences were too short. At this point, things got confused. Gorsuch announced at a House Judiciary Committee meeting that she had received "new figures" from the department—it seemed there had been some error in computing the current average sentences. The new figures upped the presumptive sentences for class 2 felonies by six months and for class 3 felonies by eighteen months.

With the new numbers, the bill passed the house and was sent to the senate, and the odd coalition supporting the bill began to fall apart. In the Senate Judiciary Committee, the director of research and planning for the Department of Corrections testified that his department opposed the bill on a number of grounds, including its lack of provision for parole supervision. He also observed that the bill, if passed, would likely increase average daily population by 93 inmates. Gorsuch was furious. There followed a heated exchange, at the end of which the witness insisted that the original figures had been more nearly accurate than the "adjusted" figures that now appeared in the bill.

Empowering Parole. By this point doubt had grown about the reliability of either set of figures. When the bill was reported out with some amendments and reached the senate floor, support had weakened enough to attract an amendment. Senator Ralph Cole, decidedly no criminal coddler, proposed that the parole board be given back some of its authority by reinstituting the old practice of allowing it to award earned time over and above the good time granted or withheld by the prison authorities, and (perhaps more significantly) allowing it to parole an inmate after the service of one-half of his or her sentence, irrespective of good time.

The parole amendments (which the senate adopted) would have restored a good deal of the parole board's back-end discretion that the legislation's original purpose was to abolish, and it would have significantly shortened the time served by many prisoners. This was too much for the district attorneys, who withdrew their support. Even without their support, however, earned-time amendments survived the conference committee and both houses agreed to the compromise hammered out in conference. The bill had some ambiguous provisions intended to allow inmates to elect whether their time credits were to be computed under the pre- or post-Gorsuch formula, and it was susceptible to the interpretation that they were to enjoy the benefits of both. This drafting problem became important later, but it didn't seem to bother anyone too much at this point. The bill was sent to Governor Lamm on June 15, 1977, to take effect on July 1, 1978.

The Governor's Veto.   Governor Lamm vetoed the Gorsuch Bill on July 15, 1977, citing the lenience of the presumptive sentences, uncertainty about the calculation of good time and earned time, and opposition from the district attorneys. Some have suggested that his real objection was more nuanced: not that he actually objected to sentence lengths, but that he feared they looked so short that they would attract Draconian amendments, which he would have to veto on fiscal grounds, setting himself up for the charge that he was soft on crime.

To understand what followed, one must know that in Colorado the governor may veto a bill within ten days during a legislative session or within thirty days if the legislature is out of session. If he does not act on time, the bill becomes law without his signature. The legislature had adjourned five days after passing the Gorsuch Bill. Lamm believed that his July 15 veto of a bill delivered to him on June 15 was timely, but many of the bill's disappointed supporters did not, and some asked the Colorado Supreme Court to declare his veto untimely and hence invalid. While this suit was pending, proponents of sentencing reform tried to persuade Governor Lamm to place sentencing reform on his "call" for the 1978 legislative session; in Colorado, sessions in even-numbered years may consider only budget matters and matters on the governor's call. Nine months later, on April 10, 1978, the Colorado Supreme Court ruled that Governor Lamm's veto was invalid. Since the legislature, although in session, could not enact any repeal or any new sentencing legislation in its regular session that year because no such item was on the call, it appeared that the Gorsuch Law would go into effect on July 1.

The Good-Time Problem.   As mentioned previously, the bill passed in 1977 had a poorly drafted provision that might have been interpreted as giving some the right to cumulate good time and earned time under both the old and new laws. Attorney General J. D. MacFarlane predicted that 400 to 500 inmates might become eligible for early release. This prospect was worse from a public relations standpoint than any Governor Lamm had sought to avoid by vetoing the legislation, and Lamm decided to seek a special session of the legislature. The legislature's leadership agreed, and in a brief special session the effective date was postponed until April 1, 1979.

There was now a little less than a year to resolve the problems in the Gorsuch Bill. A committee created by the Colorado Legislative Council, which the legislature had directed to study classification of offenses, parole, and good time, also tried to clarify the confusion over existing average sentences—reportedly with at best partial success. This committee eventually recommended a return to the regime of sentencing ranges imposed by judges, but with a new set of ranges. Perhaps because this committee hoped to avoid a fiscal impact, its proposal contained such presumptive ranges as three years, seven months, and six days to five years, four months, and twenty-four days for a class 3 felony. Things were getting silly, and it looked as though the consensus supporting reform was falling apart.

A Smoke-Filled Room.   Governor Lamm, unwilling to watch passively while the Gorsuch Law went into effect, convened important actors for a one-day conference in February 1979. Those in attendance included the governor, Representative Gorsuch, Denver District Attorney Dale Tooley, Public Defender Greg Walta, and a number of well-respected judges. At one point Gorsuch, Tooley, and Walta caucused apart from the others. Out of that caucus came an agreement that, with minor modification, became the final version of the Gorsuch Bill.

The numbers were settled. Each class of felony other than class 1 was assigned a sentencing range: for class 2, eight to twelve years; for class 3, four to eight years; for class 4, two to four years; and for class 5, one to two years. Each sentence would include in addition one year of parole supervision. Terms outside the presumptive range could be handed down within what might be called the "expanded presumptive range," which would range from half the minimum to twice the maximum. Thus for class 3 felonies, for example, since the presumptive range was four to eight years, the actual permissible range was two to sixteen years, but any sentence greater than eight or smaller than four years had to be justified in a written statement and automatically reviewed by the court of appeals.

The new law provided for good time, which if fully awarded by the prison authorities could cut a sentence in half, and earned time at a rate of up to one month per six, which would be administered by the parole board. Thus both forms of back-end discretion were preserved, and in the case of good time expanded: the formula of 2:1 was much more generous than the pre-Gorsuch formula of 3:1. And judges who could choose a sentence from within a range that might cover as long as twenty years were hardly shorn of discretion, either. Moreover, judges faced no constraints on their discretion to impose probation (except in capital cases) and very few constraints on their discretion to sentence concurrently or consecutively for multiple offenses.

Thus the legislation was of questionable value in achieving its original goals of equalizing sentences and divesting judges and administrators of discretion. Truth in sentencing was another casualty. Public Defender Greg Walta favored the final bill because its generous good-time and earned-time provisions constituted what he called a "shell game" that would "protect us from the rage of the general public." The Gorsuch Bill as it emerged from caucus was introduced on March 5, and with little debate by March 29, 1979, had been passed with minor amendments, reconciled in both chambers, and signed by Governor Lamm. Except for certain sentencing laws that were unaffected—lengthy mandatory prison sentences for habitual offenders and indeterminate sentencing for sex offenders—the Gorsuch Bill became the law of sentencing.

Changes in the 1980s Almost uniformly, amendments and additions in the 1980s led to longer and harsher sentences. A 1981 amendment created classes of offenders who were required to be sentenced in the so-called aggravated range; that is, to a sentence at least one day longer than the presumptive maximum. They included those who were on bond, probation, parole, deferred judgment, or in confinement at the time of their crime. Even this law, however, did not preclude probation for such offenders—it merely required that if they were sentenced to incarceration, it had to be in the aggravated range.

One sort of offender was required, under the 1981 laws, to be sentenced to incarceration, and for a term of years in the aggravated range: those convicted of "crimes of violence." This all-purpose phrase was redefined and broadened nearly every year; roughly, it meant any of a large number of serious crimes committed with the use of a deadly weapon or resulting in death or serious bodily injury. For these crimes, the judge's discretion was more severely constrained than otherwise, since probation was ruled out and the sentence had to be longer than the presumptive maximum.

Perhaps more important was a little-noticed change that reinvented parole release. The Gorsuch Law had given the parole board discretion only to grant or withhold earned time. Otherwise it was limited to parole supervision and revocation. But the scheme changed in 1985 so that an offender, rather than being entitled to release at the end of his or her determinate sentence, minus presentence credits, good time, and earned time, merely became eligible for parole on that date. Release could be granted or denied between the date determined by these deductions and the straight-time expiration of the inmate's sentence. This might have made little difference had the parole board routinely granted parole at the first possible date as it had pre-Gorsuch; it did not do so, instead it granted parole release at first eligibility in only about one-third of cases.

A far more visible change in 1985 doubled the term of years that formed the top of the presumptive range for each class of felony; hence, the range for class 2 felonies grew from eight to twenty years. Keeping in mind that judges could, in what they thought were extraordinary cases, sentence in a range that varied from half the minimum to twice the maximum, the domain of judicial discretion became even larger, for example, four to forty-eight years in the case of a class 2 felony.

The sponsors predicted that judges would reserve sentences in the top of the newly expanded presumptive range for unusual cases, but this did not prove to be the case. A 1987 study by Mary Mande of the Colorado Department of Public Safety concluded that the average sentence imposed after 1985 was far above the midpoint of the presumptive range. Moreover, Mande concluded that the higher sentences did not seem to be reserved for the most serious offenders; the relation between lengths of sentence and severity of offenses was weak.

By the time the 1985 amendments took effect, most of the objectives of the 1970s' sentencing reforms had been undermined. Judges had enormous discretion over who went to prison and for how long. The parole board could release the inmate at any time during the second half of the sentence imposed. Equally important, these alterations to the Gorsuch scheme contributed to the out-of-control growth of the state's prison population in the 1980s. So did mandatory minimum sentences that were enacted during the 1980s for crime-of-the-week-type offenses ranging from habitual sex offenses against children to habitual shoplifting to distribution of twenty-eight or more grams of cocaine.

Proposed Sentencing Guidelines   An attempt was made in the mid-1980s to develop and enact sentencing guidelines. Senator Jeff Wells of Colorado Springs, assisted by Mary Mande and Bill Woodward of the Colorado Division of Criminal Justice, developed legislation that would have created a guidelines sentence for each offense classification. The bill was a serious attempt to link sentencing policy to corrections capacity—each sentencing cell was shown to legislators with an "impact statement" explaining the effects of any change on the prison population. The proposed guidelines used the existing offense classifications but divided them into violent and nonviolent instances. The guidelines took account only of offense severity and prior convictions (again distinguishing violent from nonviolent); each cell contained not a single term of years but a very narrow range. The guidelines would have been advisory; a judge could, by giving reasons on the record, depart from the range. Changes had been made to satisfy prosecutors, in particular to make both prior misdemeanor and prior felony convictions "count." In the end, however, the district attorneys persuaded Governor Lamm to veto the bill.

Halfhearted Changes in the Late 1980s   In the still later 1980s, two changes were enacted that constituted a halfhearted effort to bring the prison population under control. A new felony classification was created; its presumptive sentencing range is from one year to two years. Various offenses were reclassified downward. And an effort was made to lighten the impact of the mandatory aggravated sentences required for some crimes; rather than requiring a sentence in the aggravated range in such cases, the law was changed in 1988 to require only that a sentence of at least the midpoint of the presumptive range be imposed. These ameliorative measures did not make a dent in the prison population, although the rate of growth apparently slowed somewhat.

And that brings matters back to where I began. The optimistic view of hard choices is that they represent opportunities. Colorado has an opportunity now, albeit one that was forced on it, to decide what is really important in its criminal justice system. As I write, the choices are being made; the history is being written. Many resist the idea that sentencing policy should be driven by economic considerations, but if there is a moral in this tale, it seems to be for them. It is the same message my mechanic gives me when I want to procrastinate about dealing with some incipient automotive malfunction: pay me now or pay me later.

Utah's Conjoint Guidelines for Sentencing and Parole
Richard J. Oldroyd

A sentencing commission is at work in Utah, as in many other states, charged by the legislature to prepare proposals for consideration in 1995. Six months into the process when this is written, the commission has adopted a policy statement that reflects a complex mix of views.

The policy statement provides, as in most states, for visibility and accountability in decision making, for recognition of victims' interests, for a continuum of punishments, and for incarceration for violent and dangerous offenders.

However, unlike more just-deserts-based approaches, this one also calls for individualized consideration of each offender's circumstances, for presumptions in favor of nonprison sentences and against incarceration, for acknowledgment of offenders' efforts at self-improvement, and for distinguishing for many crimes between short sentences for "situational" offenders and longer ones for repeat offenders.

The commission can build on nearly two decades of guidelines experience, which has included felony guidelines for sentencing and parole and, for a time, pilot guidelines for misdemeanors. Two generations of felony guidelines have been developed—in 1979 and 1985. This article recounts Utah's experience with those guidelines and describes the foundations on which the new commission's proposals will stand.

Utah in 1977 started a sentencing reform process based on a unique collaboration among the affected agencies. In other guidelines states, sentencing commissions developed guidelines and then tried to sell them to other agencies. In Utah, by contrast, sentencing and parole release guidelines were jointly developed by the courts, the parole authority, corrections, the prosecutors' association, and the state criminal justice coordinating body. The process has not been without difficulty, but it has continued through major evolutions. Utah's prisons are full, but they are within their capacities. Only Maine, Minnesota, North Dakota, and West Virginia have lower incarceration rates. Tensions exist between governmental entities, but the spirit of cooperation remains alive.

The Setting   Utah has a population of 1.8 million, 80 percent of whom live in a narrow corridor ninety miles long known as the Wasatch Front. Utah is one of the most urban states, and one of the fastest growing.

Utah may be most distinctive because of its large number of children. The fertility rate (number of women having children) is the highest of any state and is 37 percent higher than the national average. As a result, Utah's median age of 25.7 years is the youngest in the nation by nearly four years. The percentage of school-age children is 50 percent higher than the national average.

Since young people aged fifteen to nineteen are arrested for most crimes at much higher rates than those older or younger, it would not be unreasonable to expect Utah's crime problem to be more severe than those of most other states. Reported rates for most violent and property crimes are, however, substantially below national averages. Reported rates of the commonly underreported crimes of rape and child sexual abuse are higher. It is unclear whether this is because these offenses are more common or because people in Utah are readier to report them.

Utah's is an indeterminate sentencing system that retains parole release. The penal code was passed in 1973. Felonies are divided into four categories bearing different authorized sanctions: capital—punishment by death or life without parole; first degree—punishable by five years to life in prison; second degree—punishable by one to fifteen years in prison; and third degree—punishable by zero to five years in prison. The court decides whether a convicted felon should be imprisoned. The Board of Pardons decides for how long.

In 1977 the legislature created a Blue Ribbon Task Force on Criminal Justice. This happened about the time when the rehabilitative philosophy was being discredited and other approaches were being implemented around the country. California adopted determinate sentencing, in which the legislature specified sentences for every crime. Many states were considering the abolition of parole. The Utah legislature was concerned about inequities in sentencing but conservatively chose not to make major changes. The courts were asked to formalize their policies by developing sentencing guidelines, and the Board of Pardons was asked to develop release guidelines.

The 1979 Guidelines   The courts and the board began separately to develop guidelines. Corrections officials watched with interest but became concerned when they realized that offenders might be sentenced under one philosophy and released under another. Corrections officials, the courts, and the board decided to develop conjoint guidelines based on agreed premises.

First, sentencing and parole guidelines for felonies and misdemeanors should be developed and overseen within the framework of the existing indeterminate sentencing law by a committee representing the state's judiciary and the Board of Pardons.

Second, the guidelines, based on the severity of crimes and offenders' risks of continued criminality, should guide rather than dictate sentencing.

Third, guidelines should be developed to be consistent with the current practices of the Utah courts and Board of Pardons and might later be modified to be more prescriptive.

The above postulates were accepted and the process of developing guidelines began. The first task was to develop a "risk assessment instrument." Since no appreciable financial resources had been allocated for this project, staff selected the best variables from the prediction literature and worked with the judges and the board to make them appropriate and useful for the settings in which they would be used. The variables ultimately included age at first arrest, prior juvenile record, prior adult arrests, correctional supervision history, supervision risk, percentage of time employed in the last twelve months, alcohol usage problems, other drug usage problems, attitude, address changes in the last twelve months, and family support. In addition, points were added if there had been an adjudication for an assaultive offense within the preceding five years.

The next task was to measure crime severity. The simple approach was chosen of using the degree of the charged crime as established in the Utah code. First- and second-degree offenses were divided into two categories: serious and moderate.

A matrix was constructed and data were collected to see how well the two dimensions of risk and crime severity accounted for current sentencing practices. The fit was not perfect, but a strong general pattern existed. Staff developed the dispositional recommendations within the matrix and established five risk assessment categories.

The final task was to fix recommended times to be served in prison. A recent study of time served in the Utah state prison provided some basic direction. A tentative sentence range was put in a matrix cell. The prison population was sampled and the ranges adjusted so that the average length of stay would be approximately twenty-eight months if the midpoints of each range were used.

The matrix was approved by both the judiciary and board for a six-month trial. Adult probation and parole staff were trained, and in January 1979 the guidelines were implemented. After six months, there were enough data to validate the risk assessment instrument and to see if the guidelines were being followed. Through random sampling procedures, fifty felons were selected who had successfully completed probation and another fifty were selected who had failed. Similar samples were selected for misdemeanants. The risk assessment appeared to be functioning as anticipated for felons, but not for misdemeanants. For this and other reasons, the guidelines did not continue being used for misdemeanor cases.

1979 to 1985   Between 1979 and 1985, several major events occurred with significant implications for sentencing. In the late 1970s and early 1980s, Utah became terrified by the disappearance of young women and children. Eventually most of these cases were resolved through the arrests of Ted Bundy, who sexually molested and murdered many young women, and another offender, Arthur Gary Bishop, who had molested and murdered the young men. The alarm at these and some high-profile child sexual abuse cases led to passage of mandatory minimum sentencing provisions for sexual offenses against children and for rape. Another notorious incident occurred one Christmas when a prisoner released to a halfway house absconded and then kidnapped a young woman with the intent of sexually assaulting her. While attempting escape, she was shot in the back and became paralyzed.

The resulting media attention and public outcry produced a major change in criminal justice focus in Utah. Prevailing sentiment, long oriented to rehabilitation, shifted toward punishment and incapacitation. The director of the Division of Corrections was asked to resign, and the division was removed from the Department of Social Services and made into an independent Department of Corrections. All subsequent directors have had backgrounds in law enforcement rather than in social work, as had previously been the case.

The 1985 Guidelines   In 1983 the Utah legislature created the seventeen-member Commission on Criminal and Juvenile Justice, whose principal purpose was "to ensure broad philosophical agreement concerning the objectives of the criminal and juvenile justice system in Utah and to provide a mechanism" to achieve those objectives (Utah Stats. s63-25-1). Part of the statutory charge was "to develop, monitor, and evaluate sentencing and release guidelines for adults and juveniles; and forecast future demands on the criminal justice system, including specific projections for secure bed space" (s63-25-4[7,8]).

The commission started work on revision of the guidelines. Its first premise was that to enhance truth in sentencing, the guidelines should be broadened to implement and guide law enforcement, prosecution, and defense decisions.

The Utah guidelines have never been statutory but have been endorsed by the agencies that use them. They have become part of the judicial rule-making process, but they have not gone through any other rule-making procedure. Their purpose has never been to dictate sentencing practice but to institutionalize a process and provide a standard to compare sentencing-related decisions against. Prosecutors and defense attorneys view the guidelines as the most likely outcome as they explore charges and plea negotiations. Presentence investigators prepare the guidelines forms as part of their investigation and must justify recommendations that vary from the guidelines' suggested disposition. Similarly, the judges have been asked to give reasons for varying from the guidelines on the court record. The Board of Pardons considers the guidelines in release decisions, stating reasons for departure.

The board also tries to achieve equity between judges and various geographical regions in setting release dates.

There have been several evaluations of guidelines compliance. Overall compliance has been quite high (over 90 percent); however, that should be expected since sentencing in the majority of cases is clear-cut, with about 70 percent receiving probation. Guidelines compliance is much lower when the focus is on those who are recommended for or receive prison. Because the Utah guidelines are intended to guide rather than dictate sentences, and there is encouragement to deviate for appropriate aggravating or mitigating circumstances, compliance has been monitored more for purposes of refining the guidelines than for evaluating their success or failure.

A major challenge in the 1985 guidelines was to take account of mandatory minimums enacted in 1983 for offenses against children. The legislation mandated a ten-year prison sentence for a sex offense where the victim was a child. The minimum sentence could be reduced to five years if mitigating circumstances existed or increased to fifteen years for aggravating circumstances. The legislation bound both the courts and the Board of Pardons. It created particular problems for the board since the board tried to reduce disparity in sentencing. The vast majority of sexual abuse cases are plea bargained to offenses that do not carry the mandatory minimum term, leaving the board the dilemma of trying to balance justice in situations where a person convicted of a third-degree felony really committed a crime much more egregious than a person sentenced to ten or fifteen years. No good solutions have been found. Another problem is that in 1983, second-degree murderers were serving an average of eight years in prison. The board felt that murder was more serious than child sexual abuse and that time to be served for those offenses should increase dramatically to maintain a sense of equity. This had substantial impact on the number of prisoners and on the costs of incarceration.

In 1985, the social and political climate was one of getting tough on crime. The emphasis in corrections had shifted from trying to support and be helpful to offenders to a law enforcement approach in which the goal was to ensure that offenders who committed crimes or violated probation or parole agreements were incarcerated. The guidelines were adjusted to incarcerate more offenders. The times to be served in the guidelines matrix were originally conceived to represent the typical sentence. The revised guidelines proposed these times as minimums.

Guidelines for Misdemeanors   After a format for felony guidelines was developed in 1979, interest shifted to misdemeanor guidelines. Ultimately nothing was implemented, but some of the ideas might be valuable as the guidelines are reconsidered and revised.

One of the major difficulties was that resources to respond to misdemeanor offenses vary greatly from jurisdiction to jurisdiction. In the urban areas there were community service programs, restitution programs, residential treatment programs for alcohol and drug offenders, and numerous other treatments or services. Some of the rural areas had only a jail and nonresidential mental health services.

To attempt to deal with this problem, the concept of calculating "criminal justice points" was developed. The basic notion was that a day in jail could be equated with so many hours of community service, a fine or restitution amount, or so many hours of residential or outpatient treatment.

Guidelines for 1995   There is general agreement that the 1985 guidelines need major revision. In 1992, a special sentencing study committee was established to make recommendations. It concluded that the major threat to the cooperative guidelines process was the tendency of the legislature to create mandatory minimum sentences in response to high-profile crimes, and it felt a need to integrate guidelines more closely with the legislative process. As a result, the study committee proposed, and the legislature enacted, legislation to create a twenty-member sentencing commission attached to the Commission on Criminal and Juvenile Justice. The principal purpose of the sentencing commission is to make recommendations to all three branches of state government concerning sentencing.

The sentencing commission is about six months into its work. Its goal is to have a proposed structured sentencing system ready for consideration in the 1995 legislative session. At this point, it has drafted a policy statement. Some of the policy premises expressed in its mission statement were described in the opening paragraphs of this article. What the commission's recommendations will be remains to be seen.

Kansas Adopts Sentencing Guidelines
David Gottlieb

On July 1, 1993, Kansas will add its name to the list of states that have scrapped an indeterminate sentencing system and replaced it with a determinate system of presumptive sentencing guidelines.

Under the new guidelines, parole release has been eliminated and possible time off for good behavior has been reduced. A guidelines matrix establishes presumptive prison or nonprison sentences for most convicted felons and, for a small number, allows the judge to choose between prison and noncustodial penalties. The guidelines were precipitated by prison overcrowding and related federal court orders. The governing legislation directed the commission to take correctional resources into account in developing guidelines. The latest estimates are that the guidelines will not reduce prison numbers but may slightly reduce admissions and stabilize the population.

The Background   The guidelines were written by a sentencing commission created by the legislature in 1989. The commission's initial recommendations were transmitted to the legislature in 1991. After debate and amendment, the bill was deferred for a year. During the 1992 session, a guidelines bill passed both houses and was signed into law to take effect July 1, 1993. In the legislative session just concluded, the legislature resisted efforts to change the new system substantially or eliminate it, although a number of technical changes were approved.

The move toward sentencing guidelines in Kansas is tied directly to prison overcrowding. During the early to mid 1980s there was little political support for revamping sentencing. Although the attorney general, among others, had been arguing in favor of determinate sentencing as a way of getting "tough on crime," the initial determinate sentencing proposals received only lukewarm support. On two occasions, bills were proposed but failed to reach the floor of either house.

The state's prison overcrowding problem then worked an abrupt change in the politics of sentencing reform. From 2,416 in 1979, the prison population in Kansas nearly tripled in ten years, reaching 6,172 inmates in 1989.

This increase had predictable results. By the mid-1980s the prisons were operating at double their capacity. Almost inevitably, a federal judge declared the major institutions in violation of the Eighth Amendment's prohibition of cruel and unusual punishment. Population caps were imposed at the two major prisons. The state's initial response was a major and expensive building program that slowed, but did not stop, the overcrowding.

In 1989, a general consensus was taking shape that something more had to be done. After holding hearings on prison overcrowding, the legislature created the Kansas Sentencing Commission. The commission was charged with developing sentencing guidelines based on "fairness and equity" that would provide a mechanism to "link justice and corrections policies." The commission was directed to take into account "correctional resources, including but not limited to the capacities of local and state correctional facilities" (Kans. Stats. §74-9101 [b] [2]). Thus, like Oregon's guidelines system, the system in Kansas was an attempt to write capacity-based guidelines for an already overcrowded system.

At least at the beginning, the move toward guidelines sentencing had broad support. The attorney general, already on record as supporting determinate sentencing, was made a member of the commission and supported its work. Other law enforcement organizations expressed support, as did the major public defender offices and a number of academics. Civil rights organizations supported the move to guidelines, particularly after the commission published a report documenting significant racial disparities at every stage of the current system. Opposition at the beginning consisted of those who had the most to lose by the commission's work—the judiciary and parole board.

The Guidelines   The system fundamentally alters the purposes and structure of punishment in Kansas. Kansas has now explicitly embraced retribution and incapacitation as principal reasons for punishment and imprisonment. Thus, the crime of conviction, rather than the nature of the offender, is now the primary determinant of the sentence.

Purposes. The move to a just-deserts orientation required the commission to weigh the relative harms resulting from different crimes. The commission concluded that society's principal interest was in protecting individuals from physical and emotional injury; it ranked harms to private property rights as secondary and harms to governmental institutions as least important.

The commission, in drafting its guidelines, also attempted to achieve "equity" and "truth in sentencing." Equity is to be achieved by providing similar punishments for defendants who are convicted of the same crime with similar criminal records. Truth in sentencing will be achieved by requiring that the actual term fixed by the judge will match, more or less, the actual sentence served.

Category	A	B	C	D	E	F	G	H	I
Severity Level	3+ Person Felonies	2 Person Felonies	1 Person & 1 Nonperson Felonies	1 Person Felony	3+ Nonperson Felonies	2 Nonperson Felonies	1 Nonperson Felony	2+ Misdemeanors	1 Misdemeanor No Record
I	204 / 194 / 185	193 / 183 / 173	178 / 170 / 161	167 / 158 / 150	154 / 146 / 138	141 / 134 / 127	127 / 122 / 115	116 / 110 / 104	103 / 97 / 92
II	154 / 146 / 138	144 / 137 / 130	135 / 128 / 121	125 / 119 / 113	115 / 109 / 103	105 / 100 / 95	96 / 91 / 86	86 / 82 / 77	77 / 73 / 68
III	103 / 97 / 92	95 / 90 / 86	89 / 85 / 80	83 / 78 / 74	77 / 73 / 68	69 / 66 / 62	64 / 60 / 57	59 / 55 / 51	51 / 49 / 46
IV	86 / 81 / 77	81 / 77 / 72	75 / 71 / 68	69 / 66 / 62	64 / 60 / 57	59 / 56 / 52	52 / 50 / 47	48 / 45 / 42	43 / 41 / 38
V	68 / 65 / 61	64 / 60 / 57	60 / 57 / 53	55 / 52 / 50	51 / 49 / 46	47 / 44 / 41	43 / 41 / 38	38 / 36 / 34	34 / 32 / 31
VI	46 / 43 / 40	41 / 39 / 37	38 / 36 / 34	36 / 34 / 32	32 / 30 / 28	29 / 27 / 25	26 / 24 / 22	21 / 20 / 19	19 / 18 / 17
VII	34 / 32 / 30	31 / 29 / 27	29 / 27 / 25	26 / 24 / 22	23 / 21 / 19	19 / 18 / 17	17 / 16 / 15	14 / 13 / 12	13 / 12 / 11
VIII	23 / 21 / 19	20 / 19 / 18	19 / 18 / 17	17 / 16 / 15	15 / 14 / 13	13 / 12 / 11	11 / 10 / 9	11 / 10 / 9	9 / 8 / 7
IX	17 / 16 / 15	15 / 14 / 13	13 / 12 / 11	13 / 12 / 11	11 / 10 / 9	10 / 9 / 8	9 / 8 / 7	8 / 7 / 6	7 / 6 / 5
X	13 / 12 / 11	12 / 11 / 10	11 / 10 / 9	10 / 9 / 8	9 / 8 / 7	8 / 7 / 6	7 / 6 / 5	7 / 6 / 5	7 / 6 / 5

Presumptive Imprisonment (boxes above black line)   Border Boxes   Presumptive Probation (boxes below black line)

Figure 2.17. Presumptive Sentences for Nondrug Felonies. *Source:* Kansas Sentencing Commission (1992).

The Grid. As in many other guidelines states, Kansas will use a grid to specify presumptive sentences for "ordinary" cases. As is shown in figure 2.17, virtually all felony offenses have been ranked at one of ten severity levels, with level 1 crimes, such as aggravated kidnapping, the most severe, and level 10 crimes, such as pirating a sound recording, the least. A separate grid (see figure 2.18) creates four rankings for drug offenses.

The second measure used to compute the sentence is a criminal history scale. The Kansas scale consists of nine different categories, using the number and nature of prior convictions as markers of "increased culpability." Felonies are treated more severely than misdemeanors, and "person crimes" are treated more seriously than nonperson crimes. The scale used in Kansas is virtually identical to that first implemented in Oregon.

The two measures form the axes of the matrix used to compute most sentences. At the intersection of each crime severity and criminal history a narrow sentencing range is set forth. The grid itself is divided by a "dispositional line." The boxes below the bold black line establish presumptive probation sentences. The boxes above the line presume imprisonment. The commission also established a few "border boxes" that allow judges to choose between imprisonment or a community-based sanction.

Category	A	B	C	D	E	F	G	H	I
Severity Level	3+ Person Felonies	2 Person Felonies	1 Person & 1 Nonperson Felonies	1 Person Felony	3+ Nonperson Felonies	2 Nonperson Felonies	1 Nonperson Felony	2+ Misdemeanors	1 Misdemeanor No Record
I	204 / 194 / 185	196 / 186 / 176	187 / 178 / 169	179 / 170 / 161	170 / 162 / 154	167 / 158 / 150	162 / 154 / 146	161 / 150 / 142	154 / 146 / 138
II	83 / 78 / 74	77 / 73 / 68	72 / 68 / 65	68 / 64 / 60	62 / 59 / 55	59 / 56 / 52	57 / 54 / 51	54 / 51 / 49	51 / 49 / 46
III	51 / 49 / 46	47 / 44 / 41	42 / 40 / 37	36 / 34 / 32	32 / 30 / 28	26 / 24 / 23	23 / 22 / 20	19 / 18 / 17	16 / 15 / 14
IV	42 / 40 / 37	36 / 34 / 32	32 / 30 / 28	26 / 24 / 23	22 / 20 / 18	18 / 17 / 16	16 / 15 / 14	14 / 13 / 12	12 / 11 / 10

 Presumptive Probation   Presumptive Imprisonment

Figure 2.18. Presumptive Sentences for Drug Felonies. *Source:* Kansas Sentencing Commission (1992).

The judge may depart from the presumptive sentence only for "substantial and compelling reasons." The statute lists five mitigating and four aggravating factors. The list is explicitly stated as "nonexclusive." If the judge decides to depart upward from the presumptive sentence, the increased sentence cannot be more than double the presumptive prison term. The statute also contains criteria for departures from probation to imprisonment, and vice versa. A departure may only be imposed upon notice and after a hearing.

Sentence Severity. The sentence imposed will closely correspond to the sentence served. Reductions for good time, previously as much as 50 percent, have been reduced to 20 percent. The legislation also abolishes the current parole system. Inmates will receive a short period of "postrelease supervision" that will be imposed as part of each felony sentence.

The commission anticipates that fewer property offenders will be going to prison after the implementation of the guidelines. The likelihood of drug offenders receiving imprisonment will increase, and the lengths of sentences served by many violent and sex offenders will also increase.

Impact Projections What is more difficult to predict is the effect that the guidelines will have on prison population. The estimate made in 1991 when the initial guidelines were proposed was that the guidelines would stabilize the rate of imprisonment but that the population would continue to increase. That prediction disappointed some legislators and led to changes in the matrix. Between 1991 and the final bill in 1992, the legislation was changed to include border boxes in what had previously been imprisonment boxes; it also incorporated the 20 percent reduction for good time and eliminated a competing proposal to add 20 percent as "bad time" for troublesome inmates. Members of the commission predicted that these changes would produce a modest short-term drop in sentenced inmates and stability over the longer term.

However, virtually everyone recognized that there was no way to predict the effect of the new system with great assurance. A number of safety mechanisms were suggested. The legislature opted to provide for a "trigger mechanism." The secretary of corrections must notify the commission at any time that prisons in the state have been filled to 90 percent or more of capacity. The commission is then obliged to propose modifications to the sentencing guidelines "as deemed necessary to maintain the prison population within the reasonable management capacity of the prisons as determined after consultation with the secretary of corrections." The commission's recommendation must then be submitted to the legislature, which is free to accept or reject it. However, the legislation does, for the first time, provide accountability for prison population. If population pressure occurs, the legislature must alter the guidelines or increase prison capacity. The concentration of responsibility will be a marked change from the mid-1980s, when the legislature, the judiciary, law enforcement, and the parole board, among other agencies, debated who was responsible for Kansas's massive increase in prison population.

Retroactivity Perhaps the most controversial aspect of the guidelines in the past year has been the question of their retroactivity. Throughout the process, a number of different constituencies have argued for making the guidelines retroactive. (Of course, for ex post facto reasons, guidelines can be applied retroactively only if they will benefit the particular defendant.) Civil rights leaders, concerned that the current system treats minority group members unfairly, have urged retroactive treatment. Law enforcement officials have generally opposed retroactive treatment. Articles in some newspapers have warned of a flood of dangerous criminals upon implementation of a guidelines system. Controversy over retroactivity spurred some local law enforcement officials to oppose the guidelines. That opposition, while not sufficient to block the guidelines as a package, did produce a compromise on retroactivity.

The current retroactivity provision authorizes case-by-case review of inmates convicted of lower severity crimes. If the inmate would have been entitled to presumptive probation under the guidelines, the inmate will have the opportunity to seek to have his or her sentence reduced to probation. If the inmate would have had a presumptive sentence of imprisonment under the guidelines, the sentence will be administered under the "old law."

During the 1980s, Kansas experienced an enormous increase in its prison population. The guidelines do not purport to turn back the clock to the incarceration rates of the 1970s. Instead, they are an effort to control the increases, to regard prison as a scarce resource, to attempt to reserve that resource for the most culpable offenders, and, most of all, to make the legislature accountable for whatever is done.

Ohio Adopts Determinate Sentencing
David Diroll

Ohio's new truth-in-sentencing law takes effect in July 1996. It abolishes parole, virtually eliminates good time, and will assure that most prisoners serve 95 to 100 percent of their announced sentences. Its passage derailed three-strikes proposals and gives judges considerably greater discretion than before over sentences affected by mandatory minimums. It should divert many currently prison-bound offenders to community penalties and should increase the resources and programs available to carry them out. And all without a sentencing guidelines grid.

To Tell the Truth   Truth wasn't so pretty at the summer 1995 meeting of the National Conference of Sentencing Commissions. Sentencing experts discussed how to "deal with" truth in sentencing. Would Congress continue to mandate truth (or at least 85 percent of truth) to receive federal aid? Would states mortgage their futures for federal funds? Would indefinite sentences, parole, and good time survive? Could some of these half-truths in sentencing be accepted as truth by the public? Would truth in budgeting follow truth in sentencing? Is honesty the best policy?

Into this context come the recommendations of the Ohio Criminal Sentencing Commission, which were adopted by the Ohio General Assembly in mid-1995, effective July 1996. Rather than resist or finesse truth in sentencing, the commission led the truth-in-sentencing movement in Ohio.

The commission was concerned that a six-to-twenty-five-year sentence never meant twenty-five, rarely meant six, and usually meant four or five years. The commission was concerned that this uncertainty undermined public trust and was unfair to both victims and offenders. The commission was concerned that the actual length of a prison sentence was determined by the parole board—an unelected body meeting in private—rather than by the sentencing judge. The commission was concerned that good time, which by statute was to be earned by good behavior, was actually earned by breathing. These concerns were shared by the commission's judges, prosecutors, and defense attorneys.

Under the commission's plan, most offenders will serve the exact time imposed by the sentencing judge. There is no funny math. Three years equals three years. Some offenders will serve less than 100 percent, but only because the judge grants the release ("shock" probation was retained but, in the interest of truth, renamed "judicial release") or acquiesces in the release (by not vetoing a furlough or boot camp placement). The only exception to judicial control of the sentence is a small earned-credit program—one day per month—during meaningful participation in school, work, or treatment programs.

In short, this is pretty honest truth in sentencing. Most offenders will serve 100 percent of their sentences, and the average will be over 95 percent of the term originally imposed.

The Whole Truth   So far, this sounds like what troubles many sentencing commissions. Often, legislators achieve truth in sentencing, or substantial truth in sentencing, simply by requiring that offenders serve all or a greater portion of the term imposed by the judge. This major adjustment may occur without concomitant changes in correctional resources or in the pool of offenders diverted from prison. Truth-in-sentencing laws may force careful, thought-out guidelines to yield in some cases or, in practice, to cover only nonviolent offenders who are not covered by the new truth. Many sentencing commissions see merit in good time and parole board reviews and are uncomfortable eliminating these tools.

Did Ohio's commission simply sing with the tempo of the times? Perhaps, but there is more to the song. The commission's plan authorizes a full continuum of community sanctions, makes almost all felons eligible for them, steers judges toward using them, and provides funding for planning and implementing these sanctions. New sanctions are formally authorized and further invention encouraged.

The plan applies truth in sentencing to nonprison terms by ending the fiction of the suspended sentence. Judges will no longer impose a prison term on an obviously probation-bound offender, only to suspend it. The plan gives community corrections greater status. Judges will sentence appropriate offenders directly to local sanctions, while warning offenders that failure will result in ratcheting up the continuum and could lead to a prison term.

The plan also builds on the boot camp model for some prison-bound offenders, but it goes beyond the paramilitary theme. It authorizes shortened terms in return for successful completion of intensive school, work, and vocational regimens. In the interest of truth in sentencing, the sentencing judge would have to approve any reduction.

The commission's plan will entail longer sentences for violent offenders but ease the growth in Ohio's prison population by diverting more drug users and property offenders into local sanctions. The onus will be on community programs—residential and nonresidential—to salvage lives and shorten criminal careers. Truth may set some free.

The commission's truth-in-sentencing philosophy had another, less tangible, benefit. Many Draconian measures were delayed in the legislature pending the commission's recommendations. Once the commission's plan emerged, its tougher measures for repeat violent offenders and certain others allowed legislators to put other matters in context. We recognize this deference may be temporary.

Ohio's Plan in a Nutshell   Ohio did not adopt a sentencing grid. Instead, the commission attempts to foster predictability, fairness, and reasonable uniformity by guiding judges with presumptions and policing them with appellate review.

Purposes and Principles. The commission's plan begins by setting forth basic sentencing purposes and principles to be considered by judges in all felony cases. Punishment and public protection are the overriding purposes. The judge must consider the need for incapacitation, rehabilitation, deterrence, and restitution. The judge also must consider a list of factors that indicate the offender's conduct is more or less serious than conduct normally constituting the offense. And the judge must consider a list of statutory factors indicating whether recidivism is more or less likely. Sentences cannot be based on race, ethnicity, gender, or religion.

Presumptions and Appellate Review. The commission placed all felonies into five classes. For first- and second-degree felonies (the highest levels), the plan instructs judges to impose a prison sentence, unless the plan's purposes and principles justify a nonprison sentence. The prosecutor could appeal any nonprison sentence imposed on a first- or second-degree felon.

In sentencing a fourth- or fifth-degree felon (the lowest levels), the judge must consider whether the offender has been to prison before, caused harm to a person, threatened harm with a weapon, violated a position of trust, or acted for hire, among other factors. If any of these factors is present, the judge must determine if the offender is amenable to an available local sanction. If not, the judge must send the offender to prison. This sentence would not be appealable.

If, however, the fourth- or fifth-degree felon had not been to prison before, did not cause harm to a person, and so forth, the judge would have to sentence the offender to a nonprison sanction, unless the purposes and principles of sentencing justify a prison term. The defendant could appeal any prison term under this provision.

Drug sentences follow a slightly different version of this model. And Ohio created a distinction between powder and crack cocaine. Because crack is more addictive, associated with more violence, and the subject of more calls to police, it is punished more severely. But the differences are less dramatic than in federal law.

Guidance as to Prison Terms. The plan sets forth a range of definite sentences for each felony level. The judge must select the shortest term from the range for any offender who has not been to prison before, unless the court finds on the record that the shortest term will demean the seriousness of the offender's conduct or not adequately protect the public.

The plan states that the longest sentence in the range should be reserved for the worst forms of the offense or for offenders with the greatest likelihood of committing future crimes. The defendant can appeal any time the longest sentence is imposed.

The ranges of prison terms are based on the indefinite minimum sentences, previously in place, minus good time. Under the old system, if a judge wanted to guarantee a four-year prison sentence, the judge would impose "6 to 25." With good time virtually automatic (at 30 to 33 percent) and applying to both the minimum and maximum sentences, the offender would reach the parole board in four years. Under the old system, first-degree felony "minimums" ranged from four to fifteen years. Minus good time, offenders served three to ten years. With this in mind, the commission set the new range at three to ten years.

This "shortening" was criticized by some prosecutors. But it is balanced by removing caps that artificially limited consecutive sentences and by a repeat violent offender (RVO) surpenalty. The RVO enhancement allows judges to double the normal range for offenders who have been to prison for a high-level felony before and who commit another high-level felony that involves actual or threatened serious physical harm. Inclusion of the RVO penalty in the plan fouled off the pitch for a three-strikes provision.

Mandatory Prison Terms. Since most mandatories were added in the last fifteen years, the commission deferred to these recent decisions. However, rather than impose specific three-, five-, seven-, or fifteen-year terms for these offenses, judges were told that a prison term was required but were allowed to select the length from the basic ranges for the level of offense. Thus, some prison terms remain mandatory, but judges were given more discretion to determine their length.

An exception is Ohio's three-year enhanced penalty for having a firearm in commission of a felony. The commission retained the three-year term but changed the penalty to one year when the gun was not used, brandished, or otherwise a factor in the offense.

Bad Time and Postrelease Control. By shifting to a more determinate system, Ohio gains greater certainty. But a purely determinate plan would leave the state with little control over potentially dangerous offenders. Thus, prison misconduct that is tantamount to a crime could result in the addition of "bad time." And every felon was made eligible for some period of postrelease control.

The parole board will no longer make release decisions, since the time served is set by the judge. The board will, however, review the records of all inmates sometime before their release dates. The board will determine whether postrelease supervision is warranted and the level of supervision.

Resources will not allow meaningful supervision for everyone released from prison. But the plan allows targeting of inmates who are likely to offend again. Supervisors will have the entire continuum of sanctions to use for initial placements and for punishing violators short of prison.

Victims' Rights. The commission recommended comprehensive revisions to Ohio's victims' rights laws. These changes put victims in position better to understand the process and to be heard.

Other Truths   Without doubt, the General Assembly adopted the commission's plan in part because of the integrity of the chief justice of the Ohio Supreme Court, who chairs the commission, and the support for community corrections of a very popular governor. But the plan's perceived balance and comprehensiveness had broad appeal, too. It is an honest truth-in-sentencing law that is mindful of public safety but attentive to cost and crowding issues.

Will it work? We don't know. Ohio's last comprehensive criminal code revision occurred twenty years ago, when there were about 8,500 inmates in Ohio prisons. With over 43,000 inmates in 1995, the state has five times as many inmates as when those changes were made. This plan does not reverse the trend; it flattens the growth rate. It is not a crash diet for prison crowding. It is weight management.

We hope the commission's plan will guide sentencing in a predictable, fair, and somewhat uniform manner, without a sentencing matrix, while providing truth in sentencing. Meanwhile, the commission has turned its attention to something really hard: misdemeanor sentencing.

Massachusetts, Missouri, and Oklahoma Establish Sentencing Commissions
Judy Greene

Massachusetts, Missouri, and Oklahoma have joined the lengthening list of states that have established sentencing commissions. In each, the goal is to design structured sentencing systems that incorporate the concept of truth in sentencing and reserve sufficient prison space for the most serious, violent offenders, while guiding judges toward broader use of intermediate sanctions. All three commissions are working in a context of sharply increased demands on jail and prison space over the past year.

Massachusetts's Sentencing Commission   Chapter 432 of the Acts of 1993 established a fifteen-member sentencing commission within the judicial branch. The commissioners were named in the spring of 1994. Robert A. Mulligan, the chief justice of the Superior Court, was appointed by Governor Weld to serve as chair.

Nine—three judges, three prosecutors, and three defense counsel—were appointed to hold voting powers.

The commission is to recommend sentencing policies and practices to the legislature. These are to include guidelines for felonies and misdemeanors, which must be enacted by the legislature to take effect. Sentencing policies must be keyed to corrections resources, and the commission is to ration resources to prevent prison population levels that exceed capacity. The guidelines are to be designed to avoid unwarranted disparity and to be "entirely neutral as to race, sex, national origin, creed, religion, and socioeconomic status of offenders" (Mass. Stats. 279, App. §3 [f]).

The Task Force on Justice, convened in 1990 by the Boston Bar Association and the Crime and Justice Foundation, proposed establishment of a sentencing commission. Finding that every part of the justice system was overwhelmed—clogged court dockets; an understaffed probation system; overcrowded jails and prisons—the task force urged repeal of mandatory sentencing laws and creation of a sentencing commission.

Prison overcrowding has not abated since the report was produced. Massachusetts's prisons operate at 137 percent of design capacity. But the guidelines commission proposal sat on a back burner until 1993, when Governor William Weld's truth-in-sentencing proposals were introduced and the two proposals were fused into legislation that passed that year.

John Larivee, executive director of the Crime and Justice Foundation, stresses the need for guidelines. "The problems we cited have not improved since we wrote our report. Since the shift to truth in sentencing, population pressures have begun to build—particularly at the local level. The counties are getting hammered!"

Chapter 432 also contains truth-in-sentencing provisions that took effect in July of 1994. The "Concord Sentence" was eliminated. This was a purely indeterminate term of imprisonment; the judge set only a maximum sentence. Parole eligibility ripened after a fraction of the term was served (e.g., an offender with no prior imprisonment was eligible for release after six months of a five-year term).

Under the 1994 legislation, parole is retained but eligibility standards were changed for normal prison sentences. Previously, prisoners became eligible for release *before* completing their minimum terms. Nonviolent offenders could go before the board when one-third of the minimum was served; violent offenders at two-thirds. Now the full minimum term must be served. The legislation also repealed "statutory good time" of twelve and a half days per month off the maximum sentence. (Earned good time—up to seven and a half days per month off the maximum *and* the minimum—was retained.) The new law eliminated judicial authority to suspend sentences and prohibited split sentences to prison (a judge may still impose a short term of local incarceration followed by probation).

The legislature provided a detailed format for presumptive guidelines. The guidelines are to incorporate an array of community sanctions—standard probation, intensive supervision, community service, home confinement, day reporting, residential programming, restitution, and day fines. The commission is to study these sanctions and to submit recommendations to the legislature regarding their development and funding.

The commission is to fix "target sentences" within durational ranges. If the target sentence is two years or more, the ranges may extend only 20 percent above or below the target. Below two years, the court may have more flexibility in setting a term of incarceration or sentencing to intermediate sanctions. The durational ranges are to be set within current statutory minimums and maximums. For offenders who receive imprisonment, a judge would set a maximum sentence within the range. The minimum sentence (which triggers parole eligibility) is to be two-thirds of the maximum. Judges may depart from these ranges, giving written reasons for doing so. Both the offender and the commonwealth can appeal. The legislation directs, however, that (excepting first- and second-degree murder) a judge—by departure—may set a minimum sentence below any applicable mandatory minimum sentence, citing mitigating circumstances that justify doing so. Such a sentence could involve a shortened term of imprisonment or an intermediate sanction.

Four committees have been organized to direct and oversee key elements of the process.

Outreach and Education. The inclusion of all interested constituencies in guidelines development will enhance the chances of successful implementation. The commission has planned an intensive series of focus groups and public hearings over the design period. Focus groups for victims have gathered their concerns and recommendations. A meeting has been conducted with chiefs of police, and focus groups are underway for prosecutors and defense counsel.

Research. Information from law enforcement and criminal justice data bases is being gathered and integrated to allow the commission to study existing sentencing practices. Simulations will be used to project the effects of the guidelines.

Legislation. This committee is to articulate the guiding principles that will shape the guidelines. It has decided that a grid will be used to classify offenses according to seriousness and criminal history by severity. Work is proceeding to rank order offenses and formulate policy statements to guide judges in application of guidelines.

Intermediate Sanctions. The commission plans to integrate a broad array of intermediate sanctions into the system and is compiling an inventory of existing programs. A more detailed inquiry including site visits is underway.

Proposals were to have been submitted to the legislature by April 1995, but the commission requested a one-year extension. "Everybody knew that wasn't a realistic deadline," says R. J. Cinquegrana, a commissioner who serves as chief trial counsel for the Suffolk County District Attorney's Office, "but of course the extension means we'll be throwing our results on the table during an election year." Cinquegrana thinks that the provision authorizing judges to depart below existing mandatory minimums was too broadly written and predicts it will not survive in that form. "The governor supports mandatory sentencing; that's a small but extremely important constituency. Wholesale avoidance of these laws would doom the guidelines to defeat. The real issue is to determine just how *much* time should be suffered by those who violate our drug laws." Commissioner Jane Haggerty, chief of the Appeals Division for the Essex County District Attorney's Office, agrees that some accommodations may be in store to preserve mandatory minimums. "These issues haven't really been discussed yet, but I don't think anybody is going to take the position that drunk drivers or violators of the gun laws should not get some jail time. It's at the high end of the drug mandatories where changes may be needed. Those who traffic 200 or more grams of cocaine now face the same fifteen-year mandatory term as those convicted of second-degree murder."

Justice Mulligan agrees that the issue of mandatory sentences will not be easy to resolve but is convinced the departure provision must be retained. "For every case I see that represents the problem a mandatory term was meant to address, I see one that doesn't. Sure, we've got some big-time drug traffickers who should be put away for a long-term stay in prison. But there are many others who are just what I call 'go-alongs.' I've seen women who were just holding drugs, or acting as a go-between for their husband—and they sit in jail facing ten years, while the guy has made bail and is gone."

Mulligan is concerned that the guidelines won't have the desired effect on prison and jail populations if judges are not given broader discretion to depart below statutory minimums. "We want to make rational policy decisions. Some prison expansion may be needed but we don't want to go the way of most other states into a building boom."

Missouri's Sentencing Advisory Commission In 1994 the Missouri legislature enacted Section 558.019 of the Missouri Crimes and Punishment Code, which established a new schedule of mandatory minimum prison terms for recidivists and those convicted of dangerous felonies and also created an eleven-member Sentencing Advisory Commission charged with development of a voluntary system of recommended sentences for felonies. "The impetus for the sentencing commission was that we were busy passing a truth-in-sentencing bill," says state senator Joseph Moseley, who sponsored the bill and chairs the commission. "As we approached changing the statutes to require minimum terms of up to 85 percent for the most serious violent offenders, we realized that we needed to balance it by including proactive measures for the less serious, nonviolent offenders—such as release of drug abusers to treatment. We knew that the best approach would be to develop a sentencing grid."

The commission is to work within existing statutory minimums and maximums to develop a system that will take account of the nature and relative severity of offenses; the criminal histories of offenders; current sentencing patterns; and available resources. As the guidelines are to be advisory only, they will require no approval by the legislature. The guidelines—which are to be completed by July 1, 1995—will simply be "published and distributed." The commission is to study their use through June 1998, when a final report will be submitted to the governor and the legislature.

The commission set to work in January 1995 and produced a draft statement of purpose: "The purpose of the Missouri Sentencing Guidelines is to recommend a uniform sanctioning policy which will ensure certainty, uniformity, consistency, and proportionality of punishment, impact on victims, and the protection of society. It is intended that the articulation and implementation of sanctioning standards will result in minimal sentencing disparity and a rational use of correctional resources to achieve the longest lasting public safety."

The commission formed two subcommittees: one to rank offenses and account for an offender's prior criminal history; the other to develop a way to handle aggravating and mitigating circumstances. By the beginning of May they had reviewed materials from states with sentencing guidelines and undertaken a preliminary analysis of Missouri sentencing data. They had also devised and circulated a questionnaire among commissioners to explore views concerning major issues. The results reveal a wide degree of consensus, including agreement to adopt a two-axis grid as the basic structure for expressing standards to guide sentencing discretion. They also agreed that the parole system should be retained.

The Missouri code categorizes most felonies into four classes. The commission intends to follow that ranking system closely and to establish a separate grid for each class. The commission has considered creation of a separate grid for drug crimes that would scale relative gravity of these offenses separately from other crimes and facilitate referral to treatment. They are agreed that code classifications should not be used in weighing an offender's criminal history, instead assigning weight according to whether the offense was committed against a person or against property, or was a "nonvictim" crime. Adult felonies and class A misdemeanors will count, but by early May it was not decided how juvenile records would be handled.

Probation is precluded for many charges (e.g., some drug offenses; "armed criminal action") and the 1994 crime bill included a new schedule of truth-in-sentencing minimum prison terms that further restrict parole eligibility under a sliding scale that sets minimums from 40 percent of the sentence imposed (for those with one prior prison commitment) to 80 percent (for those with three or more priors), and requires that every offender convicted of seven "dangerous felony" offenses serve 85 percent of the term imposed. "The task of creating a structure compatible with statutory truth in sentencing and mandatory minimum provisions is not as hard as it might seem at first glance," according to commissioner Dora Schriro, director of the Department of Corrections. "The schedule of mandatory penalties passed during the 1994 legislative session is triggered by prior *prison commitments*, not prior convictions, and—except for the requirement that those convicted of 'dangerous felony' offenses serve 85 percent of their time—the mandatory sentencing requirements are pretty much in line with parole practices before they were enacted."

Schriro contends that judicial discretion must be guided toward more use of treatment-oriented sanctions in light of the commission's determination that crime control is a primary purpose of sentencing. "Our guidelines will reflect the advantages that may be won—in terms of the *longest-lasting* prospects for improving public safety—through sentencing offenders to an array of DOC-sponsored treatment programs." These include:

post-conviction drug treatment—community- and institution-based short-term relapse prevention programs for nonviolent first offenders placed on felony probation;

120-day institutional treatment—a 120-day prison-based assessment program that will provide the court with a treatment needs and community risk report;

shock incarceration—assessment, drug education, skills training, and release planning in a secure setting for felony offenders;

boot camp—community restitution, education, and discipline for seventeen- to twenty-five-year-old nonviolent first offenders;

long-term cocaine treatment—a two-year prison-based program for nonviolent repeat offenders with cocaine dependence.

By early May the commission found itself facing the fast-looming July 1 deadline with major design and drafting work still ahead. With such a tight schedule it does not plan public hearings. Instead, the commission is concentrating on completing its work, and it plans to take advantage of regularly scheduled judicial conferences and training sessions in the summer to disseminate the results.

Prison population pressures are rising fast, and this has added urgency to the commission's work. Schriro points out that the prison population growth rate for many years was relatively stable at 1.44 per day, about 500 more prisoners each year. During the 1994 election season the growth rate shot up to more than five per day and has since continued to increase to nine inmates per day by May. The prison system has a capacity of 18,500, with a population of about 18,300. In response to the alarming trend, the legislature has approved $367 million to finance a 9,481-bed expansion.

Successful implementation of guidelines will require the support of the bench. The system will not include appellate review of sentences, and it is not clear whether judges will be required by the state court administration (which will track compliance) to give written justification for sentences that depart from the guidelines recommendations. "The primary obstacle may be the courts' unwillingness to comply with voluntary guidelines, given that many judges feel their discretion has already been seriously eroded," says commissioner Gail Hughes. "The great majority of cases are settled with plea bargains which include stipulations as to sentence. Missouri's prosecutors enjoy *enormous* discretion."

Senator Moseley says, "We chose a voluntary system to increase the comfort level of those whose support we need. Once we have the data and can demonstrate that we're doing the right thing, I think we can enact them into the state codes. Right now there is consensus on all the broad parameters. Of course, when we get down to putting numbers in the boxes we may run into difficulty, but I think we can get it done by the deadline."

Oklahoma's Truth-in-Sentencing Policy Advisory Commission Current efforts to reform Oklahoma sentencing can be traced to 1989 when a Sentencing and Release Policy Committee was created to assess sentencing policies and recommend improvements. After three years' work (during which the notion of sentencing guidelines was rejected), the committee submitted recommendations to abolish parole and establish a determinate sentencing system in which offenders would serve 80 percent of time imposed. Offenses were ranked in ten "schedules" with very broad sentence ranges (e.g., fifteen to sixty years for Schedule I). The scheme provided a catalogue of "enhancements" to account for previous felony convictions and factors that relate to offense severity. Legislative action was delayed through 1993, however, because the committee was unable to project the impact of its proposals on the prison population. After a simulation model constructed by the National Council on Crime and Delinquency produced impact estimates, the legislature rejected the proposal.

Instead, the Oklahoma Truth-in-Sentencing Policy Advisory Commission was created. Chaired by state senator Larry A. Dickerson, it has fifteen members (including three judges) and is to develop a truth-in-sentencing act that will "adequately reflect the offense committed, reasonably safeguard society as a whole, provide an opportunity for rehabilitation of the offender, insure the incarceration of the most violent offenders, and truthfully present the punishment actually to be executed." Sentencing criteria are to take the resources and constitutional capacity of the Department of Corrections into account, and recommendations are to be accompanied by an estimate of fiscal and population effects of new sentencing criteria on DOC and local facilities. If the proposed criteria will push populations in excess of capacity, the commission "shall present an additional set of criteria that are consistent with that capacity." The commission is to provide an interim report during the 1995 session and to submit final recommendations prior to the 1996 session.

Key commission members were involved in the previous design effort, and they remain committed to its basic features. They were initially determined to avoid a long process that would culminate during an election year. Commissioner Robert Ravitz, the Oklahoma City Public Defender, expressed this view: "Truth in sentencing is the only salable concept around. We have *done* it already. With a few changes to meet the objections raised before, we can resubmit this again in 1995."

Set on this accelerated schedule, the commission adopted two guiding principles to shape its deliberations: that the effect of its recommendations not be limited to existing resources, and that imprisonment be reserved for the most violent and habitual criminals. Drawing on many features of the previous proposals, work began on a matrix sentencing structure for felony offenses that would establish very broad ranges of punishment, coupled with a system of enhancers that would increase sentences on the basis of the offender's prior criminal history and the current crime's severity.

A draft grid has nine offense schedules on one axis and nine enhancement levels on the other. Wide durational ranges in offense level 1 represent "base ranges of punishment"—for example, 30 to 300 months for a Schedule B crime (a violent act that is committed with intent to kill or with reckless disregard for human life); 6 to 36 months for Schedule E offenses (a nonviolent act that creates a risk of injury to a person or a risk of harm to property). Each enhancement level would increase the time to be imposed. Separate matrices are planned for sex offenses, DUI, and drug offenses.

The commission has approved language describing offense enhancers. They include such factors as whether the offender committed the act while in "active criminal justice status" (a one-level increase); engaged in a pattern or scheme of criminal activity (one level); used a firearm (two levels); or preyed on an especially vulnerable victim (two levels). Enhancers for property offenses are keyed to dollar value. Prior record enhancers would produce increases proportional to the severity of the prior offense—six levels for a Schedule A felony, down to one level for a misdemeanor.

A system being considered to structure use of alternatives to incarceration involves sentencing presumptions. The cells on the grid are divided into four zones, with seven (those involving the least serious offenses and two or fewer enhancers) designated as "presumed probation." Thirteen cells are designated "presumed community incarceration" (a jail or correctional center term, inpatient treatment, or house arrest). Forty-three cells are designated as "presumed imprisonment," and the remaining eighteen (those containing Schedule A and B offenses) are designated "mandatory imprisonment." The drafters intend this scheme to supplant current minimum mandatory sentences. If this provision is adopted by the commission and enacted by the legislature, a judge could not depart from the presumptions without giving reasons on the record.

The commission has adopted a recommendation abolishing parole, with offenders sentenced to incarceration serving at least 80 percent of the sentence imposed. Twenty percent would be reducible through earned credits. The state's prison population has been legislatively capped and is being managed through a variety of back-end mechanisms, including up to forty-four days earned credit per month of incarceration and "preparole" community supervision. DOC estimates that average offenders now serve less than one-quarter of their terms. The commission wants to scale back credits to a maximum of ten days per month, capping the total at 20 percent of the sentence.

In April the commission reversed its plan to submit recommendations during the 1995 legislative session. Executive director Paul O'Connell believes that the final product will be completed by the fall, in time for thorough review prior to the 1996 legislative session. In submitting an interim progress report, the commission warned that the planned reforms would require significant funding increases. Commissioner Paul Anderson, district attorney in Payne County, says that is the primary difficulty. "Our draft structure is a well thought-out scheme, but the problem is the long ranges of imprisonment for many offenses coupled with the 80 percent requirement. As best we can tell, this will result in a 30 to 40 percent increase in the state prison population—and that is unacceptable at the legislature. No one wants to raise taxes and this has stymied the process." Anderson believes that support for community-based sanctions is growing. He has suggested that the commission submit three versions of their matrix with punishment ranges proportionately scaled to different impact levels on the prison population, but he says that reliable estimates of impact on prison population cannot be produced before December.

Concern about prison population levels is likely to increase. Cliff Sandell, DOC director of research, warns that his department is operating above "crisis capacity" and is facing sharp new population pressures: "Our April receptions were the largest in history."

THREE

Sentencing Reform outside the United States

Hubris often characterizes the large and the powerful. Sentencing policy, domestically and comparatively, offers examples. A delusion often afflicts policy makers in states like New York and California, or in Washington, D.C., that their problems are unique, that the often formidable brainpower that can be focused on solutions is enough, and that little useful can be gleaned from the experiences of others. The New York State Committee on Sentencing Guidelines (1985) offers one example. That committee, the largest and best-funded state sentencing commission, made little effort to draw on the experiences of earlier guidelines commissions and proposed guidelines the legislature would not accept.

The extreme case is the U.S. Sentencing Commission. None of the initial seven commissioners had any prior guidelines experience. Moreover, the commission, which began serious work in 1985, made no effort to learn from the experiences of the successful commissions in Minnesota, Pennsylvania, and Washington (except that Minnesota's director, Kay Knapp, was hired as the commission's first staff director, but she was forced out within a year by internal commission politics). Except for Kay Knapp, neither any staffer or member of the successful state commissions nor any of the sizable national pool of sentencing consultants and researchers were involved in the commission's work. The guidelines adopted, easily the most unpopular American sentencing reform initiative of this century (see, e.g., Reitz and Reitz 1995; Tonry 1996, chap. 3), include many unsuccessful features that could have been avoided had the experiences of state sentencing commissions been drawn on.

The same lack of willingness to learn from others operates at a national level. It appears never to occur to American policy makers that much could be learned from other countries. American policy makers and scholars know astonishingly little about sentencing practices and innovations in other countries. With one exception—day-reporting centers based on English models (Larivee 1991)—no significant sentencing policy innovations from other countries have been widely adopted.

Transfers in sentencing technology are common among European countries and members of the British Commonwealth. Day fines, penalties that can simultaneously be scaled to an offender's assets and income (by calculating daily income) and the severity of the crime (expressed in the number of units), began in Scandinavia early in this century, were adopted successfully in Germany in the 1970s, were pilot tested in England in the 1980s and for a short time implemented, and are in 1996 being tested in a number of European countries. Community service, though the first organized programs were established in California in the 1960s, was initially implemented nationally in England in the 1970s, followed by the Scots and the Dutch in the 1980s and the Swiss in several cantons in the 1990s. When Sweden recast its sentencing laws in 1988, Finland's 1976 reforms served as the primary model.

Some American innovations have been emulated elsewhere. Although organized community service programs were never widely established as prison alternatives, the early California programs were a catalyst for the later English success. Proponents of truth-in-sentencing laws adopted in several Australian states beginning in 1989 claimed to be inspired by American sentencing reforms. National law reform commissions in Canada and Australia proposed creation of sentencing guidelines systems patterned on American state guidelines. Punitive intensive supervision probation programs (commonly referred to as ISP), patterned on American programs, were established in South Africa, New Zealand, and some Australian states, and pilot projects were established in England and the Netherlands. England's home secretary in the mid-1990s proposed establishment of American-style correctional boot camps and three-strikes laws.

So far, little sentencing technology has crossed oceans to the United States. Pilot community service and day-fine programs have been established. A model community service program in New York City accomplished all of its goals (McDonald 1986) and continues in 1996 to operate successfully, but it was not replicated elsewhere. Day-fine projects were established in at least eight jurisdictions in the 1980s and 1990s, but most were abandoned and none succeeded in operating as a prison alternative (Tonry and Hamilton 1995, chap. 1).

The only successful import, day-reporting centers, was based on English programs (Mair 1993; Tonry and Hamilton 1995, chap. 4). These programs, which allow offenders to sleep in their homes but require their presence in a facility during the day, are generally used as early release programs from county jails. They offer various combinations of drug testing, supervision, counseling, and treatment programs.

A number of European sentencing practices and policies warrant serious consideration by American policy makers. One is the Scandinavian approach to sentencing reform (Jareborg 1995); rather than create numerical sentencing guidelines expressed in grids, the new Finnish and Swedish laws establish clear principles to guide sentencers and provide presumptions for what kinds of sentences should be imposed when. The policy guidance is clear, but the guidelines are verbal, not mathematical.

Another is a German and Dutch procedure known as "conditional dismissal" or "prosecutorial fines." In cases where guilt is clear, prosecutors allow offenders to pay a fine, in the amount that would have been ordered were the offender convicted; no conviction is entered and the charges are dismissed after passage of a designated crime-free period.

A third and a fourth are community service and day fines as prison alternatives. Although they have not been notably successful to date, they continue to have promise as credible, mid-level punishments for moderately serious crimes. The major problems to date (except with New York City's community service program) have been refusals by policy makers to provide enough money to pay for credible, well-managed programs and to authorize their use for other than minor offenses (in Phoenix, Arizona, which has one of the few surviving day-fine programs, for example, they are used in place of unsupervised probation for trifling crimes).

A fifth, no doubt utopian in America's current political climate, is the view that sentencing policy is a nonpolitical subject in which knowledge, good faith, and fairness to offenders are key elements. From the 1930s through the early 1970s, that view was predominant in the United States. Remarkable as it seems in the 1990s, the U.S. Congress in 1970 repealed most mandatory minimum sentence provisions in federal law because they often resulted in unjustly severe punishments and sometimes led to use of cynical evasion techniques by judges and lawyers (Tonry 1996, p. 142). It is hard in 1996 to imagine that sentencing policies will not be entangled in ideological and partisan politics in twenty or thirty years. Few in 1970, however, could have foreseen things as they are in 1996. Time will tell.

Americans should want to learn from the experiences of other countries as they wrestle with problems of crime and punishment that affect every developed country. This chapter contains articles on all the countries mentioned. That by André Kuhn describes recent prison population trends in seven countries and discusses reasons for significant changes. Some offer overviews of sentencing practices and policies in particular countries, together with explanations of recent changes and the reasons for them. Other articles describe the implementation and evaluation of new programs and laws.

Prison Populations in Western Europe
André Kuhn

Prison overcrowding is an important concern in all Western countries. The undesirable effects of crowding are legion, including increases in corrections costs, delays before imprisonment, and deterioration in living and working conditions for inmates and corrections officers.

This article examines prison population trends in seven European countries—the Netherlands, Switzerland, Italy, France, Germany, Austria, and Portugal—and the effectiveness of various efforts they have made to reduce prison populations and crowding. The most common methods, occasional amnesties and efforts to reduce use of short prison sentences, are often successful in the short term but not in the long term. The only strategy likely in the long term to control population growth is to reduce the lengths of long sentences.

The reasons for overcrowding are not clear. Increases in crime are often invoked to explain increasing prison populations, but this cannot be the primary explanation because there is no general relationship between crime rates trends and prison populations. Independent of crime rates, an increasing incarceration rate can also result either from increasing numbers of persons receiving confinement sentences rather than fines or other noncustodial penalties or from longer sentences being imposed.

Changes in prison use at least to some degree result from changes in punitiveness. Punitiveness can be thought of subjectively as being a characteristic of public opinion (the attitudes of the public toward severity of sanctions), or objectively (a social characteristic such as the severity of the sentences imposed by the judges). The objective sense partly affects the incarceration rate. All societies differ from each other in this respect and are in constant evolution. Severity of sanctions and the number of convicted persons thus vary through time and space.

It is difficult to make international comparisons of corrections practices, partly because cultural ideas about crime seriousness and punishment severity are affected by social, economic, and political conditions, and these vary between countries. Such comparisons are made more difficult by the diversity of criminal justice systems and repressive institutions. But important efforts have been made since 1983, under the aegis of the Council of Europe, to develop and disseminate periodic statistics about prison populations of member countries.

The incarceration rate is the main indicator. It is obtained by relating the number of prisoners on a specific date or as an annual average to the number of inhabitants. Such data are called statistics of "stock populations." Statistics of "flow" ("admission" or "committal" rates) document shifts in prison population. These two measures, which are often confused, involve very different notions and processes.

Objective indicators of punishment in a country can be assessed from the stock population or from the flow of prison populations. It is not clear which indicator is a better measure of the objective punitiveness of a society. The question is whether a country in which a relatively high portion of its population is confined for a limited time (significant flow) is more or less repressive than a country in which a small proportion of the population is sentenced to prison but for longer times (significant stock).

Every sentence imposed on an offender is a harm inflicted by the society. Every day in prison is a harm inflicted. Because the number of harms inflicted is an objective measure of aggregate severity, the number of days of imprisonment imposed per 100,000 inhabitants is a reasonable measure of punitiveness. Stock statistics thus may be more appropriate indicators of objective punitiveness than flow statistics. This view is reinforced by the finding that the average term of imprisonment is highly correlated with the incarceration rate per 100,000 inhabitants.

Comparisons of the punitiveness of different countries from prison statistics is not an easy task. Both the prison population and changes in it result from complex processes that are affected by the frequency and the seriousness of offenses, police efficiency, the strictness of the law and the way judges deal with it, and the modes of carrying out sentences (stay of sentence, amnesty, release on parole, etc.).

The Netherlands The Netherlands illustrates recent increases in incarceration rates in Europe. Although the Netherlands is well known for its low incarceration rate, its prison population has increased significantly.

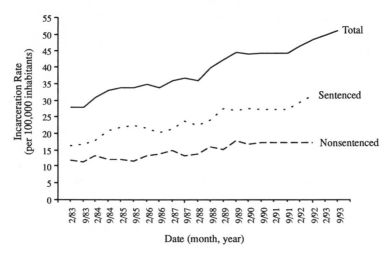

Figure 3.1. Incarceration Rate per 100,000 Inhabitants in the Netherlands, 1983–1993. *Source:* Council of Europe (1983–1993).

Figure 3.1 is derived from Council of Europe prison population statistics on incarceration rates per 100,000 Dutch inhabitants every February 1 and September 1 (only September 1 since 1991) for sentenced and nonsentenced detainees. In the Council of Europe data, nonsentenced detainees include all inmates awaiting trial (a first trial or a trial after appeal). The overall increase results more from increases in the number of sentenced offenders than from increases in pretrial detainees, especially for the 1991 to 1992 increase. Unfortunately, it is unclear why overall incarceration rates rose in 1993 because the data are not yet available. The increase is not, however, due to an increase in the use of prison sentences but in the lengths of sentences pronounced (Tubex and Snacken 1995).

Switzerland   Prison population data are available for Switzerland from 1890 to 1941 in the *Statistical Yearbook*. Prisoners are divided into two groups: sentenced and unsentenced. Since 1982, the Federal Statistics Agency (Office fédéral de la statistique) has maintained a central data bank about the correctional system that shows more recent trends.

Figure 3.2 shows the effects of economic conditions on incarceration. During the economic crisis of the 1930s, incarceration rates rose, not because of the number of sentenced offenders but because of the number of pretrial detainees. The increase in rates during periods of high unemployment was presumably caused by a change in attitudes of police, examining magistrates, and administrative authorities rather than by increases in crime, which would bring about an increase in sentenced offenders. Thus it appears that an important number of accuseds are imprisoned during economic crises because of their unemployment status who would otherwise keep their freedom (see also Killias 1991, p. 378, and Godefroy and Laffargue 1991).

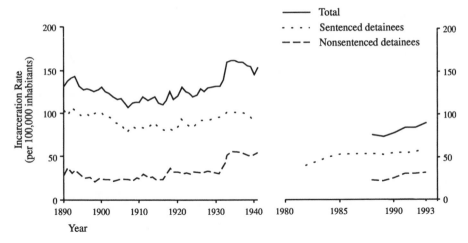

Figure 3.2. Incarceration Rate per 100,000 Inhabitants in Switzerland, 1890–1940 and 1980–1993. *Sources:* Killias and Grandjean (1986); Office fédéral de la statistique (unpublished data provided to author).

Figure 3.2 also shows that in Switzerland, as in France (Tournier 1984, p. 523), the incarceration rate fell by half between the 1930s and the 1980s, suggesting that the contemporary criminal justice system is less severe than previously. But even if the median term of imprisonment served has long remained stable at around one month, table 3.1 shows that during the last ten years the average length of sentences increased by approximately 74 percent (for more details and explanations, see Killias, Kuhn, and Rônez 1995).

Table 3.1. Average Length (in days) of Imprisonment in Switzerland between 1982 and 1992

Year	Average Length	Median
1982	73.9	28
1983	82.1	27
1984	97.0	30
1985	99.6	29
1986	110.0	29
1987	108.3	29
1988	109.2	28
1989	103.8	29
1990	111.8	30
1991	119.8	30
1992	128.5	30

Source: Unpublished data from the Office fédéral de la statistique.

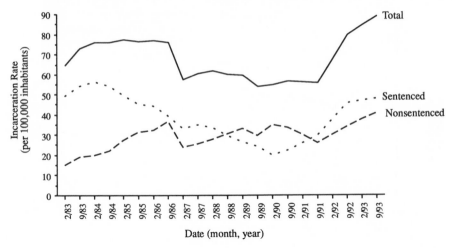

Figure 3.3. Incarceration Rate per 100,000 Inhabitants in Italy, 1983–1993. *Source:* Council of Europe (1983–1993).

Italy   Figure 3.3 shows that the overall incarceration rate in Italy dropped greatly between September 1, 1986, and February 1, 1987, from 76.3 to 57.4 detainees per 100,000 inhabitants. This reduction occurred mainly because of a December 16, 1986, amnesty.

Italy's example shows that an amnesty can significantly reduce the incarceration rate in the short term. The question, though, is whether this is an appropriate way to solve the prison overcrowding problem in the middle and long terms. Will one effect of this type of measure not likely be a rapid increase in prison population that will regain its previous level?

The Italian data suggest that an amnesty is incapable of reducing prison population for a long period. The sentenced offenders pattern shows that the incarceration rate quickly returns to the preamnesty level. The fall in corrections populations resulting from amnesties is therefore temporary and is not capable of solving over-crowding problems in the long term. The overall incarceration rate did not rise after the amnesty mainly because the pretrial detainees rate had been decreasing since 1984. This decrease was partly due to a change in pretrial detention law, which abolished compulsory arrest and introduced stricter conditions for pretrial detention.

Since February 1990 for pretrial detainees and September 1991 for sentenced offenders, the incarceration rate has increased. This seems to be an effect of illegal Albanian immigration, enlargement of the anti-Mafia fight after the assassinations of judges, and the anticorruption operation led by the magistrates.

France   Figure 3.4 on French trends has been constructed from the SEPT data base (séries pénitentiaires temporelles) belonging to CESDIP (Centre de recherches sociologiques sur le droit et les institutions pénales). This data base records incarceration rates on January 1 for every year since 1968.

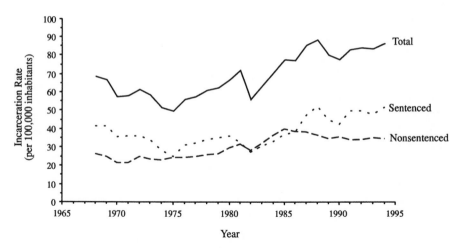

Figure 3.4. Incarceration Rate per 100,000 Inhabitants in France, 1968–1994. *Sources:* CESDIP, unpublished data base SEPT.

From 1956 to 1968, the prison population in France increased annually at an average rate of 5 percent (Tournier 1984, p. 524). From January 1, 1968, to January 1, 1975, there was a reduction of approximately 25 percent. Legal changes are responsible for the decline: June 30, 1969, and July 16, 1974, amnesty laws, introduction of partially suspended sentences on July 17, 1970, a December 29, 1972, law instituting stays of sentences and giving the judge charged with the implementation of sentences the power to grant a release on parole to those sentenced for less than three years, and an October 3, 1974, decree that granted sentence reductions to prisoners who were not part of spring 1974 insubordination movements.

The number of prisoners increased continuously from 1975 to 1981. A July 14, 1981, presidential pardon and an August 4, 1981, amnesty law temporarily stopped the "correctional inflation." But the numbers increased even more from 1982 until 1988 (the stabilization between January 1, 1985, and January 1, 1986, was due to a reduction in the pretrial detainee rate; the sentenced prisoners rate continued to grow). Following the presidential election of 1988, general pardons and one more amnesty were pronounced. A general pardon was granted on the occasion of the French Revolution bicentennial in 1989. Despite all those measures, prison populations have increased since 1990.

The French experience suggests that amnesties and other stays of sentences temporarily mask upward structural trends in prison populations. They do not solve prison overcrowding.

Germany   Figure 3.5 is constructed from data published by the German national statistics agency (Statistisches Bundesamt Wiesbaden) and shows incarceration rates in Germany per 100,000 inhabitants on January 1 every year from 1961 to 1991.

Following the reform of Germany's criminal law in 1969, use of short-term

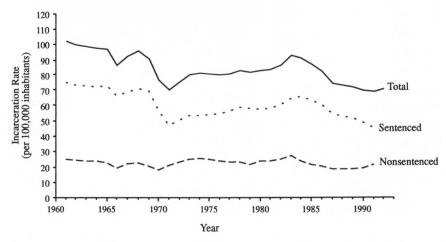

Figure 3.5. Incarceration Rate per 100,000 Inhabitants in Germany, 1961–1991. *Sources:* Statistisches Bundesamt, various years.

imprisonment was limited (short terms were first limited in 1969; in 1975, sentences of less than one month were abolished and sentences of less than six months were limited to exceptional cases). These restrictions had a temporary reductive effect on prison populations. Prison populations quickly grew, however, because of an increase in the lengths of some sentences (in 1973 there were twice as many sentences between six and twelve months as in 1968). If the German example seems to show that efforts to reduce short-term imprisonment can produce a middle-term decrease in prison population, it seems also to demonstrate that such measures do not solve the problem of prison overcrowding in the long term.

The German incarceration rate per 100,000 inhabitants decreased between 1983 and 1991. This phenomenon remains unexplained among criminal policy specialists, especially in Germany where this question has not really been explored. It appears, however, that it could be explained in part by a change in judges' and prosecutors' attitudes (see Graham 1990, Feest 1991, and Kuhn 1996).

Nevertheless, the statistics published by the Council of Europe show that the German incarceration rate has been increasing since 1990 and that the German prison overcrowding problem is perhaps not definitively solved.

Austria Figure 3.6 is constructed from data from the Ministry of Justice of the Austrian Republic. Those data represent the annual average incarceration rates per 100,000 inhabitants.

The interest of this figure lies in two points. First, a decrease in the incarceration rate in 1975 was quickly offset. Second, there was a second decrease since 1985 followed by another increase since 1990. What are the explanations for these fluctuations? An analysis of legislative changes in Austria gives a possible explanation. The use of sentences of less than six months was limited in 1975, and in 1988 partly sus-

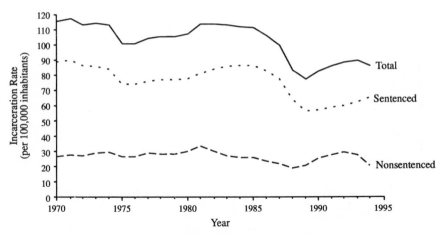

Figure 3.6. Incarceration Rate per 100,000 Inhabitants in Austria, 1970–1994. *Note:* Data
for 1994 include only the first six months of the year. *Source:* Bundesregierung Österreich;
unpublished data from the Austrian Ministry of Justice.

pended sentences were established and the eligibility date for parole release was
reduced from two-thirds of the sentence to half.

As in Germany, abolition of short-term imprisonment does not seem to have
been a long-term solution to prison overcrowding. However, extension of parole
and the establishment of the partly suspended sentence did have a significant effect
on prison populations.

Council of Europe statistics show that the total incarceration rate was relatively
stable until it fell from 96 per 100,000 inhabitants in February 1988 to 77 in September 1988. This is mainly because the law reducing the parole eligibility date
came into force on March 1, 1988. Nearly 1,500 prisoners were released (those
who had served more than half of their sentences but less than two-thirds). The
effect is therefore similar, in the short term, to the effect of the amnesty in Italy in
December 1986.

In Austria, the extension of parole release seems to have had a perverse effect;
parole use has become more restrictive. The relative stability—with increases that
seem more related to an increase in nonsentenced prisoners than in sentenced
ones—in the incarceration rate following its drop in 1988 is mainly due to the
introduction of the partly suspended sentence.

Portugal   The new Portuguese Penal Code came into force on January 1, 1983.
The new legislation—largely inspired by German law—limited the use of short-
term imprisonment and replaced most short-term sentences with other sanctions.
These amendments generated an increase of the incarceration rate per 100,000
inhabitants from 53 to 96 between 1983 and 1986, as shown in figure 3.7.

Here again, a limitation in the use of short terms of imprisonment did not
reduce the prison population. The 1986 reduction occurred mainly because of a

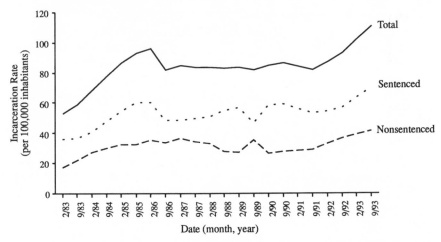

Figure 3.7. Incarceration Rate per 100,000 Inhabitants in Portugal, 1983–1993. *Source:* Council of Europe (1983–1993).

June 11, 1986, amnesty (Lei no. 16/86, June 11, 1986). The stability of the overall incarceration rate between 1986 and 1990 involved an offset between the increasing rate of sentenced offenders and a decreasing rate of pretrial detainees. The latter is the result of the new Procedure Act of January 1, 1988, which limited use of pretrial detention.

In 1991, a further amnesty (Lei no. 23/91, July 4, 1991) affected the incarceration rate, but it continued to grow after 1992. A further decrease can be predicted after 1993 because a new amnesty was enacted on May 11, 1994 (Lei no. 15/94, May 11, 1994).

Effects of Sentence Length   Does the length of sentences have an impact on prison crowding? It seems logical that the longer a sentence, the greater the burden on the corrections system. Knowing that many sentences in Europe are for less than six months, many think that prison populations could be reduced by introducing noncustodial sentences. However, such policies in Austria, Germany, and Portugal did not have that effect. Why?

In Switzerland in 1993, 85.3 percent (11,282) of imposed firm terms of imprisonment were shorter than six months. They represent 13,483 months of imprisonment. However, they represent only 18.5 percent of the 72,763 months of detention ordered by Swiss judges. The 14.7 percent of sentenced prisoners who received six months or more will serve 81.5 percent of total prison time (unpublished data from the Office fédéral de la statistique).

The effect of long prison terms on the corrections system is thus much more important than the effect of short sentences. The substitution of short terms of imprisonment is not an effective measure for reducing prison populations significantly. This conclusion has been confirmed by several studies showing that this measure may increase the severity of sanctions imposed by judges. It is now possible for a judge to sentence an offender who is not eligible for parole to four months; after the eventual abolition of short-term sentences, the judge would have to impose a six-month sentence in order to make sure that the offender goes to prison. Furthermore, imprisonment will be unavoidable when the offender fails to carry out a specific alternative sanction (generally fines or community service).

Conclusion Prison crowding is linked more to the lengths of prison sentences than to the number of persons who enter the prison system. If the goal is to reduce prison overcrowding, means must be found to reduce long-term imprisonment. That could be tried by an extension of parole release and introduction of partly suspended sentences (as in Austria); it could also be done by a change of the attitudes and the punitiveness of judges (as in Germany); or it could be done by a general reduction in the terms of imprisonment imposed. All sentences would become shorter and would then weigh less on prison populations, which would consequently be reduced.

With the scaling down of sanctions, two problems cannot, however, be avoided. First, the reduction of sentence length would have an effect on prison population only in the long term, since lengthy sentences already imposed will long affect prison populations. Second, there is the risk of alienating that part of the public that feels that criminals are not treated severely enough. But should we not be working toward a general evolution of attitudes if we really want to do something new in penal policy?

England

New Sentencing Laws Take Effect in England
Andrew Ashworth

Prison overcrowding has been a dominant feature of the English criminal justice system for the last twenty-five years. It is not merely that there have been insufficient prison places for each prisoner to have an individual cell, so that in 1991, with a prison population of over 45,000, some 13,000 were held either two or three to a cell. There has also been a growing acceptance that some offenders are sent to prison unnecessarily, and others for unnecessarily long. Even before riots at Strangeways Prison in Manchester and in other prisons in April 1990 brought the world's attention to English prison problems, strategies aimed at reduced prison use were announced by the government in its 1990 White Paper, *Crime, Justice and Protecting the Public.* These strategies were incorporated in the Criminal Justice Act 1991, most of which came into force on October 1, 1992.

The Policy The official aim is to move toward a twin-track policy of sentencing—dealing severely with people who commit serious offenses involving violence, sex, or drugs, and lowering the level of penal response to those who commit less serious offenses. Offenders sentenced to four years imprisonment or longer will be eligible

Figure 3.8. Hierarchy of Sanctions.

for parole: if they are not paroled, they will be released conditionally after serving three-quarters of the sentence, with a liability to serve any unexpired portion if they reoffend before the end of the full term. Offenders serving less than four years' imprisonment will benefit from automatic conditional release after serving one-half. However, a major element in the strategy is its "lower track"—that a higher proportion of offenders should be "punished in the community" instead of being sent to prison. To help induce courts to use community sanctions, the act introduces a new, tougher form of sentence called the "combination order" (a mixture of probation and community service), and probation orders themselves are made more rigorous by the adoption of new "National Standards" for their content and enforcement.

The Sentencing Framework The Criminal Justice Act 1991 introduces a new sentencing framework, which can perhaps best be visualized as the pyramid shown in figure 3.8.

In all but the most serious group of cases, courts are expected to start at the base of the pyramid and work upward. If the features of the case do not indicate that the offender should be given an absolute or conditional discharge, the penalty is likely to be a fine. Around 40 percent of indictable offenses (roughly equivalent to American felonies) now result in a fine. That figure once was closer to 60 percent, and government policy is to persuade courts to impose fines more often. Magistrates' courts, which deal with less serious crimes, are required to adopt a form of day-fine system called "unit fines," aimed at achieving a fairer relation between the amount to be paid and the financial resources of the offender.

Section 6(1) of the 1991 act is designed to ensure that a court moves up from a fine to a community sentence only if it is satisfied that the offense is sufficiently serious to warrant this. Once the court crosses this threshold, it may choose between a probation order, a community service order, or a combination order. In making this choice and in deciding the length of the order, it must not only choose the one most suitable to the needs of the offender but also ensure that the "restrictions on liberty" involved are "commensurate with the seriousness of the offense."

The final step up the pyramid, from a community sentence to prison, should only be taken if the court is satisfied that the offense is "so serious that only a custodial sentence can be justified" (sec. 1(2)). If the court is of that opinion, the length of the prison sentence must be "commensurate with the seriousness of the offence" (sec. 2(2)). These provisions rule out disproportionately long sentences based on individual deterrence or general deterrence. However, there is a limited exception for incapacitative sentences. A court may impose a prison sentence for a sexual or violent offense if it believes that only such a sentence would be adequate to protect the public from serious harm from the offender; such a sentence may be longer than would be proportionate to the seriousness of the offense committed.

The Common Law of Sentencing   The 1991 act does not introduce a complete new code of sentencing laws. Instead it superimposes a framework on the existing common law of sentencing, developed by the Court of Appeal over the last eighty years. The White Paper of 1990 envisaged a partnership between the legislature and the courts, in which the Court of Appeal through its judgments would give guidance that would put "the flesh" on the "bones" established by Parliament. During the 1980s the Court of Appeal handed down about a dozen "guideline judgments," each of which includes some sentencing standards for a major offense such as drug trafficking, rape, causing death by reckless driving, and child abuse. The government expressed the "hope" that the court would continue to fulfill this function in relation to the 1991 act.

The practical impact of the Court of Appeal will probably be twofold. First, it will continue to develop a kind of tariff, for the more serious types of offense, using appellate judgments to decide what sentence levels are proportionate or disproportionate. Second, it may give an authoritative interpretation of the key concept of "seriousness of offense," which determines virtually every step up the pyramid of sanctions. The practical frequency of problems at the "in/out" borderline means that the court must give guidance on how to evaluate thefts, burglaries, and deceptions if the 1991 act is to achieve even a modest success. In the past, the court has shown far more confidence in its judgments on very serious crimes. Its jurisprudence on the more mundane crimes that are the daily fare of most judges and magistrates remains underdeveloped.

Repeat and Multiple Offenders   One unusual and controversial feature of the 1991 act is its approach to prior record and to multiple offenses, both of which tend to occur in a majority of Crown Court cases. Section 29(1) states that an offense "shall not be regarded as more serious . . . by reason of any previous convictions of the offender." While section 28 allows courts to take account of a good previous record as a mitigating factor, section 29 means that the seriousness of the current offense establishes a kind of ceiling beyond which the sentence may not go. The policy behind this is to restrain courts from imposing severe sentences on repeat small-time offenders—and, in fact, to remove many petty thieves and property offenders from prison. The same policy underlies the provision that, when a court is deciding whether an offense is so serious that only a custodial sentence can be justified, and the offender stands convicted of several offenses, the court may take account only of two of them in deciding whether the case is serious enough. Even if the offender is being sentenced for 20 or 120 check frauds, the two-offense rule restricts the court to aggregating any two of the offenses in order to gauge the seriousness of the case.

The policy behind this is to keep small-time criminals out of prison, but many sentencers find the provisions artificial and unacceptable. Some judicial circumvention is always a possibility, and it remains to be seen whether the essence of the new policy will survive.

The Likely Impact of the 1991 Act   Will the effect of introducing the new act in October 1992 be to reduce, or even to control, the English prison population? The British government itself seemed unclear about this when the 1990 White Paper was issued, but its latest estimate is that the prison population will decline by 3,500, or around 10 percent, by 1995. Many hope that they are right: a concerted effort has been put into the development of community sanctions, and in the 1980s there were successes in reducing the use of custody for juveniles and young adults.

Unfortunately, at least five sources of difficulty appear. First, the 1991 act abolishes remission (good time) and alters the provisions for early release: this will result in offenders staying longer in prison unless the courts voluntarily reduce their sentencing levels in the middle range of current sentences, from one to four years. No move in this direction has yet been announced. Second, between 1985 and 1990, the number of prisoners serving sentences of four years or longer doubled, and under the twin-track strategy this is set to continue. Third, if more offenders are given more demanding community sanctions, they may well violate them in larger numbers and thus enter prison by that route. Fourth, much will depend on how the Court of Appeal, under the new Lord Chief Justice (Lord Taylor), approaches its task: the court should give an early and positive lead to other judges.

This leads us to the fifth question, which may or may not prove to be a difficulty. Many judges and magistrates seem skeptical about the new act. Wholesale change in one's working practices is rarely welcome to anyone, and the obscurity of some language in the new act hardly encourages a sympathetic response. But the 1991 act does leave a considerable degree of discretion to the courts. Will they use this to neutralize the spirit of the legislation? Or will they use it to advance the primary aim of proportionality in sentencing, and to move toward greater use of community sentences? The new English statute leaves far more power in the hands of the judiciary than do sentencing reforms in most other jurisdictions. Criminological research might suggest that judges will alter their approach as little as possible, co-opting the new act into established working practices. Whether they will confound their critics and show that the judiciary can participate in the process of reform remains to be seen.

England Repeals Key 1991 Sentencing Reforms
Martin Wasik

Three key provisions of the Criminal Justice Act 1991, which was intended to effect major changes in English sentencing policy and practice, were repealed to take effect this summer of 1993. One change, abandonment of unit fines, an English variation on day fines, is discussed at greater length elsewhere in this chapter. Two of the repealed provisions, which limited the number of current convictions to be taken into account in ordering imprisonment, and limited the relevance of prior records to current sentences, were widely seen as shifts toward "just deserts" in national sentencing policy.

The Criminal Justice Act 1991 was brought into effect in England and Wales only in October 1992. The 1991 act was, however, subjected to a campaign of strident and often ill-informed criticism in the media. Several prominent individuals expressed opposition to the reduction in sentencing discretion the act brought about.

The Lord Chief Justice, Lord Taylor, in a widely reported speech, said that the act's provisions forced sentencers into an "ill-fitting straitjacket," and that "penologists, criminologists and bureaucrats" had created a sentencing regime in the act "incomprehensible to right thinking people generally." Magistrates expressed opposition to the act and several resigned in protest.

The home secretary, after meeting with sentencers, declared that certain aspects of the act were "not working." Changes to it were hastily assembled and inserted into a criminal justice bill, dealing with quite different matters, which happened to be going through Parliament at the time. The 1993 act has 79 sections and 6 schedules. Only sections 65 and 66 and schedule 3 impinge on the 1991 act. The changes brought about by it are, however, considerable.

The so-called two-offense rule has been abolished. The rule in the 1991 act was that a sentencer considering whether to impose a custodial sentence on an offender whose case involved multiple offenses should consider the relative seriousness of up to two of these offenses. If, having done so, it appeared that custody was justified, then the totality of the offending could be taken into account when fixing the length of sentence. If, however, the sentencer concluded that no two offenses taken together justified custody, then custody could not be imposed at all.

The purpose of this rule was to provide a check against the incarceration of offenders who committed large numbers of petty crimes. Sentencers objected to what they saw as an artificial mental exercise and many clearly took the view that persistence, in itself, was sometimes enough to justify custody. Lord Taylor said that the two-offense rule "defied common sense."

The rule in section 29(1) of the 1991 act, which restricted the relevance of an offender's previous convictions to the sentence to be imposed for the latest offense, has been replaced by a much more loosely worded provision. The original section reflected the theory of "progressive loss of mitigation"; while a clean record could provide mitigation, a poor record or a poor response to earlier sentences was not in itself deemed relevant to the seriousness of the current offense or whether a custodial sentence should be ordered. There were exceptions to this rule.

First, it did not apply when a violent or sexual crime had been committed. Then the court could have full regard to previous convictions. There was a second limited (and rather obscure) exception, provided for in section 29(2), which allowed a court to infer that the latest offense was more serious than it looked, by having regard to the circumstances surrounding other crimes committed by that offender.

The basic principle in section 29(1) attracted most of the criticism. It was (mis)interpreted by the media, and by some practitioners who should have been better informed, to mean that sentencers should simply disregard all matters relating to previous convictions. Section 29 as originally drafted was also criticized as "offending against common sense," and one magistrate blamed "the pro-criminal lobby groups" for the thinking behind it. It has been replaced by a new version, which seems to allow sentencers to take account of the offender's previous convictions and his or her response to previous sentences whenever, and however, the sentencer wishes.

The 1991 act introduced a system of unit fines to magistrates' courts in England and Wales. Lay magistrates (who are legally unqualified, unpaid people) deal with the vast majority of criminal cases in England and Wales, and most of their cases are disposed of by way of fine. Traditionally, fines were fixed largely by tariff principles: a particular offense attracted a fine of a certain set amount. Magistrates were required by statute to reduce the level of the fine when the offender lacked the means to pay, but all too often ability to pay was not taken fully into account, producing a high rate of incarceration through fine default.

The unit-fine system mirrored similar schemes in other countries. It required magistrates to work through their sentencing decision in three stages. The first was to form an assessment of the seriousness of the offense and express that in terms of a number of units. The second was to determine the offender's disposable income, after making deductions for housing, subsistence, and other essential expenses. The third was to compute the fine by multiplying the number of units by the unit value in the offender's case.

The unit-fine system ran into serious trouble, however, soon after October 1992. The main problem was of the government's own making. The 1991 act increased the maximum fines that could be imposed in magistrates' courts from £2,000 to £5,000, and the lower and upper unit values were set at £4 and £100. So, for a twenty-five-unit offense, the unit fine now varied between £100 and £2,500, depending on the offender's means. People on average incomes found themselves in the top band, paying much more than they would have under the old system. The result was a middle-class backlash.

There are two likely effects of these sentencing reversals. The first is an increase in the custodial population. It is not clear what the overall effect of the unamended 1991 act would have been on the prison population. Some people forecast a downturn as a result of tighter restrictions on the use of custody, but this was offset to an unknown extent by changes to the early release rules (most prisoners serving terms between twelve months and four years will spend longer in custody under the 1991 act than they did formerly) and the effective abolition of the suspended prison sentence. The reversals made by the Criminal Justice Act 1993 will all tend to increase use of custody. Abolition of the two-offense rule will result in more minor property offenders receiving custody. Abolition of the restriction on use of prior record will accentuate the same trend, in that offenders will tend to be sentenced on the basis of their poor records, rather than on the seriousness of their latest offense. The abolition of unit fines seems bound to result in an increase in fine default.

On June 12, 1993, it was reported in *The Times* newspaper that prison service managers were preparing emergency plans involving the use of former military camps to cope with a predicted upsurge in the prison population as a result of the 1993 act changes. Home Office officials are said to have predicted privately that the changes will lead to an increase of between 5,000 and 10,000 in the prison population (currently it stands at just under 44,000).

The second effect of the changes made by the 1993 act is to create a setback for the rational development of principled sentencing in England and Wales. The 1991 act represented the first real attempt to introduce a coherent set of principles for the imposition of custodial, community, and financial penalties. Important parts of that edifice remain in place, but several of the key provisions have been abandoned in a remarkably short space of time.

England Abandons Unit Fines
David Moxon

After only seven months, the English government has abandoned its unit-fine initiative and repealed the enabling legislation. For the present, at least, the effort to incorporate means-based fines akin to day fines into the system of penal sanctions is at an end. A number of problems contributed to the government's decision, not least the extensive media accounts of bizarrely harsh fines for trivial crimes, the most celebrated being a littering fine of £1,200 for tossing a potato chip bag on the ground. This article tells the story of England's brief and troubled experience with unit fines.

Background   In the June 1992 edition of *Overcrowded Times* I wrote about the success of pilot projects designed to test the feasibility of unit fines—similar to day fines but based on a week's income rather than a day's. The basic principle is that the *number* of units imposed for an offense is determined by the seriousness of the offense, and the *amount* of each unit is determined by the offender's weekly disposable income—the income left over after allowing for basic expenditures.

The pilot projects were carried out in magistrates' courts, which have full jurisdiction over minor crimes and concurrent jurisdiction with higher courts over some more serious crimes. The pilot projects had the support of the magistrates and court staff where they had been introduced, not least because they were felt to have been fairer than the old system; a result was greater consistency between courts when fining people of similar means than had been the case previously. The scheme made enforcement easier, with fines paid more quickly and with less imprisonment for default. It was hoped that the ability to match fines to offenders' circumstances in a more systematic way would help to reverse the long-term decline in the use of the fine.

Encouraged by the evaluation research, the government introduced legislation that gave statutory force to unit fines with effect from October 1992. There were important differences between the national scheme and the pilot schemes. Perhaps the most crucial was that, when the pilots were conducted, fines could be reduced for lack of means but could not be increased for the better-off. Pilot project courts set the range for units of disposable income locally, so that well-off offenders paid roughly the same fines as they would have under the previous system. In practice, the maximum sum for each unit ranged from £10 sterling to £20 (at current exchange rates, $15 to $30), reflecting perceived differences in average local income. A nominal minimum unit value was set at £3, but in some courts this was reduced still further for the very poorest offenders.

By contrast, under the unit-fine legislation, the range of unit values widened to £4 and £100. As a result, fines could be much larger, particularly as the maximum fine that magistrates have authority to impose was raised from £2,000 to £5,000. Means assessment therefore assumed much greater importance than in the pilot schemes, particularly as all types of offense were included.

Assessing Means   Each court set local allowances based on family composition and estimates of typical local housing costs. The form was kept simple since it was felt that a complicated form was less likely to be filled in and a detailed breakdown of expenditure would be largely irrelevant, though allowance could be made for exceptional obligations. The unit value was assessed as one-third of the residual income after taking account of allowances. However, if there were few allowances to set against income, the maximum £100 unit value could be reached on quite modest incomes. It was felt by many magistrates that the scheme failed to acknowledge that, as income rises, so do financial commitments that cannot easily be reduced in the short term. Before the scheme was scrapped, many of those involved had urged changes that would have permitted a slower increase in fines as incomes rose, for example, by defining the unit value as a proportion of total income.

The increase in maximum fines did, of course, make it much more important to have good means information in every case. All defendants were sent a means form, but there was no legal obligation for them to complete it in advance of the case, partly because it was felt that it would be wrong to impose such a requirement on people who might at that stage be protesting their innocence. If the offender was in court, he or she could be asked to complete a form while waiting for the case to be heard, or asked questions in court. However, in minor (and particularly minor motor vehicle) cases, offenders commonly plead guilty by mail. If the means form was not returned with the guilty plea, the court then either had to adjourn the case for means information to be supplied or to make a judgment based on little or no information.

If the court had some information, for example, if it was known that the offender was unemployed or alternatively what the offender's job was, a reasonable estimate could often be made. However, if there was no information, after the offender had been given the opportunity of providing it, what should the court do? Some fell back on an arbitrary figure of, say, £20, which they felt would be fairly realistic in the general run of cases and would avoid further delay or complication. Others took the view that this approach was unjust, in that someone who was relatively well-off and took the trouble to tell the court would be fined more severely than someone who simply kept quiet about their means. The maximum unit value of £100 was then likely to be imposed, leaving it to the offender to challenge this assessment later on if he or she felt it was excessive.

Scope of Unit Fines   In contrast to other jurisdictions that have adopted day fines, the unit-fine system covered all offenses that reached magistrates' courts (though the scheme did not apply in the higher courts). It was felt that drawing a distinction between offenses liable to unit fines and those that were not would inevitably be somewhat arbitrary, and that a common system for all offenses would be simpler to administer. Calculating the fine would otherwise be especially awkward when a mix of offenses was dealt with on the same occasion. Some doubts were expressed, however, over whether the complication of a means inquiry was appropriate for very trivial matters. Minor motor vehicle offenses posed special difficulties because they are usually dealt with by means of fixed penalties but can go to court for reasons that have nothing to do with seriousness. To take one example, by law an American serviceman stationed in England cannot be dealt with by way of fixed penalty for a driving offense and so must go to court even on a very minor matter. This could result in a fine of several hundred pounds instead of the £36 fixed penalty that most other people would pay. The potential anomaly was recognized insofar as the legislation provided for fines to be increased to the level of the fixed penalty so that the poorest offenders would not have an incentive to take trivial cases to court, but there was no corresponding power to reduce the fine to avoid unfairness.

In practice, some courts took the view—with the approval of other local criminal justice agencies—that where a fixed penalty would have been appropriate but could not be applied for technical reasons, the unit-fines provisions should be overridden. This could be done by arbitrarily reducing the number of units or fixing the unit sum at a low level, or by giving a discharge but requiring the offender to pay costs of an amount similar to the fixed penalty.

Problem Cases   The foreseeable result of raising the maximum unit value to £100, of adopting the maximum in the absence of means information, and of including all offenses in the scheme was that large fines were sometimes imposed for very minor offenses. Because such cases and the publicity they attracted contributed strongly to the scheme's downfall, it is instructive to look at a couple of the numerous examples that caught the headlines. In the first, magistrates fined a man £500 for illegal parking after his car broke down on a road where parking was prohibited. He had exercised his right to take the case to court, rather than pay the fixed penalty by mail, because he thought he had a legitimate defense. A flawed means assessment resulted in the unit rate being set at the £100 maximum, yielding a fine that according to press reports amounted to more than twice the value of the vehicle. On appeal, the Crown Court (quite sensibly) reduced the fine to the level of a fixed penalty.

Cases of this sort highlighted two anomalies. First, a unit fine could be totally disproportionate to the fixed penalty that would normally apply, making it very difficult for anyone who felt they had a defense to challenge it. (Automobile clubs were quick to advise their members to pay up, even if they thought they were innocent, rather than risk losing in court.) Second, magistrates' authority and the credibility of unit fines were undermined by the fact that the unit-fine principle was applicable only in magistrates' courts but not in the higher Crown Court; on appeal, it was up to the individual judge. In the particular instance of the parking fine, the magistrate's view of the seriousness of the offense—awarding five units on a scale of one to ten—was out of step with the public and media perception of the offense. So what was widely perceived as an unjust decision regarding the nature and seriousness of the alleged offense was unfairly—but understandably—blamed on the unit-fine system. In this way, factors outside the unit-fine scheme contributed to its downfall.

In the second example a man was fined £1,200 for throwing a potato chip bag onto the ground instead of placing it in a nearby litter bin. The penalty in terms of the number of units was higher than usual in such cases because the offender refused to pick the litter up and was cheeky to the policeman who witnessed the incident. The unit value was fixed at £100 because the offender failed to provide any means information. However, later on, evidence was received that he was unemployed and the fine was reduced to the minimum level for twelve units, £48. The system worked effectively and fairly, but not before a great deal of publicity had been given in the national press to the curiosity that an unemployed man had been fined £1,200 for carelessly discarding a potato chip bag.

The Lessons　The English experience clearly has implications for any jurisdiction contemplating the introduction of day or unit fines, bearing in mind that acceptance by both the public and sentencers is crucial to success.

If the unit values span too wide a range, the amount of the fine is influenced much more by assessment of means than by offense seriousness. Sentencers are often uneasy at the contrast between small fines for relatively serious matters and large fines for minor offenses. This was a major issue in England and Wales, although it does not appear to have troubled sentencers in other jurisdictions that use day fines, even when much higher maximums are often available. One reason may be that minor offenses, of the sort that caused some of the most damaging criticism, are not normally included in day-fine schemes.

Quite apart from the problem of large fines for minor offenses, inclusion of all offenses can create anomalies when compared with any fixed penalty system (which by definition takes no account of means). The kinds of examples cited here raise the question of whether trivial incidents justify an investigation into income and expenses, with all the difficulties that arise when means information is not forthcoming. If the system is confined to more serious matters, the offender will usually be in court and questions about means can be raised if no form has been provided.

It is important that those operating the scheme understand and accept both its principles and its practical implications. Support for the principles was in many cases undermined because of perceived injustices in certain types of case. The need to ensure that the principles are applied consistently should not result in such rigid rules and loss of discretion that sentencers are faced either with breaking the rules or imposing sentences that they feel are unfair. (A few magistrates were so worried by the decisions they felt the legislation was obliging them to take that they resigned.)

If fines are subject to review on appeal, the appellate judges should be subject to the criteria that applied in the lower court that set the original fine, although clearly both the number of units and the amount of each unit are proper matters for review by the appellate court.

During the seven months in which the unit-fine scheme was in operation, these points received a good deal of attention and consideration and there were signs of a consensus developing in some quarters as to how to resolve them. However, unit fines came under such heavy and sustained attack from large sections of the media—and often from magistrates—that government ministers felt it better to scrap what was felt to be too mechanistic and rigid a scheme. What has replaced it does not take England and Wales back to the preexisting situation; courts are now able to increase fines for the better-off and to reduce them for the poor. The legislation also makes it clear that means are still relevant when setting fines, and courts can evolve whatever measures they think most appropriate for taking means into account. It will be interesting to see whether the experience of relating fines to means in a systematic way will have a lasting impact on the way fines are set.

Punitive Policies and Politics Crowding English Prisons
Rod Morgan

Such is the level of official uncertainty about the future trajectory of the prison population in England and Wales that this spring the Home Office Research and Statistics Department decided not to publish its usual population projection for the next decade. The department judged that the factors influencing prison admissions are too volatile to make the exercise worthwhile.

The prison population is rising rapidly—more than 10 percent since March 1993—as a result of recent fundamental changes in crime control politics and policies. The Criminal Justice Act 1991, which took effect in October 1992, was intended to reduce reliance on incarceration, and official projections called for long-term reduction or only modest growth in numbers in confinement. During the last year, however, the premises of the 1991 act have been rejected by the current home secretary, and projected increases in overcrowding and in population are the result. Before I examine these political and legal factors, perspective may be gained from a look at the two most recent official population projections.

In May 1992, the prison population was approximately 47,000 (including both convicted offenders and those confined while awaiting trial; figure 3.9 shows population figures for 1987 to 1994). The *Statistical Bulletin* published that month by the Home Office projected an average daily prison population (ADP) of 57,500 for the year 2000. The projection took into account the estimated effects of the Criminal Justice Act 1991 and was said to be more sophisticated than any of its predecessors. It took both demographic and economic factors into account, research having shown that in England the number of property offenders received into custody has tended to rise more rapidly when personal consumption declines. It was widely expected that the provisions in the 1991 act would by 1995 reduce the sentenced custodial population by approximately 3,500 from what would otherwise have been the case. The assumptions on which the projections of the pretrial population were based remained broadly unchanged. Overall the future was said to look slightly better than it did the previous year: there would still be a steady growth in numbers—a reflection of the upward trend in the number of serious offenses being recorded and offenders being convicted by the courts—but the increase would be more modest than previously predicted.

In March 1993, the picture looked even rosier. Instead of rising, the prison pop-

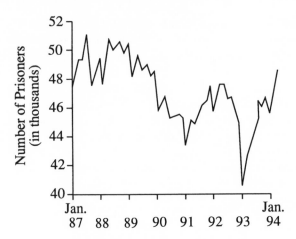

Figure 3.9. British Prison Population, January 1987–
January 1994. *Source:* Prison Service (1994).

ulation had fallen significantly since spring 1992. It stood at a little over 43,000 and
a new *Statistical Bulletin* projected an ADP of 50,400 for the year 2000, a reduction of 12.4 percent on that projected the previous year. Though no data were yet
available on the impact of the 1991 act, the trend apparent in 1991–1992 was
judged to presage a new climate in sentencing practice. This was a reasonable
proposition given the widespread public consultation during consideration and passage of the act and the extensive training provided to sentencers before implementation. It seemed as if the principles central to the legislation were already leading
to more parsimonious use of custody, particularly for young offenders.

Things look very different today. The prison population now stands well above
48,000, the number of pretrial prisoners is at an all-time high—an ominous sign—
and the current rate of increase is such that there is a very real likelihood that the
ADP of 50,400 predicted for the year 2000 will be exceeded by December 1994.
No further official population projection has appeared, but the Prison Service's
Corporate Plan 1994–97, published in April, mentions a population projection of
55,700 for the year 2000, a figure said to be based on an "interim" assessment of
the likely impact of the Criminal Justice Act 1993. This statute among other
things reversed key provisions of the 1991 act. The director general of the Prison
Service is using all his persuasive skills to maintain an optimistic mood about the
future quality of life for prisoners and prison staff, but there are signs that all the
ills of yesteryear—serious overcrowding, perennial disturbances, and poor labor
relations—are returning.

Though some overcrowding in particular institutions continued even during the
period 1992–1993, there was excess capacity in the prison system for the first time
since 1945. Today, despite new prisons continuing to come on stream, there is once
again system overcrowding. The government has announced that four new prisons
are to be added to the building program, but this additional accommodation will not

outpace the predicted population growth. System overcrowding seems destined, all else being equal, to return to the 10 to 15 percent levels that prevailed in the early 1980s and to continue at that level into the twenty-first century. This inevitably means that some facilities will be 50 percent overcrowded or more. Indeed it is notable that the specifications prepared for the latest facilities to be contracted out to private-sector management request potential contractors to anticipate overcrowding up to 33 percent. There is once again an overflow of several hundred prisoners backed up in police cells. Last September there occurred—at Wymott, the largest medium-security prison in the country—the most serious riot since the spate of disturbances in April 1990: the prison was closed for several months to repair the damage. And last year, as the Prison Officers' Association has emphasized with increasing bitterness, there was the largest number of assaults by prisoners on prison officers ever recorded, up 33 percent over 1992–1993 and 83 percent over 1989–1990. Only a court order has prevented the union from taking industrial action.

There is only one explanation for this change of fortune. The prison population has been talked up by a tide of punitive rhetoric that politicians have employed since the beginning of 1993 when criticism of the early working of the 1991 act, combined with several notorious crimes, caused a good deal of adverse media publicity and led Home Office ministers to reach for the populist phrase books. In his August 1993 *Overcrowded Times* article on the 1993 act, Martin Wasik predicted that all of its reversals of the 1991 act would tend to increase the use of custody. That much is now clear. But it is also apparent that the upward trend was well established six months before the 1993 act was passed and implemented. In the same way that sentencers anticipated the principles, and their prevailing interpretation, embodied in the 1991 act prior to their implementation, so also did they anticipate the countervailing trend represented by the 1993 act. It is worth illustrating the extent of the reversal that created this response.

Prior to the passage of the 1991 act, the government published two major consultative documents setting out the premises on which the act was based. They were a set of principles and objectives about which politicians, civil servants, and academic researchers were apparently agreed. The 1991 act and its provisions was based on three primary assumptions. First, prisons are not good places in which to encourage citizens to act responsibly and live in a law-abiding manner. "It is better that people . . . exercise self-control than have controls imposed upon them . . . however much prison staff try to inject a positive purpose into the regime . . . prison is a society which requires virtually no sense of personal responsibility from prisoners" (Home Office 1990, paras 2.6–2.7).

Second, the deterrent value of imprisonment—in its individualistic calculative sense—no longer has much crime preventive plausibility. "Deterrence is a principle with much immediate appeal . . . but much crime is committed on impulse . . . it is unrealistic to construct sentencing arrangements on the assumption that most offenders will weigh up the possibilities in advance and base their conduct on rational calculation" (Home Office 1990, para. 2.8).

Further, the document could have continued, Beccaria's four certainties of offense, detection, conviction, and punishment are now so belied by the small proportion of offenses that are reported and recorded, lead to a conviction, and result in a particular sentence, that those offenders who do base their behavior on rational calculation are likely to go ahead and do it.

Third, the use of imprisonment as a sentence is never justified in order to make offenders better in some sense. "It was once believed that prison, properly used, could encourage a high proportion of offenders to start an honest life on their release. Nobody now regards imprisonment, in itself, as an effective means of reform for most prisoners" (Home Office 1990, para. 2.8). "Offenders are not given sentences of imprisonment by the courts for the purpose of ensuring their rehabilitation. Most offenders are usually likely to have a better prospect of reform if they stay in the community" (Home Office 1991, para. 1.28).

This was government policy. It underpinned the emphasis on just deserts and the use of custody only when the seriousness of the offense meant that denunciation and retribution could not otherwise be satisfied or, in the case of violent and sexual offenses, that public protection required it. In all other cases, "community" or financial penalties were to be employed.

If ministerial rhetoric is to be believed, the "new realism" said to be embodied by the 1991 act has been substantially swept away. "Prison works," the home secretary, Mr. Michael Howard, told the Conservative Party Annual Conference in October 1993. It prevents prisoners from committing the crimes that they would commit were they not incarcerated, and he does not accept that the prospect and experience of imprisonment does not deter. Indeed, to ensure that prisons do deter, Mr. Howard has instructed the Prison Service to make the character of prison regimes "more austere": it is not yet clear what this is to mean operationally, but the decision to provide electrical outlets in cells so that prisoners can have their own television sets has already been reversed. Further, the minister is not persuaded that offenders will not be improved while in custody. Thus the Criminal Justice and Public Order Bill currently before Parliament provides for a new sentence, a secure training order, for twelve- to fourteen-year-olds. This provision, for which repeat juvenile offenders will be eligible, is an expression of the minister's faith that the research conducted in the 1960s and 70s by his own Home Office Research Unit on the damaging failure of custodial sentences for juveniles is no longer relevant. Now, he avers, we can do it better.

There are other provisions in the new Criminal Justice Bill that, like the 1993 act, will also increase the use of custody. But we do not have to wait for the bill to receive the Royal Assent. Sentencers are already responding to the new tough climate and sentencing accordingly: the large measure of discretion provided for them by the 1991 act to interpret offense seriousness, and thus "just desert," gives them ample scope. The Prison Service is preparing for a hot summer.

English Sentencing since the Criminal Justice Act 1991
Andrew Ashworth

The Criminal Justice Act 1991 represented the most far-reaching sentencing legislation in England and Wales for many a decade. It came into force in October 1992. The act sought to impose a structure on aspects of sentencing that had previously been left to the discretion of the courts. The government's goal was that Parliament would establish the structure and the courts (under the guidance of the Court of Appeal) would then refine it and apply it to particular classes of case. How have these ideals worked in practice? This question may be answered in relation to four of the leading aims of the act.

Clearer Rationale One of the aims of the legislation was to ensure that, in general, sentences should be proportionate to the seriousness of the offense. This was to apply not only to the choice of community sanctions and financial penalties but also to the length of custodial sentences. Thus section 2(2)(a) of the act provides that a custodial sentence shall be "commensurate with the seriousness of the offense." However, within months of the act's implementation, the Court of Appeal, in the leading case of *Cunningham*, (1993) 14 Cr.App.R.(S) 444, held that the key words should be interpreted as "commensurate with the punishment *and deterrence* which the seriousness of the offense requires"—a flagrant deviation from the aims of the act, allowing the courts to revert to "business as usual" and to add a deterrent supplement to sentences for certain types of offense.

Twin-Track Sentencing The government's intention was that sentence levels for nonserious crimes should be lowered, while serious offenses and dangerous offenders should continue to attract long sentences. The upper track of sentencing has certainly slotted into place. Sentences for serious violence, armed robbery, drug dealing, and other serious crimes, remain high. The statutory exception to proportionality in section 2(2)(b) of the act, allowing courts to impose longer-than-proportionate sentences where it is necessary to protect the public from serious harm from a violent or sexual offender, has been invoked many times. But perhaps the main cause of the 20 percent increase in the prison population between early 1993 and late 1995 can be traced to a judicial response to public alarm at certain notorious crimes in 1993, the catalyst being the murder of two-year-old James Bulger by two young boys. The courts are now imposing more and longer custodial sentences, but it is doubtful whether the act has influenced this.

Fewer Short Sentences The lower track of sentencing was intended to replace custodial sentences for many nonserious offenders with "punishment in the community." To this end the act provides that no offender should receive a custodial sentence unless the offense is so serious that only such a sentence can be justified: this formula was left to be interpreted by the courts, and again the Court of Appeal's response has been disappointing. All it has said, in *Cox*, (1993) 14 Cr.App.R.(S) 479, and subsequent decisions, is that the test should be that of "right-thinking members of the public."

The act also provided that sentences for offenders with a prior record should be kept in proportion to the seriousness of the offense. Many magistrates and politicians misinterpreted this as forbidding courts from imposing longer sentences on those with a prior record than on first offenders. As a result, the Criminal Justice Act 1993 replaced that provision with new language providing that "the court may take into account any previous convictions of the offender." It appears that courts are now adding greater enhancements for prior record.

The 1991 act also sought to promote community sentences for nonserious offenders: a higher proportion of community sentences has been imposed (from 23 percent before the act to 36 percent), but the proportionate use of custody has also risen from 44 to 47 percent. Use of other penalties, notably suspended sentences (down from 18 to 2 percent of the total), has fallen.

Financial Penalties   The act introduced unit fines, based on the day fine, as a means of calculating fines more fairly. The aim was to ensure that fines could be imposed on rich and poor alike, depriving them of a proportionate amount of their disposable income. The system was widely misunderstood, newspapers campaigned strongly against it, and rather than adjusting the method of calculation the government, in an extravagant political gesture, abolished the whole system in the Criminal Justice Act 1993. Courts are again left to their own devices, fewer fines are being imposed, and the amount of fines imposed on unemployed offenders has increased.

In this short article I have charted some of the many changes in English sentencing since 1992. To what extent those changes can be traced to the 1991 act is controversial. Many judges and some magistrates were unsympathetic to the act from the outset, resenting Parliament's intrusion into "their" discretion. In many respects the act was poorly drafted. Only months after the act's implementation, the British media focused on law and order, as a result of some dreadful crimes, and called for tougher sentences. And then, during 1994 and 1995, punishment policy became one of the few areas in which a weak government could hope for popular acclaim. The Criminal Justice and Public Order Act 1994 introduced tougher sentences for young offenders. At the Conservative Party conference in 1995, the home secretary announced that proposals would be brought forward for tougher approaches to repeat offenders, resembling the three-strikes laws in the United States. The Lord Chief Justice immediately made a public statement denouncing this approach (*The Times*, September 29, 1995). Thus the struggle for sentencing supremacy between the courts and the government or legislature has intensified since the 1991 act. For further discussion see Ashworth (1995).

Australia

Sentencing Reform in Victoria
  Arie Freiberg

Recent sentencing reforms in the Australian state of Victoria have shown that well-implemented sentencing policy changes—the abolition of good time—can take place without increasing prison crowding. More recent 1993 legislation, however, premised on the mistaken belief that tougher penalties reduce crime, are likely to lead to overcrowded times in Victorian prisons.

In April 1991, the Victorian Parliament enacted the Sentencing Act 1991 (Vic), the third major legislative renovation in a decade. One of the major features of the sentencing reform package was the immediate and complete abolition of remissions (good time), the effect of which had been to reduce the time served in custody by one-third. However, the abolition was accompanied by a legislative direction to the courts to decrease commensurately the lengths of sentences imposed to ensure that the period in custody served by offenders was no longer after the act than it was before. The act came into operation in April 1992. An evaluation after nearly a year indicates that the legislation has been successful in ensuring that neither the overall prison population nor the average time served by offenders has increased because of the legislation.

Background Sentencing reform during the 1970s and early 1980s in Victoria focused primarily on increasing the range of intermediate sanctions available to the courts. In 1981 the Penalties and Sentences Act consolidated a number of disparate sentencing powers in a single act and created the community service order, which required the performance of community work as a means of punishment and restitution. In 1985 a new Penalties and Sentences Act introduced suspended sentences into Victoria, permitting courts to suspend a sentence of imprisonment of up to one year, either totally or partially, and introduced a new measure, the community-based order, which combined probation orders, attendance center orders, and community service orders. Finally, the act required the courts in sentencing an offender to have regard to that person's plea of guilty, whether or not it was indicative of remorse.

The genesis of the Sentencing Act 1991 (Vic) can be traced back to one controversial case decided by the Supreme Court of Victoria in 1985 (*Yates v. V. R.* 41 [1985]), which centered on truth in sentencing, in particular, the question whether courts should take into account the effects of good time when imposing sentence. In delivering a negative response, the court trenchantly observed that the sentencing process had been reduced to a "charade" because of significant differences between nominal sentences imposed by the courts and the actual times served by offenders in custody.

Following this decision the Victorian government appointed a committee to review Victoria's sentencing laws. Reporting in 1988, the committee recommended the abolition of remissions and suggested a host of other reforms relating to statutory maximum sentences, new and modified sentencing options, enhanced guidance for sentencers, and the establishment of a Judicial Studies Board to assist the courts in sentencing and undertake research on sentencing.

Victoria has long been proud of having one of the lowest imprisonment rates in Australia. The financial implications of increasing prison populations are always a considered factor when sentencing reform is discussed. The incarceration rate per 100,000 of adult population has held steady over the last decade, at around 57 to 58. In February 1993 the prison population stood at about 2,236. By comparison, the New South Wales prison population stands at about 6,100 (about 110 per 100,000 adult population), having increased by about 42 percent over the same period.

The Sentencing Act 1991 When the Victorian government decided to abolish good time, it was concerned that the prison population not increase as a result, as had occurred earlier in New South Wales when a similar move was made. The policy adopted in Victoria was that adjustments would be made to require courts to take the abolition of good time into account when setting sentences (see generally Freiberg 1992). Section 10 of the act accordingly provides:

> (1) When sentencing an offender to a term of imprisonment a court must consider whether the sentence it proposes would result in the offender spending more time in custody, only because of the abolition of remission entitlements . . . than he or she would have spent had he or she been sentenced before the commencement of that section for a similar offense in similar circumstances.
>
> (2) If the court considers that the sentence it proposes would have the result referred to in sub-section (1) it must reduce the proposed sentence in accordance with sub-section (3).

Thus "truth," under the Sentencing Act 1991 (Vic), does not mean that the nominal sentencing patterns prevailing prior to the proclamation of the act continue so that prisoners serve the whole of the court's announced term of imprisonment, a term that was effectively never served in practice. It does mean that the perceived lack of correspondence between sentences imposed by the courts and periods served by prisoners will diminish.

Evaluation of the Legislation   In order to assess the impact of the Sentencing Act 1991 (Vic) a study was undertaken by the Victorian Criminal Justice Statistics Planning Unit and the University of Melbourne. A preliminary evaluation has examined both the total data for prison sentences in Victoria and a selection of four offenses for which a sufficiently large data base is available (armed robbery, burglary, theft, and driving while disqualified). If section 10 of the Sentencing Act 1991 (Vic) is working as it should, then the mean time for the aggregate sentence and the nonparole period after the introduction of the Sentencing Act 1991 (Vic) should be about 67 percent of that for sentences imposed before the act. The effective time in custody before and after the act should be identical.

The aggregate data so far obtained reveal that the average aggregate prison term for all prison receptions dropped from 14.7 months in the twenty-four months prior to the act to 10.8 months in the six months after the act, a drop of 27 percent. The average period before parole release eligibility decreased from 10.4 months to 8.0 months, a reduction of about 20 percent. The average estimated time in custody for all offenses has remained about the same.

Given the difficulty of policing the internal thought processes of the judiciary and the opportunities for evasion of the legislation afforded by use of rules relating to concurrent and consecutive sentences, it appears that the Victorian legislature's requirement of a mandatory adjustment to sentence lengths has been highly successful in achieving the intended outcome. The results of the study are supported by the aggregate reception data for the period of twelve months before the act and nine months after. The total number of prisoners has remained almost identical (2,222 in January 1992; 2,243 in January 1993). During that period, however, the number of prisoners received into prison fell from about 1,700 to 1,407 in a six-month period.

The implementation of the Sentencing Act 1991 (Vic) has taken place generally as predicted. The smooth transition was assisted by the long lead time between the introduction of the act and its proclamation. Information about the act and its implications was widely distributed. Comprehensive briefing notes on the act were prepared, particularly by the magistracy (which deals with over 90 percent of criminal cases), and all courts were assisted by the timely publication of a sentencing manual written in anticipation of the new legislation.

The successful outcome of this legislative intervention occurred in the face of hostility on the part of some judges and the then opposition party, which had taken the erroneous view that effective sentence lengths would decrease. The successful change was also made possible by the fact that, relative to other jurisdictions, Victorian sentencing decision making is highly centralized in Melbourne, the capital city.

Intermediate Sanctions Imprisonment is used sparingly in Victoria. Even in the higher courts, which deal with the most serious cases, fewer than 50 percent of offenders are jailed. The proportion of sentenced offenders imprisoned in higher courts has decreased from 58.3 percent of all dispositions in 1985 to 47.9 percent in 1991, a drop of nearly a fifth. In the Magistrates Court in 1991, imprisonment orders amounted to 5 percent of all principal offenses disposed of, suspended sentences amounted to 4.7 percent, community-based orders to 4.3 percent, fines to 43.8 percent, bonds to 20.3 percent, and sanctions relating to drivers' licenses, 20.9 percent.

The Sentencing Act 1991 (Vic) built on, and emphasized, a number of trends already evident. The stability of the prison population has occurred in the face of a rising crime rate and an increasing number of offenses processed by the courts. In the higher courts, from which the prison population is mainly drawn, the number of offenders sentenced jumped from 1,004 in 1985 to 1,572 in 1991, an increase of nearly 57 percent. It appears that the composition of the prison population is changing, with the courts sentencing fewer people to short terms of imprisonment. Options such as suspended sentences (introduced in 1985) and intensive correction orders (introduced in 1991 and which permit service of a term of imprisonment in the community) appear to be having some real diversionary impact, with noncustodial options displacing custody for a range of offenders.

The Sentencing (Amendment) Act 1993 In October 1992, Victoria changed governments. After ten years in office, a reformist Labour government was replaced by a party that had campaigned partly on a law-and-order platform and that was antagonistic to many of the provisions in the Sentencing Act 1991 (Vic).

In May 1993 the new government rushed through Parliament an act that could significantly increase the prison population. Responding to what it perceived to be the community's concern about the inadequacy of custodial sentences imposed on sexual and violent offenders, the legislation sought to increase custodial sentences in a number of ways. The first is to create two new classes of offenders: "serious sexual offenders" and "serious violent offenders," both of which are elaborately and widely defined. If these offenders commit certain serious sexual or violent offenses, the court is required to regard the "protection of the community" as the principal purpose of sentencing and may, "in order to achieve that purpose, impose a sentence longer than that which is proportionate to the gravity of the offense considered in the light of its objective circumstances." In addition, section 10 of the Sentencing Act 1991 (Vic) will not apply to these offenders, effectively increasing their sentences by a third. Further, sentences imposed on serious sexual offenders will be presumptively consecutive. The current legislative presumption is that all sentences are to be served concurrently.

The second method by which sentence length will be increased is the introduction of indefinite sentences. Adult offenders who are proved, to a high degree of probability, to be a "serious danger to the community" and who have been convicted of specified "serious offenses" can, at the discretion of the court, be sentenced to indefinite imprisonment. A court in such cases is required to set a "nominal sentence," equal to the previous minimum period before parole release eligibility, at the end of which a judicial review must take place. If the offender is deemed still to be a serious danger to the community, he or she must remain in custody for at least another three years, the minimum period between reviews.

It is difficult, at this early stage, to speculate about the effect of these changes on prison populations, as much will depend on the judiciary's attitude to provisions that are so contrary to the prevailing sentencing culture. In contrast to the wide and deliberate consultation that took place before the introduction and implementation of the Sentencing Act 1991 (Vic), the 1993 act was introduced into Parliament and passed within a matter of weeks with no community input at all.

Other than the provision that selectively negates the operation of section 10, the courts retain some discretion in relation to both consecutive and indefinite sentences. However, if implemented wholly as intended, it has been estimated that, on average, sentences imposed on "serious sexual offenders" will increase by 250 percent, from 5.4 years to 19 years. Assuming all other variables remain constant, such as conviction rates and sentencing practices, within ten years there may be an additional 400 such offenders in prison.

Conclusion After a decade of sentencing reform that saw imprisonment rates in Victoria remain among the lowest in the industrialized world, Victoria is now embarking on a new era of change modeled on the failed law-and-order experiments of many of the U.S. jurisdictions. Despite overwhelming evidence that such strategies increase prison populations, increase the rate of not guilty pleas by persons in the relevant categories of offenses, increase the cost of trials as all legal issues are vigorously contested, increase case processing times and judicial work loads, increase demands for legal aid, and increase plea bargaining and other techniques to avoid manifestly unjust and disproportionate sentences, the government has proceeded with legislation whose only certain outcome is that overcrowded times are just around the corner for Victoria.

Truth in Sentencing in New South Wales
  Angela Gorta

In late 1989 New South Wales made radical changes to its sentencing laws. Before the changes, New South Wales had an indeterminate sentencing system consisting of minimum and maximum prison terms, good time (called remission in Australia) that reduced both maximums and minimums, discretionary parole release, and lengthy postrelease supervision. Under the 1989 legislation, good time was eliminated and the length of the community supervision component of the sentence became a fixed fraction of the length of time in custody.

Evaluations show that the legislation has increased average time in custody by about 20 percent, thereby adding to prison crowding, and greatly diminished (by nearly half) the use of postrelease supervision and (by nearly three-quarters) the duration of postrelease supervision. This article describes the legislative changes and their effects on custodial sentencing patterns and on the size of the New South Wales prison population.

Background   Australia is divided into six states and two territories. Residents are subject to both state and federal legislation. Although New South Wales (NSW) comprises only 10.4 percent of the total area of Australia, over one-third of Australian residents live there. NSW has a resident population of approximately six million people. Its current prison population is approximately 6,100.

Before the new legislation took effect, the Probation and Parole Act 1983 governed custodial sentencing. Under this act, approximately one-third good time was deducted from both the maximum sentence and the minimum sentence before parole eligibility. For example, an offender sentenced to nine years, including three years before parole eligibility, would serve two years in custody if his parole were granted at the earliest opportunity or six years if released with full credit for good time.

Under the Sentencing Act 1989, often described as "Truth in Sentencing," no good time is deducted from the maximum sentence or from the minimum. Furthermore, the length of the community supervision component of a sentence is a fixed fraction of the length of time in custody.

The disparity between the length of the maximum sentence and the time spent in custody under the Probation and Parole Act was, at times, highlighted by both the media and politicians. While well understood by those familiar with the criminal justice system, it was difficult for the general public to understand how a prisoner sentenced to, say, nine years could be released after only two years (the minimum less one-third good time). Furthermore, calculation of good time was subject to complex guidelines.

The Sentencing Act   The Sentencing Act, which took effect on September 25, 1989, was introduced, according to the legislative history, to "restore truth in the sentencing system in New South Wales" and "to bring certainty to sentencing in the State . . . (and) . . . to ensure that the public and prisoners know exactly when a sentence shall commence and exactly when a prisoner will be eligible for consideration for parole."

The objects of the act are stated to be: "(a) to promote truth in sentencing by requiring convicted offenders to serve in prison (without any reduction) the minimum or fixed term of imprisonment set by the court; and (b) to provide that prisoners who have served their minimum terms of imprisonment may be considered for release on parole for the residue of their sentences."

The new act provides for a "fixed term," a "minimum term," and an "additional term." The minimum term must be served in custody. The additional term is that part of the sentence during which the person may be released on parole. The fixed term is similar to the minimum term in that it is the period that must be served in custody; unlike the minimum term, however, no additional term is specified, and postrelease supervision is not applicable. Sentences of six months or less are required, under the act, to be fixed terms. The "sentence," thus, consists of the time in custody plus the period of community supervision, if any.

Under previous legislation a court was required to set the maximum sentence (which would reflect the offender's criminality together with the objective and subjective factors in the case) before determining the proportion that should be served in custody. Under the Sentencing Act, the time to be spent in custody is decided first, with the length of the total sentence decided subsequently.

The Sentencing Act abolished all forms of good time, established a 1:3 ratio of the additional term to the minimum term, and removed a presumption in favor of parole for certain prisoners.

In place of good time, the act authorizes visiting justices dealing with major breaches of prison discipline to increase the period of confinement by up to twenty-eight days. Prisoners whose minimum sentences plus additional terms total more than three years are not released automatically to parole at the end of the minimum period but are considered for release by the Offenders Review Board.

Effects on Custodial Sentencing   Because the Sentencing Act simultaneously changed so many aspects of custodial sentencing, patterns and practices necessarily changed. Before the introduction of the Sentencing Act no one was sure what its impact would be. The Court of Criminal Appeal in *R. v. Maclay*, 19 NSWLR 112 (1990), judged that a "fresh approach" must be made, "giving appropriate weight to well-established principles of sentencing."

In order to determine the act's effects on the size of the NSW prison population and the sentencing practices of judicial officers, sentences imposed and time served for a sample of those sentenced before the change in legislation (specifically those discharged from NSW institutions during the six-month period January 1, 1989, to June 30, 1989) were compared with those sentenced after the change (new sentenced receptions after October 1, 1989). The initial analysis included those received up to March 31, 1990 (Gorta and Eyland 1990). A second analysis included those received up to June 30, 1990 (Gorta 1990, 1992). The most recent analysis included those whose sentences began between October 1, 1989, and August 31, 1992 (Eyland 1992). The details of the analysis are published elsewhere (Gorta and Eyland 1990).

In summary, as table 3.2 indicates, the evaluations revealed these patterns under the new legislation: prisoners were serving longer periods in custody; fewer were being given sentences that included periods of conditional release to community supervision; and those with additional terms were spending much less potential time on community supervision.

Judicial officers have tended significantly to reduce the total sentence ordered from the average maximum sentence that would have been handed down under

Table 3.2. Truth-in-Sentencing Evaluation Results

Sentencing Measures	Before Group (n = 1,832)	After Group (n = 2,908)
Average time in custody[1]	244 days	290 days[4]
Percentage whose sentences included a conditional release to community supervision component[2]	56.0%	30.9%[4]
Average maximum potential period of community supervision[3]	799 days	208 days[4]
Average ratio of period on community supervision to period in custody	2.40	0.39
Average aggregate sentence	738 days	360 days

Source: Gorta (1992).

Notes:

1. Time in custody for the Before Group was the nonparole period less good time for maximum sentence less good time; for the After Group, it was the minimum or fixed term.

2. Community supervision component refers to parole or after-care probation for those in the Before Group and to an additional term for those in the After Group.

3. In practice, an offender may not have been supervised for this entire period as the Probation and Parole Service (now known as the Community Corrections Service) maintained the right of discretionary or early termination.

4. In a more recent analysis incorporating 35 months of data after the change in legislation (Eyland 1992), the same general trends were found. The average time in custody was calculated to be 280.7 days; 27 percent had a sentence.

the Probation and Parole Act 1983, largely by reducing the length of postprison community supervision. Minimum terms tend to be shorter than nonparole periods set under previous legislation. Despite these reductions, however, the average time to be spent in custody is longer under the Sentencing Act than under the Probation and Parole Act.

Effects on Prison Population   When the Sentencing Act was introduced, its potential effect on the size of the NSW prison population was of concern. The minister for Corrective Services emphasized that "the Government is not seeking to make sentences longer." A guide published by the Department of Corrective Services noted that it was "not the Government's intention that, as a consequence of the *Sentencing Act,* longer sentences be served." Others, however, argued that the new legislation was likely to lead to increased prison sentences in NSW with an associated massive growth in the number of people in prison.

The NSW prison population was increasing prior to the introduction of the Sentencing Act. The daily average number of prisoners in NSW has been increasing since the 1984/1985 financial year when the average prisoner population was 3,473 (equivalent to an imprisonment rate of 63.8 per 100,000 NSW population). In 1991/1992, the daily average had reached 6,056 (equivalent to an imprisonment rate of 102.1). In 1991/1992, there were 1,698 prisoners more on average each day than in 1988/1989. The prison population may have continued to increase, as a result of factors such as increasing crime and policing clearance rates, had there been no legislative changes. The question is how much, if any, of the increase can be attributed to the effects of the Sentencing Act.

By comparing the average time served before and after the 1989 legislation and by assuming that the reception rate during the period under study was representative of the rate of receptions in subsequent months, it appears that the overall increase of 46 days (290 – 244) in the average time to be served is equivalent to an eventual overall increase in the NSW prison population of approximately 490 additional sentenced prisoners on any day. By examining differences in the distribution of time spent in custody before and after the legislative change and assuming that the sentencing patterns of prisoners received between October 1989 and June 1990 were representative of the sentencing patterns of prisoners received in the following months, it was estimated that the increase in the prison population would be most marked four to seventeen months after the introduction of the legislation. This estimated population increase is probably an underestimate in that it is based on an assumption that all prisoners in the After Group are released at the ends of their minimum periods.

Other Effects   While the increase in prisoner numbers at a time when the NSW prison population was already overcrowded was a serious consequence of the Sentencing Act, there are also more positive consequences. Chan (1992) has identified five "windows of opportunity for turning the tide." They include: (1) the public acceptance of changes in established sentencing practices; (2) not all sentencers have ignored the effect of the abolition of good time; (3) even though the Court of Criminal Appeal in *Maclay* did not tell judges and magistrates to reduce their sentences to correct for the absence of good time, "it did not tell them that they could not do so either"; (4) the Court of Criminal Appeal has made a longer parole period a possibility for some prisoners; and (5) the continual monitoring of sentencing trends and imprisonment rates and the publication of these results have highlighted the connection between sentencing policy and the penal system.

Sentencing and Punishment in Australia in the 1990s
Arie Freiberg

Unevenly, but seemingly inexorably, Australia is being swept up in the international movement to "get tough" on crime in the hope that such policies will restore communal peace and harmony. Prison populations, which were steady or declining, have begun to grow in a number of jurisdictions.

Predominant concerns about sentencing have fundamentally changed in twenty years. In the 1970s, prison conditions, development of community-based penalties, and reduction in sentencing disparities were the major goals of sentencing reform. In the mid-1990s these problems—though certainly not solved—have been displaced by calls for tougher penalties, incapacitative sentencing, and vindication of the interests of victims.

Responsibility for sentencing in Australia is divided among six state governments, two self-governing territories, and the federal government. Federations inevitably tend to diversity, and Australia is no exception. Recognizable similarities in policy and legislation most commonly emerge from "the temper of the times," sometimes from conscious emulation, and rarely from concerted action. It is very difficult to do justice to the changes in all the jurisdictions and to describe each of their histories and legislation in minute detail. Rather I have tried to capture the general flavor of the changes and speculate on their likely impact.

In sentencing, as in so many other areas of social policy, the only constant is change. Following the election of politically conservative governments in New South Wales in 1988, Victoria in 1992, and Western Australia and South Australia in 1993, and the reelection of a conservative government in the Northern Territory in 1994, the political complexion of the country has changed dramatically.

Background   Sentencing reform in Australia in the late 1960s and early 1970s reflected the times. Disillusionment with the prison led to the development of a wider range of community-based sanctions. These were primarily community work, suspension of sentence, and forms of periodic detention. Though imprisonment rates increased slowly across Australia in the 1970s, they declined in the early 1980s, possibly as a result of the proliferation of alternatives to custody, including increased use of parole (Walker 1994, p. 22).

Federal Government   The first systematic review of sentencing took place in South Australia (South Australia, Criminal Law and Penal Methods Reform Committee [Mitchell Committee] 1973) but did not bear fruit until 1988, when it appeared in a greatly attenuated form in the Criminal Procedure (Sentencing) Act 1988 (SA). More influential in sentencing policy, although not in practice, was a comprehensive inquiry into sentencing undertaken by the Australia Law Reform Commission (Australia Law Reform Commission 1988). Although the federal government plays only a small part in the administration of criminal justice in Australia, the eight-year inquiry helped shape the sentencing debate. It identified a long list of problems: "lack of sentencing policy; lack of guidance as to criteria for determining sentence; lack of guidance in relation to the procedure for determining sentence; unsatisfactory penalty structure and lack of satisfactory rules as to sanction choice; lack of sentencing information; early conditional release; excessive use of imprisonment as a sanction; under utilization of noncustodial options; unwarranted disparity; [and poor] public perceptions of the sentencing process" (Zdenkowski 1994, p. 192).

The commission's final report, delivered in 1988, precipitated a number of changes to federal legislation, not all of which were in accordance with its recommendations. The changes, however, are relatively minor—concerning parole release and eligibility and some specification of factors courts should consider in sentencing—and do not constitute an integrated package. They failed to remedy the problems the commission identified, have made federal sentencing law unduly complicated and opaque, and have generated further calls for change.

New South Wales   No lengthy inquiries preceded New South Wales's sentencing reform. In 1989, responding to law-and-order concerns, Parliament abolished remissions (good time) in the name of "truth in sentencing." Together with a direction that the nonparole period must be two-thirds of the sentence imposed, and in the absence of a binding legislative direction to the courts to reduce sentence lengths to compensate for the abolition of remissions, the result was a rapid increase of 47 percent in prison population from 4,369 in 1988–1989 to 6,117 in October of 1993. Truth in sentencing in New South Wales came to mean longer sentences, rather than a sentencing system in which the time served by offenders more closely reflected the sentence imposed by the courts.

Victoria   The background to sentencing reform in Victoria has been set out in a previous article in *Overcrowded Times* (August 1993). The reforms outlined briefly there, and in more detail elsewhere (Freiberg 1994), have partially influenced the development of sentencing legislation in other jurisdictions. The Sentencing Act 1991 (Vic) provided for abolition of remission but—unlike its New South Wales predecessor— directed judges thereafter to reduce sentence lengths to adjust for absence of remission. Judges by and large complied, and the prison population held steady.

Queensland   In 1992, four years after an extensive examination of the prison system, Queensland comprehensively recast its sentencing legislation by enacting a new Penalties and Sentences Act. Adopting a similar structure to the Sentencing Act 1991 (Vic), it consolidates the sentencing powers of the court in one act. It provides for a range of identified "intermediate" orders including probation orders, community service orders, and a new order, the intensive correction order. The intensive correction order is modeled on a similar Victorian sanction that filled a perceived gap between imprisonment and community service. This order is intended to replace short terms of imprisonment (up to one year) with intensive community service work, counseling, and possibly residence in community residential facilities for treatment for periods of up to a week. In Victoria, this order has had little impact, primarily because the resources required to support its rehabilitative component have not been provided. Though drawing its clientele from the potential prison population, thus fulfilling its role as a diversionary device, the intensive nature of the supervision and the onerousness of its community work provisions have resulted in a relatively high failure rate (30 to 40 percent).

Both the Queensland and Victorian acts codify the general common law sentencing principles, identified as imposition of just punishment, deterrence, rehabilitation, denunciation, and protection of the community. No priorities among principles are established.

Queensland's legislation differed in one significant way from Victorian legislation. It contained a provision allowing for indefinite sentences to be imposed on violent offenders. This concept was imported into Victorian law in 1993 by the newly elected conservative government and, together with other measures to deal with serious violent and sexual offenders, marked a turnaround in Victorian sentencing policy.

Northern Territory   Victoria's influence is also noticeable in a major piece of sentencing legislation that was put before the Parliament of the Northern Territory prior to an election in June 1994 that saw a conservative government returned to office. The Sentencing Bill, part of a law, order, and public safety campaign, combined the form and structure of the Victorian and Queensland statutes with even "tougher" innovations, including fixed nonparole periods for sex offenders (70 percent of the term of imprisonment), the abolition of remissions with no compensatory changes in sentence length, and the introduction of indefinite sentences for violent offenders. The bill did not pass but is expected to be reintroduced with even more stringent measures such as mandatory imprisonment for repeat offenders.

Prison Populations Australia's prison population is low relative to that of the United States. In April 1994 the average imprisonment rate was 86.2 per 100,000 of the total population, or 114 per 100,000 of the adult population (seventeen years of age and over). These average figures, however, mask great variations between the states (see table 3.3) (Australian Institute of Criminology 1994).

These variations can be attributed to a number of factors. The Northern Territory and Western Australia have larger Aboriginal populations which, in turn, have a very high imprisonment rate. In the Northern Territory, their imprisonment rate is 1,431 per 100,000 of the Aboriginal population. The Territory's population is also more masculine and more youthful than in other jurisdictions (Walker 1994, p. 23). In Western Australia, approximately 35 percent of prisoners are Aboriginal, although they make up only 2.7 percent of the general population.

Another important factor is the rate of imprisonment for fine default. Although fine defaulters represent only a small percentage of prisoners in custody at any one time, in some jurisdictions they make up a large proportion of prison receptions. In 1993, fine defaulters made up 41 percent of total prisoners received in prison in South Australia, 40 percent in Western Australia and Tasmania, 28 percent in New South Wales, 26 percent in the Northern Territory, 18 percent in Queensland, and only 7 percent in Victoria.

The variations, however, cannot be attributed to differences in crime rates nor to different offense profiles. Victoria's historically low imprisonment rate, relative to other Australian states, and indeed the rest of the world, can be attributed to a number of features of its sentencing policy: an extensive range of noncustodial options that are well supported by the courts; a successful suspended sentence option, particularly for first offenders; the reluctance of the courts to impose imprisonment in default of fines and for breach of intermediate sanctions; limited use of short terms

Table 3.3. Australian Prison Populations, April 1994

Jurisdiction	Daily Average Population	Rate per 100,000 Adult Population	Rate per 100,000 Total Population
Northern Territory	484	414.9	286.1
Western Australia	2,062	163.4	121.7
New South Wales	6,354	138.5	105.0
South Australia	1,268	112.4	86.5
Queensland	2,406	101.0	75.6
Victoria	2,431	71.3	54.3
Tasmania	245	69.5	51.8

Source: Australian Institute of Criminology (1994).

of imprisonment, resulting in low prison reception rates; and restrained use of very long or indefinite sentences of imprisonment.

Overall in Australia, imprisonment is used sparingly. In Victoria, for example, in 1993, in the higher courts that deal with only 5 percent of criminal cases, sentences of imprisonment comprised only half the sentences imposed, with suspended sentences of imprisonment, introduced in 1986, accounting for nearly 30 percent of sentences. Intermediate sanctions make up 10 percent of sentences and lower order sanctions such as bonds and fines, another 10 percent. In the magistrates' courts, which sentence 90 percent of all offenders, over 50 percent of offenses are dealt with through lower order sanctions such as fines and bonds, with imprisonment accounting for around 13 percent of sentences. Here, as in the higher courts, the suspended sentence has occupied an important sentencing space, diverting offenders not only from imprisonment but also from intermediate sanctions such as the community-based order.

Until recently, in Victoria, the general judicial and political ethos was that imprisonment was a necessary and expensive evil, the use of which was to be a very last resort. However, that ethos has withered in the face of increasing public fear of crime, a fear that is paradoxically rising as the general crime rate falls.

Recent Changes in Sentencing Policy Sentencing policy in Australia has historically been governed by the concept of proportionality. However, proportionality, or just deserts, is conceived of as setting the general outer limits of punishment, rather than determining the exact type and precise quantum of the sanction. The policy has been affirmed on numerous occasions by the High Court of Australia as a primary sentencing consideration, and the recently introduced statutory codifications of principles in Victoria, Queensland, and, most recently, in Western Australia ("A sentence imposed on an offender shall be commensurate with the seriousness of the offence") are regarded as merely legislative recognition of what was decided by the High Court.

Principles of proportionality have also found their way into new juvenile justice legislation, previously dominated by considerations of welfare. Thus Queensland's Juvenile Justice Act 1992 requires that there be a "fitting proportion between sentence and offence" and contains an ordered sanctioning hierarchy to assist in calibrating offenses and penalties.

However, these developments are counterbalanced by other measures. In some jurisdictions, for limited classes of offenses, courts are required to have regard primarily to "the protection of the community" as the principal purpose of sentencing and may, in order to achieve that purpose, impose a sentence *longer* than that which is proportionate to the gravity of the offense. Proportion gives way to prophylaxis.

Preventive sentences are not unknown in Australia. Habitual offender provisions existed in the statute books of most jurisdictions but fell into desuetude in recent decades. Their revival or transformation into "dangerous" or "violent" offender laws is now a feature of most sentencing regimes, including juvenile justice statutes (see Crime [Serious and Repeat] Offenders Sentencing Act 1992 [WA]). Based on past crimes or predictions of future offending, or their combination, these laws have, as yet, been little used. However, they represent a significant, symbolic, and temptingly expandable inroad into the hitherto sacrosanct concept of proportionality. As courts are increasingly induced, or coerced, to use preventive sentences, warehoused offenders will make up a greater proportion of the stock of prisoners. Although baseball is but a minor sport in Australia, the notion of "three strikes and you're out" is becoming a familiar refrain in conservative circles.

A process of conflating the concepts of proportion and prevention is underway. Longer sentences are more deserved, it is sometimes said, because previous scales of punishment were unduly lenient and manifestly inadequate for modern notions of desert that more closely reflect the punitiveness of the populace rather than the theorizing of criminologists, sentencing commissions, or indeed, the judiciary. Longer sentences are also more incapacitative and possibly deterrent. Hence such sentences are sought for a wider range of offenses.

The pressure to increase sentence length is manifested in moves to increase statutory maximum penalties, court imposed sentences, and the actual time served by prisoners. Increasing statutory maximums has long been the first response to crises in criminal justice, and this has indeed occurred in a number of jurisdictions. In particular, penalties for sex offenses and offenses against children have been increased in response to pressures by victims' lobbies and some women's groups. In Victoria, in particular, the courts have responded by significantly increasing the length of sentences imposed on such offenders. This may also have the unintended (or perhaps intended) effect of raising sentences across the board in order to maintain some semblance of internal consistency in the judicial sentencing scale.

Another technique by which sentence length can be increased is by requiring courts to cumulate sentences for multiple offending. Principles of totality, proportionality, and rehabilitation are made subservient to the notion that each offense must be justly, proportionately, and manifestly punished. In part, this development is an outgrowth of the victims' movement, sections of which regard concurrent sentences as a denial of justice to individual victims.

Longer sentences can also be brought about by the abolition of remissions. As noted, New South Wales and Victoria did so some years ago, Queensland and South Australia have recently followed suit, and the Northern Territory is likely soon to do so, usually without adjustment of the sentence imposed by the courts.

In my 1993 *Overcrowded Times* article, I indicated that the Victorian prison population had remained stable after remission was abolished. This occurred in part because section 10 of the Sentencing Act 1991 (Vic) directed judges to reduce sentences to take account of the absence of remission and thereby to provide that offenders' actual time in prison would not be increased compared with time served by comparable offenders during the period when remission was available. Since that time, however, the situation has deteriorated markedly.

A decision by the Supreme Court of Victoria in 1994, subsequently upheld by the High Court of Australia, threw the operation of section 10 of the Sentencing Act 1991 (Vic) into doubt. Its effect was that section 10 was applicable only to offenses whose maximum penalty had not changed. Because the Sentencing Act 1991 (Vic) has altered most maximums as part of a major rationalization of the penalty structure, section 10 was effectively rendered nugatory. As a result of this decision, and pressure from government, victims' groups, and the media for longer sentences, the Victorian prison population increased from around 2,200 in early 1993 to approximately 2,500 in mid-1994. Victoria's experience now resembles that of New South Wales.

The response of government to increasing prison populations in Australia is to build new prisons, preferably private ones. In 1989 there were no private prisons in Australia. In 1994, Queensland had two, and New South Wales two. Within two years, Victoria will have three private facilities housing over half the total (and growing) prison population. Other states are actively considering this option. In a short time, Australia will probably have the highest proportion of private prisoners in the world. The modern discourse is about prison management and fiscal responsibility, not about penal philosophy. Apparently, reduction in prison use is no longer a major political aim. Prison populations can be larger, so long as the prisons are cheaper and more efficient.

The trend away from a focus on just disposition of individual cases and toward a focus on efficient management and economy is also evident in other aspects of court administration. In Victoria, guilty pleas, particularly early pleas, are encouraged by legislation that permits courts to discount sentences on account of such conduct.

In New South Wales, plea bargaining has now been formalized by the introduction of "sentence indication hearings" that permit accused persons to appear before a judge, even before entering a plea, to obtain an indication of the sentence that they would be likely to receive if they plead guilty. The accused person is able to change his or her mind as to the plea and to opt for a trial.

Future Directions   New South Wales is Australia's most populous state and goes to the polls in March 1995. In January 1994, the government instituted an inquiry into sentencing reform that produced an issues paper later that year (see New South Wales, Attorney-General's Department, *Sentencing Review* 1994). The government's stated aim was to fine-tune, rather than dramatically overhaul, its sentencing legislation, and to strengthen its commitment to truth-in-sentencing legislation.

Reflecting developments elsewhere in Australia, the reform agenda included the consolidation of legislation; codification of sentencing principles; rationalization of sanctions; a sanction hierarchy; strengthening habitual criminal legislation; the desirability of an American-style sentencing grid to provide consistency in sentences of imprisonment; the need for a Sentencing Policy Advisory Council; the giving of power to the Court of Criminal Appeal to issue sentencing guidelines judgments; and the need for minimum sentences and for a wider range of intermediate sanctions, in particular, for an intensive corrections order.

Queensland, also facing an election within the next year or so, sees all political parties vying for the "tougher than thou" crown. Proposed revisions to its criminal code contain many increases in maximum penalties, with many carrying life imprisonment. Plans are afoot to build a superjail to hold intractable prisoners, and the state is adding another 500 cells to the prison system to cope with overcrowding and to cater for expected increases resulting from longer sentences and the abolition of remissions. Even in advance of implementation of these policies, prison numbers had increased by approximately 200 in two months to 2,627 in June 1994.

Conclusion   Sentencing reform in Australia over the last five years reflects sharply disparate policies. On the one hand there has been the move to consolidate, articulate, and rationalize sentencing laws. The modern Australian sentencing statute contains statements of principle and some guidance to sentencers, and it stipulates the range of sanctions, how they are to be used, and what should happen in the event of breach. Sentencing policies should provide a rational maximum penalty structure, an internally consistent penalty scale, and a coherent set of relationships between the forms of sanctions. Many of the statutes recently introduced contain these elements.

On the other hand, the inherently political character of sentencing policy making engenders ad hocery and irrationality. Harshness replaces hope, retribution displaces rehabilitation, and prevention erodes proportionality. The constituency of legislators and judges is changing. The state of prisons is of less concern than the state of the victim's health. The victim impact statement comes to carry more weight than the presentence report. The delicate sentencing balance between the interests of the offender, the state, and the victim is shifting away from the former to the latter. In sum, the greater sensitivity to the rights and interests of victims and the protection of the community in general is being reflected in more severe sentences.

Sentencing reform builds on the economics of fear. The most recent reforms represent a poverty of the political and criminological imagination. They confirm the observation that in Australia, at least, "rates of imprisonment tend to increase during periods of conservative government and tend to reduce during periods of Labor government" (Walker 1994, p. 26).

As these negative policies fail to "control" crime, as they have failed to do wherever they have been tried, the calls for harsher penalties will likely increase, resulting in the imposition of even more punitive and ineffective sentences. The perverse result of these policies will be a cycle of repression in which crime will increase, rather than decrease. Then, a new round of sentencing reform may again begin, searching again for ways to empty the overcrowded prisons and deal with offenders in more humane and positive ways.

Other English-Speaking Countries

Sentencing and Punishment in New Zealand, 1981–1993
Justice T. M. Thorp

New Zealand since 1980 has experienced many of the crime and punishment trends of other countries. Crime rates, especially for violent crimes, are up, as are prison populations. Community sanctions have increased in scale and number.

While political pressures for harsher crime control policies have increased, New Zealand, perhaps more than most English-speaking countries, has managed a balance between policies aimed at deterrence and incapacitation and policies aimed at rehabilitation and social reintegration.

A brief chronicle of developments since 1981 may interest readers because it confirms the universality of most of the central problems facing penal reformers and the difficulties of overcoming them and achieving consistent and coherent penal policy.

New Zealand is a South Pacific country with a population of 3.5 million, approximately 20 percent of whom are indigenous Maoris or immigrant Pacific Islanders. New Zealand inherited its legal systems and penal policies from Britain in 1840, when it came under British rule, but in this century has been trying to work out penal policies that suit its particular circumstances. It dropped corporal punishment in 1941 and capital punishment in 1961.

Levels of Criminality   In the early 1980s, the crime rate in New Zealand fitted the general pattern throughout the developed Western world. The only aberrant figure was for burglary, which was among the highest, a circumstance believed to be related to the fact that most residences stand in their own grounds and shrubberies and offer burglars favorable working conditions.

At that time, over 90 percent of offenses dealt with in the Magistrates Court were punished by fines, while a little over 60 percent of those sentenced in the High Court, which then tried all serious crimes, were imprisoned. This resulted in an imprisonment rate of 90 per 100,000 population, less than those in North America and some of its Pacific Island neighbors, but higher than in neighboring Australia and most European countries.

As measured by the total number of convictions, criminality in New Zealand has increased only slightly since the early 1980s. Table 3.4, comparing convictions for nontraffic offenses in 1981 and 1992, shows an overall increase of about 16 percent, which is close to the increase in population. However, the 1991 figure hides the introduction of two diversion schemes, which are discussed below, and without which the increase might have been 10 percent higher.

Even more important, there has been a much larger increase (approximately 79 percent) in convictions for violent crime than overall. Within violent offending, the largest increases were for the most serious offenses (over 100 percent each for homicides and sexual assaults). The sexual assault increase was almost certainly affected by higher rates of reporting and by many complaints that concerned inci-

Table 3.4. Number of Convictions for Nontraffic Offenses, 1981 and 1992

Offense Type	1981	1992
Violent	5,994	10,733
Other, Against Persons	2,154	2,701
Against Property	40,665	57,859
Involving Drugs	6,363	11,458
Against Justice	3,239	11,106
Against Good Order	12,614	6,264
Miscellaneous	26,760	14,094
Total	97,789	114,215

Source: Data provided to author by New Zealand Department of Justice.

Table 3.5. Average Daily Prison Population Counts, 1981–1992

Year	Sentenced Inmates	Remand Inmates	Total* Inmates
1981	2,461	229	2,690
1982	2,378	239	2,617
1983	2,565	283	2,848
1984	2,714	288	3,002
1985	2,525	287	2,812
1986	2,310	341	2,651
1987	2,636	354	2,990
1988	2,953	364	3,316
1989	3,203	380	3,583
1990	3,498	408	3,905
1991	3,752	430	4,182
1992	3,834	444	4,278

*Figures may not always add to the total because of rounding.

Source: Data provided to author by New Zealand Department of Justice.

dents said to have occurred years earlier, during the complainant's childhood. But even with adjustments for those influences, increases in sexual assaults would still be well above the overall rate of increase.

In addition, research on violence suggests that there are now more cases of extreme violence. This is certainly the view held by the media and by a small but growing and vocal section of the public, whose response has been to petition the government for severer penalties.

Custodial and Noncustodial Sentences   Table 3.5 shows average daily prison population counts from 1981 to 1992. These have increased by 59 percent since 1981, reaching 4,694 at the end of 1993. Changes already made in average effective sentence lengths will increase the count to 6,000 within a few years unless there are further changes to the penal legislation.

The growth in prison numbers in New Zealand cannot be attributed to lack of interest in, or acceptance of, noncustodial alternatives.

While the population increases are unremarkable (indeed modest) by U.S. standards, they have occurred during a period of growth in alternatives to prison, including *periodic detention*, where the offender lives at home but does weekend work, generally for public or charitable organizations; *community service*, imposed as x hours of community service and carried out in the community under the instruction and control of a probation officer; *community care*, participation in a rehabilitative program run by a community organization; and *supervision* by a probation officer.

Table 3.6 shows that noncustodial sentences increased by 150 percent over the period, and that by 1992 five times as many offenders were serving noncustodial sentences as were in prison.

Table 3.6. Noncustodial Sentences, 1981 and 1992

Sentence Type	1981	1992
Periodic Detention	4,448	13,501
Community Care	—	869
Community Service	1,232	4,482
Probation/Supervision	2,772	2,580
Total	8,452	21,432

Source: Data provided to author by New Zealand Department of Justice.

Principal Policy Reviews, Legislative Responses   During the 1980s several advisory committees were set up which, influenced by the tempers of their times, proposed a variety of reforms. While some were implemented in total, most stopped short of establishing any clear overall new direction.

Penal Policy Review 1981. This was a comprehensive review of penal policy, carried out by a suitably qualified commission. Among its principal recommendations: (1) Prisons should be recognized as the sanction of last resort and be used as little as possible—the "frugality" principle. (2) The principal criterion for use of imprisonment should be the need to protect the public from violent offending. (3) Imprisonment should not be imposed for property offending except in special circumstances. (4) Prisons should be small regional institutions allowing the maintenance of family ties. (5) A policy of "through-care" should be developed, based on the needs of individual inmates and preparation of individual release plans. (6) There should be more involvement of Maoris and Pacific Islanders in courts and prison systems, these people being heavily overrepresented in the prisons.

Government response was not rapid, but the report substantially influenced the form of the Criminal Justice Act 1985, which incorporated the frugality principle, a presumption against imprisonment for property offenses, and a presumption in favor of imprisonment of seriously violent offenders. It also gave greater emphasis to the making of restitution orders to victims.

Ministerial Committee of Enquiry into Violence 1987. This inquiry saw the continuing increase in violent offending as the inevitable consequence of a breakdown in family and societal ties and standards: "The public now has the community it deserves." In penal policy, it recommended longer effective sentences and a strengthened presumption in favor of imprisonment for violent offending.

The report, delivered in an election year, received an almost immediate response in a series of amendments to penal legislation that provided for (1) greater use of indeterminate sentences, principally for sexual recidivists; (2) longer periods before parole eligibility for murder; (3) a strengthening of the presumption for imprisonment for violent offending; (4) removal of eligibility for parole from seriously violent offenders; and (5) authority for the parole board to cancel the statutory right to remission of one-third of the term (time off for good behavior) if it determines that the inmate is likely to reoffend in a violent way if released into the community. That jurisdiction has been sparingly exercised, a total of seven orders having been made in seven years. However, proposals have recently been made for broader incapacitative authority in the parole board.

Committee of Enquiry into the Prison System 1989. This major inquiry picked up the findings of the 1981 committee, developed them, and looked particularly at prison management and alternatives to prison.

The committee vigorously questioned the desirability of imprisonment as the appropriate sanction except where there was a significant risk of further violent offending. It argued that prisons "cannot do more than provide secure and humane containment," and that to expect them to achieve punishment and reformation at the same time and in the same institution was "imposing incompatible and unattainable goals" on them.

It accordingly recommended that the majority of offenders be placed in small regional "habilitation centers," with community residents on their governing bodies. These would provide an environment for positive change, as the offender would be able to maintain community ties and, having been required to face the reality of his or her criminal lifestyle and its consequences for victims, be readier to engage in rehabilitative programs.

As subsidiary recommendations, the committee supported a moratorium on prison building, a lid on prison populations (invoking Michigan's 1981 Overcrowding Emergency Powers Act as a model), more Maori input, and the abandonment of corrective training schemes (three-month sentences for young offenders involving tight military-style discipline and a heavy program of physical exercise).

Government reaction to the report was muted. A number of the report's secondary recommendations were adopted, but not its basic proposition that prison and habilitation functions should be separated. A 1993 amendment has, however, made legislative provision for habilitation centers and home detention units, though none has been established to date.

Diversion Schemes   Meantime, however, two schemes have been developed that divert offenders from the courts. The first is a "police pretrial diversion scheme," and 5,619 offenses were dealt with by this method in 1991. The offender must acknowledge guilt and responsibility and make good any loss to the extent he or she can do so. Most offenses diverted are for minor theft, minor assault, and cannabis possession. Although a recent examination of the scheme suggested that some diverted cases would not otherwise have come to court, overall the scheme appears to be diverting a significant proportion of lesser offenses. In most cases a charge is filed and later withdrawn with the consent of the court, but procedures vary in different police districts, and the size of the operation calls for the establishment of common guidelines and procedures.

The second diversion scheme followed the introduction of the Children and Young Persons and Their Families Act 1989, which had as one aim the diversion of "young offenders" (people under seventeen years of age) from the court process. It relies heavily on the use of court-directed family conferences to secure the involvement of the offender's family and community, and of victims who are prepared to participate. Eighty percent of cases involving young offenders are being diverted from the formal court process.

Law enforcement officials say that the scheme creates difficulties for them in achieving reasonable control of experienced young criminals, but they believe that it is worth further development. In general it is seen as an advance on previous arrangements for young offenders, principally because it involves their families and communities in a more direct and effective way. The crucial test will be whether it results in any decrease in overall juvenile and young adult criminality.

Sex Offender Treatment   Another development has been the establishment of a specialist sex offenders' treatment unit, modeled on North American institutions, with a second unit under construction. The first unit has established the superiority of specialist units in the treatment of this type of offender and probably represents the most direct benefit this country has obtained from its examination of North American penal institutions.

In 1993, in response to a petition with 300,000 signatures that followed a particularly horrible child rape/murder, legislation was passed increasing rape sentences and imposing new controls over inmates following their release from prison.

At the end of 1993, New Zealand had arrived at a position—an increase in imprisonment rates coupled with a dramatic increase in the use of noncustodial penalties—not unlike those in most American jurisdictions.

It is disappointing to a New Zealander that more has not been achieved. The country is fortunate enough never to have experienced large-scale organized crime, it has enjoyed long periods of reasonable economic conditions and generally egalitarian policies, and it has a history of willingness to endorse new policies that seem likely to enhance social justice.

It is unclear what direction further reforms are likely to take. While other jurisdictions have more advanced facilities and programs designed to solve their crime problems, New Zealand is applying a larger than average amount of its energies and resources in trying to identify and address the particular needs and potential abilities of individual offenders to assist their rehabilitation. It seems to be getting some, though not a dramatic, return on that investment, and it is placing a good deal of faith on further developments along those lines.

Sentencing Reform in Canada
Anthony N. Doob

The front-page headlines in Toronto's two legitimate newspapers told it all. "Ottawa wants crackdown on violent offenders" announced the *Globe and Mail*, Canada's self-styled "national newspaper." "Ottawa aims to put fewer in prison," reported the Toronto *Star* to its readers. These two stories were reports of the same event: the tabling in Canada's Parliament on June 13, 1994, of amendments to the criminal code.

There was no contradiction between a "crackdown on violent offenders" and a policy to "put fewer in prison." Finding a way to do both in the same bill is a skill that Canada's minister of justice, a "self-described liberal on justice matters" the *Globe and Mail* reported, seems to be perfecting in the wake of the "Just Desserts Murder" that took place in Toronto in April. For both juveniles and adults, recent proposals for policy changes, if enacted, could have the effect of increasing the severity of punishment for small numbers of offenders convicted of violent crimes while substantially reducing the use of confinement for other offenders.

The Just Desserts Murder   In April 1994, a young white woman, sitting in a fancy downtown Toronto dessert café, was shot and killed by two young black men during a robbery on a quiet weekday evening. The restaurant, Just Desserts, is situated in a largely upper-middle-class neighborhood in downtown Toronto. To almost anyone in Toronto with ten dollars to spend on a piece of cake and coffee, it is a well-known spot.

The Just Desserts Murder created a hurdle for any reasonable criminal justice policy to overcome. The spring of 1994 was a "tough on crime" time in Canada. Even before the young woman was buried, various public officials were calling for harsher sentences, a tougher young offenders law (though the accused people are all adults), restrictions on immigration (at least one of the accused is not a citizen, although he has apparently spent most of his life in Canada), tougher gun control laws, more police, harsher sentences in the criminal code, and more minimum sentences (even though Canada has a mandatory sentence of life imprisonment for murder, a maximum of life imprisonment for robbery, and mandatory minimum sentences for the offense of using a firearm during an indictable offense).

Youthful Offenders   The new national government in Ottawa made election campaign commitments for changes in certain criminal laws and apparently felt obliged to carry through on these promises quickly. In early June 1994, the minister of justice tabled a range of amendments to the Young Offenders Act that set the tone for the adult criminal code amendments that would follow a few weeks later. The young offenders amendments were almost universally seen as "toughening" the way in which serious young offenders would be treated and were, therefore, supported by some and criticized by many who felt that they "didn't go far enough" in a harsh direction. The focus of press coverage was on the way dispositions for youth charged with murder were being toughened. In 1993, forty-two young people (ages twelve through seventeen, inclusive) were charged with murder across Canada. The proposed changes to the act might also increase the likelihood that a few hundred youths charged with serious violent offenses in addition to murder would be transferred to adult court rather than have their cases heard in youth court.

The press was strangely silent about other proposed amendments to the Young Offenders Act which, for example, will require the sentencing judge to state the reasons why each and every noncustodial disposition is inappropriate whenever a sentence involving custody is being handed down. Little public attention was paid to a proposal creating a presumption, except in the case of "serious personal injury" offenses, of a noncustodial sentence for a young person. An additional "principle" contained in the amendments, that "protection of society" is best achieved "by addressing the needs and circumstances of a young person that are relevant to the young person's offending behavior," appears never to have been mentioned in the press.

The use of custodial dispositions for young persons in Canada has been growing dramatically over the past ten years. If the provinces—which administer criminal law (including the youth justice system)—act in a way that is consistent with the intent of the minister of justice, the types of kids entering custodial facilities will change. Specifically, the minister of justice, in toughening one part of the Young Offenders Act, may have dramatically lowered the number of young persons in Canada being given custodial sentences for property crimes. About eleven thousand young people in 1992–1993 found guilty of property offenses (27 percent of all young people convicted of property offenses) were given custodial sentences. About four thousand young people were sentenced to custody for violent offenses.

But keeping minor offenders out of expensive custodial facilities will depend, in large part, on ten provinces and two territories. The split between federal responsibility to make criminal law and provincial responsibility to administer it creates a myriad of problems in Canadian public policy.

Nevertheless, there is a chance that the minister of justice's "get tough on young offenders" amendments will serve to differentiate between those very few youth who have committed serious crimes and the thousands imprisoned for minor offenses. A few may spend more time in prison, but thousands could receive community sanctions instead of prison.

Adult Offenders The minister of justice then moved on to adult sentencing. Here, too, the theme was to toughen at one end but make room for change by keeping less serious offenders out of prison. Whether he has accomplished this goal in his recent bill is debatable. The bill has a couple of provisions that are, in effect, restatements of common law provisions making abuse of trust or authority and "hate motivation" aggravating factors in sentencing. The bill contains a number of provisions attempting to minimize the use of imprisonment. For example, it mandates noncriminal proceedings for minor criminal offenders where the provinces deem it appropriate. A number of provisions dealing with fines may, if used sensibly, reduce the number of offenders jailed because of their failure to pay fines. The critical issue in the success of these provisions will be how they are implemented at the provincial level.

Although the size of the impact of this bill may be debatable, what is not debatable is that, if this bill passes, the Parliament of Canada will have, for the first time, made a formal statement about the purpose and principles of sentencing. There is every reason to believe that both bills—the amendments to the Young Offenders Act and the sentencing bill—will be enacted. The government has a large majority in the House of Commons and, with a tradition of effective party discipline in votes, it is likely that the bills will become law without serious changes.

Parliament will, for the first time in Canadian history, be giving a bit of guidance to judges on sentencing. There is a slightly confusing statement of purpose: "The fundamental purpose of sentencing is to contribute, along with crime prevention initiatives, to respect for law and the maintenance of a just, peaceful and safe society by imposing just sanctions that have one or more of the following objectives." There follows a traditional list of purposes: denunciation, deterrence, incapacitation, and rehabilitation, and the politically correct objective "to promote a sense of responsibility in offenders, and acknowledgment of the harm done to victims and to the community."

There is also a strongly worded "fundamental principle": "A sentence must be proportionate to the offense and the degree of responsibility of the offender," and courts are reminded that similar cases should result in similar sentences.

Courts, unfortunately, are not given much help in how they are to accomplish proportionality. Given Canada's chaotic set of maximum sentences, courts are given no guidance even on the relative seriousness of different offenses. But it will likely be a bit harder to justify a severe sentence given purely for purposes of general deterrence.

Sentencing Reform   A cynic might suggest that Canada hasn't moved very far in reforming sentencing. That may be true; certainly it is if comparisons are made to England's Criminal Justice Act 1991 or the U.S. Sentencing Reform Act of 1984. But even if Canada hasn't accomplished very much, a great deal of time and a lot of effort have been devoted to doing so.

The Sentencing Commission   The actions and inaction of the past decade express the strength of the commitment to fundamental sentencing reform in Canada. The Canadian Sentencing Commission was established in 1984 by the then Liberal government. The commission was a "policy recommending" commission rather than an implementation commission. It proposed in its spring 1987 report that there be a meaningful statement of purpose and principles for sentencing and a set of guidelines assented to by Parliament. Its recommendations received quite positive reviews, particularly outside of Canada.

A parliamentary committee examined the commission's suggestions and recommended, in the fall of 1988, a somewhat watered down, occasionally toughened up, and sometimes contradictory set of proposals. The thrust of their recommendations was, still, that Parliament should have a role in sentencing and that the sentencing process should be structured. Nothing happened. An election was called for late 1988, and sentencing fell off the public agenda for almost two years.

In the summer of 1990, the government released a set of papers on sentencing and parole and slid even further from the commission's recommendations. The government's recommendations were generally even less clear than had been the report of the parliamentary committee almost two years earlier. The papers suggested the creation of a "sentencing and parole commission" with, as the minister of justice pointed out, "a broad purpose." However, the verbs used in its proposed mandate betrayed the Conservative government's agenda on sentencing. The commission was to "develop the work of the Canadian Sentencing Commission"; in other areas it was to advise, encourage and promote, consider, or evaluate. The one firm requirement was that it was to "make annual reports."

The 1992 Sentencing Bill   The 1990 papers properly faded into obscurity, though the information released with the Conservative government's June 1992 "sentencing bill" suggested that the amendments were a "refinement" of its 1990 proposals. The government never quite got around to passing its sentencing bill. It is possible that there simply wasn't much enthusiasm. For all of its faults, it had one very important feature: it was an attempt to give Parliament a role in sentencing policy. Judges under the 1992 bill would have had to look first to the criminal code for sentencing policy. They then could turn to their (provincial) Court of Appeal.

The sentencing bill and criminal justice policy were probably irrelevant to the outcome of the 1993 federal election. The minister of justice who introduced the bill, Kim Campbell, became prime minister in early 1993. She called an election for later that year and managed to change dramatically the political landscape of the country. She led the Conservative Party from having a comfortable majority in the 295-seat Parliament to being an also-ran nonparty with two seats.

The statement of purpose and principles contained in the bill introduced by the Liberal Party in June 1994 had an uncanny resemblance to that contained in the 1992 bill. There are differences, however, and some provisions concerning fines and the creation of a "conditional sentence" were not contained in any form in the earlier bill. But the significance of the bill lies less in its specific provisions than in the possibility that it may begin a process of structuring sentencing in Canada.

One federal official, for example, has suggested that Courts of Appeal may take a more active role in creating guidelines than they have in the past. In Canada, the Courts of Appeal have been quite active in sentencing decisions for years; however, even their so-called guidelines judgments have been heavily criticized, largely because the courts appear not to be comfortable in laying out guidelines that do a good job of structuring the decision for the "next" case that comes to them.

The optimistic view is that the Parliament of Canada, if the proposed bills are enacted, will have finally broken the stranglehold that judges have had on sentencing policy. By legislating a purpose, a fundamental principle, some other principles, and a couple of aggravating factors in sentencing, they may have started down a path from which there is no return. For example, Parliament may have laid the groundwork for arguments that Parliament now has an obligation to do something to ensure that unwarranted disparity is reduced. The movement in the direction of structured sentencing may be uneven, and we may not, at this point, know exactly where it is going. But if the Canadian Parliament does pass the bill introduced by the minister of justice in June 1994, sentencing in Canada may never be the same.

Sentencing in South Africa
Stephan Terblanche

Sentencing in South Africa, as in many countries, is changing. The peaceful revolution and establishment of an elected representative government will affect all spheres of government, including the justice system, in ways that cannot be predicted.

This article offers a snapshot of sentencing policy in South Africa and an overview of developments since 1980. Most of my comments are based on personal experience or on experience working on government committees and commissions. This is because well-maintained statistical data on the judicial system are virtually nonexistent and corrections data, though available, are limited in scope.

Basic Sentencing Policy: Individualization   The South African sentencing tradition has been shaped by English law, which was almost completely adopted for procedural purposes after the British occupation of the Cape in 1806. Nonetheless, and especially since the 1940s, sentencing in this country has developed its own character. Basic sentencing policies have changed little since the 1970s. Belief in the desirability of individualized sentencing is widespread. A substantial body of opinion holds the view that the legislature should leave the courts as much discretion as possible so that judges can determine the most suitable sentence for every individual offender.

Two factors limit this discretion. The first is statutory penal provisions. Every statutory crime is provided with a maximum term of imprisonment (which is not the case with the common law crimes). Similarly, the various levels of courts are restricted in the sentences they may impose: the district magistrates' courts may normally not impose sentences exceeding twelve months' imprisonment, and the regional magistrates' courts are limited to ten years. The supreme courts have, technically speaking, an unlimited jurisdiction, but they have decided that twenty-five years' imprisonment is to be exceeded only in exceptional circumstances.

The second limiting factor is the review and appellate functions of the supreme courts, where some consistency in sentences is sought, and where excessively harsh or mild sentences will be set aside. The appellate division is not keen to lay down general principles for sentencing but tends to limit its functions only to the case it is dealing with. Nevertheless, its decisions are important and have to be taken into account by the lower courts when imposing sentence.

One of the most influential policies was laid down by the appellate division in *S. v. Zinn* 1969, 2 SA 537 (A), where the court decided that when imposing sentence: "[W]hat has to be considered is the triad consisting of the crime, the offender and the interests of society." These considerations, namely the seriousness of the crime, the personal circumstances of the offender, and the interests of society, have become the foundation for sentencing in South Africa. In addition, courts imposing sentence are expected to take account of the four main purposes of punishment: deterrence, retribution, prevention, and rehabilitation. How the four purposes of punishment are to be integrated with the triad of aspects has never been fully explained. Although these basic factors are far from exact, they have been very important in shaping the sentencing practices we have today.

Government's Response to Overcrowding   In 1976 a Report from the Judicial Inquiry into the South African Criminal Justice System under the chairmanship of Justice G. Viljoen (hence the Viljoen Commission) was tabled in Parliament, and in 1980 its findings were as valid as they had been a few years earlier. The Viljoen Commission was only the second judicial commission that had looked at the criminal justice system as a whole since South Africa became a union in 1910. It was also the last.

As a result of prison overcrowding, and partly in response to the Viljoen Commission's report, the minister of justice in 1980 established the Working Group on the Overcrowding of Prisons. It consisted of three highly placed officials serving in the Department of Justice and the Prison Service, plus two full-time researchers. The group's work had several consequences.

First, community service legislation was enacted for the first time. Unfortunately, magistrates and judges were expected to search for placement agencies themselves, and community service has thus never been used to any significant degree.

Second, articles were published on sentencing options other than imprisonment. They included "periodical imprisonment," in which the prisoner serves a prison term periodically, typically over weekends. As in the rest of the world, this good idea has never been used to a substantial degree. Other articles stressed the use of various positive conditions for the suspension of imposed sentences, which require the offender to do something in order to prevent the sentence coming into operation. These articles were supplemented by tours throughout the country to stress the importance of imposing sentences other than imprisonment whenever possible.

Third, the Working Group's most important contribution was the introduction of a sentence called "correctional supervision." This is similar to intensive probation (and was modeled on the Georgia system), and would normally consist of house arrest, community service, monitoring, and various individualized programs.

The Working Group was instrumental in shifting sentencing policy toward greater emphasis on community-based forms of punishment. Since August 1991, when a pilot project was started in Pretoria, more than 18,000 people have been sentenced to correctional supervision. It has become the third most important form of sentence imposed in South Africa (after fines and imprisonment). In S v. R 1993, 1 SA 476 (A), the appellate division observed that correctional supervision makes it possible to approach sentencing differently, in the sense that it is now possible to impose a severe punishment without resorting to imprisonment.

Changes in Policy   Changes in sentencing policy and practices since 1980 have been subtle rather than explicit. Some crimes that might have been considered quite serious in 1980 (such as the possession of small quantities of drugs) are considered less serious today. Some aspects of an offender's personality may receive more emphasis, in accordance with current thought on the human psyche. And, of course, the perception of the courts of the interests of the community may change over time. Less serious offenses are more leniently treated, and much emphasis has been placed by the authorities, the supreme courts, and academic writers on avoidance of imprisonment if at all possible. Increased reliance on community-based forms of punishment is part of this trend. Of course, apartheid was also still in full swing in 1980, and since all the crimes that could only have been committed by black people at that stage have now been abolished, the criminal justice system has been rid of many of the petty crimes associated with apartheid.

The Current Sentencing System   One difficulty in describing sentencing in South Africa is the dearth of sentencing statistics. The Department of Correctional Services, which manages all prisons, has fairly comprehensive statistics on all prisoners, but the Department of Justice has rudimentary statistics only. Nobody, therefore, really knows how much of which type of sentence is imposed on offenders. A recent study (in preparation) shows that in 1992 fines were imposed in 84 percent of drunk driving cases, in 52 percent of assault cases, and in 41 percent of theft

Table 3.7. Reported Crime in South Africa, 1983 and 1992

Year	Murder	Rape	Robbery	Car Theft	Housebreaking	Serious Assault
1983	8,573	15,342	38,229	44,771	148,766	121,716
1992	20,135	24,812	79,927	73,619	254,941	137,800
Increase	134.9%	61.7%	109.1%	64.4%	71.4%	13.2%

Source: South African Institute of Race Relations (1994).

cases. The respective figures for suspended sentences were 8 percent for drunk driving, 27 percent for assault, and 21 percent for theft. Direct imprisonment was imposed in 2 percent of drunk driving cases, in 6 percent of assault cases, and in 19 percent of theft cases.

However, about 30 percent of offenders convicted of theft and assault wound up in prison. This is because a term of "alternative" imprisonment is added to every fine, in case of default. If the fine is not paid, the alternative comes into operation automatically and immediately.

In 1980, more than 60 percent of all people admitted into prison were there only because they did not pay their fines. Time for payment may be allowed in the discretion of the court, but this is done only in a relatively small percentage of cases, and very inconsistently. The Criminal Procedure Act provides other measures for collecting fines but they are seldom used.

When the above-mentioned study was conducted, correctional supervision was not in operation in most of South Africa, and its influence has not been measured with any accuracy.

Although the majority of crimes are dealt with in a fashion similar to that described above, South Africa has of late been subjected to a major increase in the incidence of serious crime. Table 3.7 gives an indication of this trend, which has been particularly marked over the last five years. Sentences for such serious crimes are restricted to imprisonment and the death penalty. A moratorium has been placed on executions since 1989. By December 31, 1994, there were 346 people on death row, now awaiting a decision by the Constitutional Court on the legality of the death penalty. Prior to 1990, the death penalty was compulsory if the court did not find any mitigating circumstances to be present, and the accused bore the onus to prove the presence of such circumstances. The Criminal Law Amendment Act 107 of 1990 reinstated the discretionary imposition of the death penalty, allowing it if, in all the circumstances, the court is of the opinion that the death penalty was the only proper sentence.

Imprisoning Offenders South African prisons admitted 369,045 people during 1993. Of these, 120,781 were sentenced offenders; the bulk of the rest were awaiting trial. More than 20 percent of the average daily prison population is made up of prisoners awaiting trial.

Table 3.8. Prison Population Figures in South Africa, 1979/1980 and 1993

Group	Total Population		Average Prison Population		Rate of Imprisonment (per 100,000)	
	1979/1980	1993	1979/1980	1993	1979/1980	1993
African	16,000,000	23,100,000	73,911	78,616	462	340
Colored	2,600,000	3,400,000	21,990	28,151	846	838
Indian	800,000	1,000,000	551	711	69	70
White	4,500,000	5,200,000	4,225	4,320	94	84
Total	23,800,000	32,700,000	100,677	111,798	423	361

Source: South African Department of Correctional Services (unpublished data provided to author).

The population figures of prisoners for 1979/1980 and 1993 are set out in table 3.8 (reports of state departments traditionally ran from July 1 of one year to June 30 of the next, but since 1993 the calendar year is used).

These figures show a substantial reduction in the rate of imprisonment of African people. This may partly be explained by improved methods of record keeping since 1980. The abolition of the apartheid crimes would have contributed as well, although probably not much. In my opinion, however, it should mostly be ascribed to changes in sentencing policy, especially in case of sentencing for less serious crimes. Shorter terms of imprisonment make up a decreasing portion of the prison population. This is as a result of the belief that, for purposes of rehabilitation in prison, a longer term is preferable, and that short-term imprisonment should be avoided as far as possible.

Sentences of two years and more consituted 8 percent of total sentences in 1982/1983 and 20 percent in 1991/1992; sentences of more than six months but less than two years made up 9 percent of sentences in 1982/1983 and 19 percent in 1991/1992; sentences of six months and less made up 83 percent in 1982/1983 and 60 percent in 1991/1992.

The trend toward prisoners with longer sentences taking up a larger part of the daily prison population has been evident over the past decade, and there are no signs of any change. This is a result of steadily increasing crime rates and the increase in the seriousness of crime. It is also possible that sentences for serious crimes have gotten longer in an attempt to stem the crime wave.

Most people sent to prison are sentenced to determinate imprisonment. Statutory provision is, however, made for (mainly) two forms of indeterminate sentences. Life imprisonment was explicitly added to the Criminal Procedure Act in 1990. By the end of 1993, 1,857 people were in custody for life or terms longer than twenty years. Another indeterminate sentence is imposed on offenders found to be habitual criminals. These people are normally detained for terms lasting between seven and fifteen years. On December 31, 1993, there were 3,868 such people in South African prisons.

The Future The future of any endeavor in South Africa can best be described as uncertain. A peaceful revolution has taken place here over the past four years. That we have our first fully representative government ever, and that the supreme law of the country is contained in a constitution with a chapter on fundamental human rights for the first time will influence future events. Some predictions can be made, however.

1. The criminal justice system is not high in the current government's priorities; housing and education are far higher.
2. The juvenile justice system will probably be overhauled in the near future.
3. The prison system will remain under severe financial pressure, which should prevent any large-scale building programs. The new Constitution, however, will put pressure on the authorities to find other means to reduce overcrowding.
4. Traditionally, African people dealt with crime in a fashion closely related to civil procedures. One might expect a bigger drive for various community-based measures in the future. The role of correctional supervision could well be extended. More informal measures may also be put into place.

Germany

Germany Reduces Use of Prison Sentences
Thomas Weigend

Between 1968 and 1989, the former West Germany greatly reduced the proportion of convicted offenders sentenced to prison. In 1968, roughly a quarter of convicted offenders were sentenced to imprisonment. Two years later, the size of that group had dropped from 136,000 to 42,000, and the percentage of convicted offenders who were imprisoned had fallen from 24 percent to 7 percent. In 1989 (the latest year for which data are available), only 33,000 persons, less than 6 percent of adults convicted in West Germany, were sent directly to prison.

It is important for North American readers to understand that these figures include all persons sentenced to incarceration—the equivalent in the United States of sentenced jail inmates plus prison inmates. This article tells the story of how and why West Germany has steadily reduced its reliance on prisons.

The remarkable decline in prison use is due to a determined assault on use of short-term imprisonment. At the turn of the century, more than 50 percent of offenders received prison sentences of three months' duration or less. Legislation passed in 1921 obliged the courts to impose fines instead of short prison terms whenever the purpose of punishment could as well be achieved by a fine. Even so, the portion of short prison sentences among all prison sentences remained high; 83 percent of offenders sentenced to imprisonment in 1968 received sentences of six months or less. By that time, the German legislature had embraced the idea that short-term imprisonment does more harm than good: it disrupts the offender's ties with family, job, and friends, introduces the offender into the prison subculture, and stigmatizes the offender for the rest of his or her life. It also does not allow sufficient time for promising rehabilitative measures. Moreover, the data on the deterrent effectiveness of short-term imprisonment were inconclusive at best.

As a consequence, the German legislature in 1970 enacted section 47, sub. 1 of the Penal Code: "The court shall impose imprisonment below six months only if special circumstances concerning the offense or the offender's personality make the imposition of a prison sentence indispensable for reforming the offender or for defending the legal order." That amendment meant, in effect, that prison sentences below six months could be imposed only under exceptional circumstances for purposes of rehabilitation or general prevention. The number of such sentences dropped dramatically from 184,000 (1968) to 56,000 (1970); after some ups and downs, that figure reached a low of 48,000 in 1989 (and many of these were suspended).

At the same time, the German legislature extended the possibility of suspending short-term prison sentences (suspension being the German equivalent of probation). According to section 56 of the Penal Code, the court shall suspend the execution of prison sentences of up to one year whenever the offender can be expected to refrain from further offenses without a prison experience. German law expressly prohibits split sentences: prison sentences can only be suspended or executed in full (except for the possibility of parole, which exists after the offender has served one half of the term). Sentences of more than two years' imprisonment cannot be suspended.

When the court suspends a prison sentence, it determines a probationary period of two to five years; suspension can be revoked (and the sentence executed) if the offender commits another crime during that period. The court can combine suspension with various conditions and restrictions, including the duty to make restitution to the victim or to pay a sum of money to the state or to a charitable organization, to avoid the company of certain individuals, and to report regularly to the court or to the police. The offender can also be assigned to a probation officer.

German courts have made use of the suspension option with consistently increasing frequency. In 1968, the year before the reform, only 36 percent of prison sentences were suspended. By 1979, that portion had climbed to 65 percent, and it has not significantly changed since then (1989—67 percent). Prison sentences of six months or less have been suspended even more liberally (1989—77 percent). Revocations of suspension have *diminished* despite the more generous use of suspension. Whereas 46 percent of suspensions were revoked in 1986, less than a third (29 percent) were revoked in 1989.

For minor offenses, German law since 1975 offers an additional option of informal sanctioning. According to section 153a of the Code of Criminal Procedure, either the public prosecutor or the court can "invite" a suspect to pay a sum of money to the state, the victim, or a charitable organization in exchange for dismissal of the criminal prosecution. The theory of this quasi-sanction is that the suspect, by making the payment, eliminates the public interest in prosecuting the minor offense. The payment neither requires a formal admission of guilt nor implies a criminal conviction, but the (presumed) offender must pay an amount of money roughly equivalent to the fine that might be imposed if he or she were convicted. The use of this procedural option has greatly increased since its inception; prosecutors and courts employ it not only in petty cases but also for sanctioning fairly serious, especially economic, offenses without trial. Taking the quasi-sanction

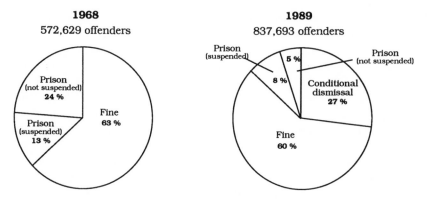

Figure 3.10. Criminal Sanctions in Germany, 1968 and 1989. *Sources:* Statistisches Bundesamt (1968), (1989).

of section 153a into account, the distribution of criminal sanctions in Germany before and after the reforms of 1970 and 1975 is shown in figure 3.10.

The de-emphasis of nonsuspended short prison sentences and the introduction of conditional dismissal produced a marked shift from custodial sentences (which, even in 1968, had a comparatively low incidence) to monetary sanctions. One might expect this shift to have led to a proportional depletion of German prisons. Curiously, that has failed to occur. Figure 3.11 shows the numbers of persons (excluding pretrial detainees) held in German prisons on March 31 of selected years.

Although the reform of 1970 brought about an immediate sharp reduction of the prison population, by 1984 the number of prisoners had surpassed the high point of 1968. Since that time there has been a slow but steady decline.

Cynics might argue that the rise in the number of prisoners in spite of the

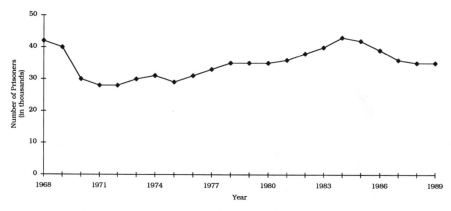

Figure 3.11. Number of Persons Held in German Prisons, 1968–1989 (excluding pretrial detainees). *Sources:* Statistiches Bundesamt (1968), (1971), (1973), (1984), (1988), (1989).

decline of imprisonment rates demonstrates that available prison space will always be filled. Yet there are rational explanations for this development. First, the overall number of convicted offenders has increased, though not dramatically, from 573,000 (1968) to 609,000 (1989). More important, those who receive nonsuspended prison sentences tend to receive longer sentences than before: within fifteen years, the share of lengthy sentences (two to fifteen years) among all nonsuspended prison sentences increased from 9 percent (1974) to 15 percent. This change may be due to the increase in drug-related offenses, which tend to draw heavy sentences.

Moreover, the initial imposition of a noncustodial sentence does not necessarily mean that the offender can avoid prison, since about one-third of suspended sentences are revoked (usually due to the commission of a new offense). Offenders who receive fines can be sent to prison for nonpayment. Under German law, nonpayment can transform a fine into a prison term; the state need not show that the offender willfully refused to pay although he or she had the means to do so (sec. 43, Penal Code). Although only 6 to 7 percent of fined offenders eventually serve a prison term because of nonpayment, this group, due to the large absolute numbers involved, imposes a heavy burden on the corrections system: each year approximately 30,000 such persons enter prison.

In recent years, the German states have increasingly attempted to reduce that number by offering destitute offenders an alternative to prison. They can enter community service programs and thereby work off the fine instead of "sitting it off." These programs, though reaching only a limited number of offenders, have been described as fairly successful, especially when they are adequately staffed and organized.

Has the reduced emphasis on imprisonment in the German sentencing system led to an avalanche of new crime? Superficially, the figures seem to support that notion, as seen in figure 3.12.

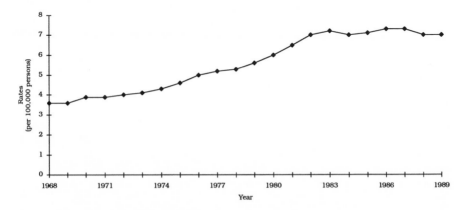

Figure 3.12. Offenses Known to Police, 1968–1989 (excluding traffic offenses). *Source:* Bundeskriminalamt, Polizeiliche Kriminalstatistik (1989), (1990).

The absolute, as well as the relative, incidence of crime known to the police has roughly doubled within the twenty years since the reform of sentencing law. However, there are several considerations that make the existence of a causal link between the two developments most unlikely. A closer look at the statistics reveals that the fairly abrupt 1970 change in sentencing practice is not reflected in the steadily rising curve of the crime rate; its increase precedes 1968. On the other hand, there is no change in sentencing practice to explain the leveling-off of the crime rate after 1983. Moreover, the police statistics tell us very little about actual changes in the occurrence of crime since they cannot provide knowledge of offenses not detected, not reported, or not recorded by the police. Finally, common criminological wisdom has it that crime rates are influenced by demographic, economic, and social factors to an overwhelmingly greater extent than by sentencing styles. (If it were otherwise, the United States should have a minuscule crime problem as compared with, say, the Netherlands or Japan.)

It is more interesting to look at the relationship between crime and criminal justice from the opposite side. What effect does the (perceived) increase of the crime problem have on sentencing? As we have seen, official sentencing policy in Germany has responded in an anticyclical fashion by discouraging the imposition of prison sentences in the face of a growing crime rate. The statistics cited above have demonstrated three features, which can possibly be interpreted as effects of the legislature's remarkable policy decision: the courts have greatly reduced the proportion of nonsuspended prison sentences; the overall number of convicted persons has remained stable; and the number of prisoners has fluctuated and has now returned to the level of the 1960s.

German policies, and the faithful implementation of their directives by prosecutors and courts, have led to a "cushioning" of the crime wave of the 1970s. A greater percentage of offenses than before were resolved without conviction, and potential overcrowding of prisons was avoided by increased use of fines and suspended sentences. The crime wave ebbed after 1983, independent of any action or inaction on the part of criminal justice policy makers. It can only be hoped that they keep as cool today as they did in the 1960s, even in view of the problems associated with reunification and with the currently publicized arrival of "organized crime" in Germany.

Sentencing and Punishment in Germany
Hans-Jörg Albrecht

Sentencing policies and practices in Germany have changed substantially since 1970 as a result of a significant shift from imprisonment to fines, probation, and various diversionary programs. The prison population declined from 56,870 in 1963 to 34,398 in 1994. Unsuspended sentences to imprisonment fell from 92,576 in 1967, fluctuated between 30,000 and 40,000 in the 1970s and 1980s, and totaled 32,359 in 1991 (5 percent of all penalties). Fines, by contrast, have been ordered in 82 to 84 percent of cases since 1970 (83.8 percent in 1991).

This article explains how and why those shifts occurred and describes modern

German case disposition from presentencing through parole, with primary emphasis on the adult system.

Sentences for Adults Each offense in the German Criminal Code (GCC) carries a minimum and a maximum penalty. The main penalties are day fines and imprisonment. The number of day fines varies with the seriousness of the offense, from 5 to 360. The size of one day-fine unit (which may vary between DM 2 and DM 10.000 or $1.30 and $6,700) must be adjusted to the offender's (daily) net income. On default, imprisonment has to be imposed (with one day equaling one day fine). Under particular circumstances (59 GCC), the court may caution the offender and postpone the imposition of a day fine. The minimum prison term is set by statute at one month and the maximum at fifteen years. Life-term imprisonment is restricted to murder. Under a 1992 amendment (42a GCC), confiscation of property may be imposed in addition to a sentence of imprisonment of more than two years and may amount to the total value of the defendant's assets.

German criminal law also allows for so-called measures of rehabilitation and security that are not dependent on personal guilt but depend solely on the degree of dangerousness exhibited by an offender and the corresponding need for preventive action. These are based on the consideration that proportional punishment may be insufficient for very dangerous offenders who are likely to recommit very serious crimes.

Under very restrictive conditions, the court may order additional detention in a detoxification center (for a maximum of up to two years), indeterminate detention in a psychiatric facility (in case of insane offenders and very serious crimes), or an additional incapacitative sentence (if ordered the first time a maximum of ten years applies, if ordered the second time detention is indeterminate). Finally, revocation of the driving license may be ordered as a measure of security.

In quantitative terms, these measures do not play a significant role, with the exception of withdrawal of the driving license. An incapacitative sentence is imposed in some thirty cases per year; approximately 500 offenders are sentenced to indeterminate detention in a psychiatric facility and 700 offenders are detained in detoxification centers. Incapacitative sentences are under heavy criticism because of problems of establishing dangerousness and the assumption of excessive punishment (Kaiser 1990).

Young Offenders Sentencing options available in German criminal law differ according to age. Juveniles (fourteen to seventeen years old) are dealt with by a system of juvenile criminal justice. Eighteen- to twenty-year-olds are presumed to be adults and therefore to be fully responsible. However, under certain conditions, defined in 105 Youth Court Law, young adults may be prosecuted and sentenced as if they had been juveniles when committing the crime. For juvenile offenders, rehabilitation must be sought, while in adult criminal law the basic statute on sentencing (46 GCC) requires punishment to be proportional with the guilt of the offender and seriousness of the offense.

In practice, treatment of older adolescents as adults is the exception, not the rule. At the beginning of the 1990s, slightly more than 60 percent of all young adults were sentenced as juveniles, including 80 percent sentenced for sexual crimes and property offenses and virtually all young adults sentenced for robbery and homicide.

The adult Penal Code penalty ranges do not apply to juveniles. Nor do day fines or adult imprisonment. Youth court sanctions consist of educational measures (e.g., community service, participation in victim-offender mediation, etc.), "disciplinary measures" (e.g., short-term detention up to four weeks), and youth imprisonment with a minimum of six months and a maximum of five years (in cases of very serious crimes imprisonment may be extended to ten years).

Sentencing Adults   German criminal law provides for a two-step procedure in sentencing of adults. The first requires a decision on the amount of punishment that is proportionate to the offender's guilt and the seriousness of the offense (46 GCC). Rehabilitation and deterrence may be pursued only within a narrow range of penalties deemed consistent with the principle of proportionality. Although sentencing statutes are not very explicit on the relations among sentencing goals or on mitigating and aggravating circumstances, doctrine and court practice widely agree that rehabilitation and deterrence cannot justify sentences outside the range determined by the principle of proportionality. However, the problem of how proportionality can be established and how control of proportionality by superior courts can be exerted is not resolved (Albrecht 1994). Finally, sentencing is determinate (German adult criminal law, although heavily influenced by the idea of rehabilitation, never allowed for indeterminate sentencing).

The second step requires a decision on whether punishment should be imposed as a day fine, a suspended prison sentence, or an immediate sentence of imprisonment. In this decision, not individual guilt or seriousness of the offense but individual and general prevention should play the decisive role.

The most important changes occurred in 1969 and 1975 when legislation was introduced (47 GCC) ordering that (day) fines should have priority over short-term imprisonment (of less than six months) with rare exceptions. The rationale was that short-term imprisonment was incompatible with respect to rehabilitation due to the short period available for treatment and the corruptive effects of the prison environment. These policies manifest the view that first-time offenders and those who commit crimes of a nonserious nature should receive fines, while longer prison sentences should concentrate on a small group of heavy recidivists, as well as serious crimes.

Prison sentences of one year or less require regular suspension in case of low risk of recidivism (56 GCC). Sentences of up to two years may be suspended if the offender presents a low risk of recidivism and particular circumstances of the offense or the offender justify a suspended sentence. In case of suspension, a period of probation of up to five years must be determined and the offender may be placed under the supervision of a probation officer. Conditions include community service, a summary fine, or compensating the victim. If conditions are not met or if the offender commits a new crime, suspension may be revoked and the original prison sentence has to be served.

The Prosecution System Judges' sentencing decisions are but a small part of the system that determines the type and size of punishment an offender receives. The public prosecutor's office plays an important role. Since the mid-1970s, the discretionary powers of the public prosecutor to dismiss criminal cases have expanded considerably. German Procedural Code 153a empowers the prosecutor to dismiss minor cases if the offender accepts punitive conditions determined by the prosecutor (a fine, community service, compensation).

A March 1993 law empowered the prosecutor to dismiss a case if the offender's guilt does not necessitate a penalty. This was motivated by economic problems due to German reunification. As rebuilding the justice system in the East requires enormous resources, the need was felt to streamline criminal procedures further in order to reduce costs. Budget concerns have outweighed legitimate interests in maintaining proper lines between the public prosecutors' task of investigating and indicting criminal cases and the judges' task of imposing criminal punishments.

The public prosecutor can choose between two procedures in bringing a case to court. One is the regular criminal trial. However, a simplified procedure may be initiated that consists solely of written proceedings. If the prosecutor concludes that guilt can easily be proven and that a day fine is sufficient punishment, a penal order providing for a day fine may be proposed to the judge. If the judge agrees, a penal order is mailed to the suspect, who is entitled to appeal. In the case of ordinary crimes (approximately 1,300,000 cases per year), 30 percent are dismissed, 40 percent are dealt with by simplified procedures, and the rest receive a full trial (30 percent).

The procedural option of simplified procedures was extended in 1993. Now, if the offender has a defense counsel, the public prosecutor may propose a suspended sentence of imprisonment of up to one year in a simplified procedure. As only 6 percent of all criminal penalties today involve prison sentences of more than one year, in theory a full trial (and sentencing by the criminal court) could be restricted to a minor part of all offenders. Economic pressures and administrative convenience thus have supported the trend toward noncustodial sentences (especially day fines).

Parole Time served by imprisoned offenders and the type of penalty ultimately executed may deviate considerably from the disposition made by criminal courts. According to 57 GCC, a prisoner may be paroled after having served two-thirds of the prison sentence; under exceptional circumstances parole may be granted after half of the prison term. Several important changes should be mentioned. In 1977 the Prison Law established a system of correctional courts having authority over parole. In case of life-term imprisonment, the constitutional court ruled that regular statutory parole must be available for prisoners sentenced to life imprisonment (which until then was granted by way of clemency only) (BVerfGE 45, p. 187). In 1986, legislation was introduced (57a GCC) setting the minimum to be served at fifteen years. The constitutional court in 1992 (BVerfGNJW 1992, p. 2947) decided that the rule of law requires predictable periods of detention for life imprisonment. The court ruled that (with the exception of very dangerous offenders) life prisoners for average cases (of murder) should ordinarily be paroled after fifteen years. For cases with aggravating circumstances, a standard of approximately twenty years was set. Thus, in practice, the constitutional court converted life imprisonment into a fixed period of imprisonment, something critics, assuming that life imprisonment violated human rights, had demanded since the 1970s (Jung 1992).

Another major change affecting prison admission occurred around 1980 with respect to fine collection. In the early 1980s, fine defaults (and substitute imprisonment) increased, mainly due to increasing rates of unemployment (Kerner and Kästner 1986). To reduce the additional burden put on the prison system, community service was introduced as an option for fine defaulters, with overall positive results (Albrecht and Schädler 1986). Currently, the conversion rate is six hours of community service equals one day of imprisonment.

Sentencing Practice and Sentencing Outcomes   Several long-term trends concerning sentencing practices stand out. The absolute number of offenders convicted and sentenced has been stable since the early 1970s (ranging between 600,000 and 700,000 per year; 1991 was 622,390). Diversion policies have cut off steadily increasing numbers of suspects. The use of prison sentences (suspended and immediate) has also been stable since the beginning of the 1970s, with a proportion of 16 to 18 percent of all sentences (in 1991, 16.2 percent). Sentences to immediate (that is, nonsuspended) imprisonment declined until the end of the 1980s from 92,576 in 1967 to 40,270 in 1970 and varied between 30,000 and 40,000 per year in the 1970s and 1980s; (in 1991, 32,359, or 5 percent of all criminal penalties). The number of suspended prison sentences increased until the end of the 1980s (in 1991, 70 percent of all prison sentences, or 11 percent of all criminal penalties).

The use of fines increased sharply at the end of the 1960s (as a consequence of giving fines priority over short-term imprisonment) and remains stable (82 to 84 percent of all sentences during the last two decades; in 1991, 83.8 percent).

Overall, while the number of sentenced prisoners decreased considerably in the long run (from 56,870 in 1963 to 34,398 on January 30, 1994, yielding a rate of sentenced prisoners of 55 per 100,000), the number of offenders placed under probation supervision increased (from 27,000 in 1963 to 150,000 at the beginning of the 1990s).

These changes lend support to some assumptions on the roles and functions of various punishments. Imprisonment constitutes physical and immediate control over restricted parts of the population. But facilities are limited and inelastic and inadequate for dealing with growing numbers of offenders. With expanded intermediate sanctions such as probation and parole, the role of imprisonment changed from immediate physical control to a last resort, strengthening the deterrent impact of a type of sanction based on supervision and control outside the prison but nevertheless backed up by the threat of imprisonment.

While the use of imprisonment may be reduced to a relatively stable (and perhaps mere symbolic) level, the greater elasticity of probation and parole allows for a widened scope of judicial control over varying proportions of precarious populations. In the 1960s, the probation population was characterized by low risk (as measured by prior record) and a limited need for control (no prior record and a stable work history were major preconditions for granting suspension of a prison sentence). However, since the mid-1970s the target group for probation is a group that formerly had been sent to prison.

A consequence of this distribution of criminal penalties was a high average period of imprisonment compared with other European countries (approximately six months compared to an average of around two months, e.g., in the Netherlands), but a low rate of offenders receiving unconditional prison sentences and admitted to prisons. Thus, the prisoner rate at the beginning of the 1990s was around 60 per 100,000 in Germany. In the Netherlands and Sweden, the corresponding rates were well below 50 per 100,000. However, the rate of offenders receiving sentences of immediate imprisonment was 73 per 100,000 in Germany, 113 in the Netherlands, and 166 in Sweden.

Although long prison sentences may appear to be the result of recent criminal law reforms, prison sentences amounting to more than two years are rare. In 1991, of all prison sentences (immediate and suspended), 6,560, or 6 percent, exceeded two years (of these, 56 were for life imprisonment). Among all criminal penalties in 1991 only 1 percent were prison sentences of two years or longer. Although judges could impose many longer sentences, they do not. Sentencing decisions tend toward the minimum authorized penalty, displaying not a normal but a J-type distribution within the range allowed for most offense categories.

However, there was no uniform development in sentencing practice over the last two decades for most traditional crimes, such as traffic offenses, property crimes, robbery, and assault. Prosecutors and courts tended to use diversion options and to favor fines and prison sentences near the minimum allowed. For simple theft, approximately nine out of ten offenders were fined. A corresponding pattern characterizes assault cases and criminal damage. However, serious property offenses and drug offenses also result in considerable proportions of fines. In 1991, one out of four aggravated thefts was followed by a fine and approximately half of drug offenders received fines.

For other offenses, sentencing practice changed at the end of the 1970s. This was true for violent sexual offenses, especially rape and child sexual abuse (Schöch 1992). The length of prison sentences for rape cases increased significantly with prison sentences of more than five years doubling between 1978 and 1983 (up to 10 percent of all rape sentences from 5 percent). Sentences for drug offenses, especially drug trafficking, also became more severe. In a 1980 amendment of the law on illicit drugs, both minimum periods of imprisonment and maximums were raised substantially, and courts tended to impose prison sentences well above the minimum.

These changes in sentencing practice became evident between 1980 and 1984 when the prison population increased significantly because of increasing proportions of drug and violent offenders in the prison population and because of a growing number of admissions with long-term prison sentences.

A turnaround took place again around 1985, which partially may also be explained by demographic changes. In the second half of the 1980s, pretrial detention rose, with growing numbers of offenders of foreign nationality being placed in pretrial detention (they make up more than half of the pretrial detention population in the 1990s).

What Lies Ahead? German criminal law provides for a simple system of penalties. Day fines and (suspended or immediate) imprisonment serve as the main penalties. Community service and probation are not available as penalties in their own right. The last decade saw vivid debates on whether other penalties should be added. Particular interest was voiced for the development of alternative penalties (community sanctions) and for strengthening the roles of restitution and victim-offender mediation. The issues discussed involved proposals to extend the range of prison sentences eligible for suspension from two to three years, to allow partial suspension of prison sentences, to transform suspended prison sentences into a criminal penalty of its own (comparable with probation and stressing the punitive impact of supervision, as well as the conditions), to introduce community service and revocation of the drivers' license as sole sanctions, and, most important, to establish out-of-court restitution procedures as a major device for dealing in a principled way with all types of criminal offenders (Schöch 1992). Some of these proposals have been taken up in a formal draft brought into the Federal Parliament at the end of 1993 by the Social Democratic Party (Entwurf eines Gesetzes zur Reform des strafrechtlichen Sanktionensystems, Bundestagsdrucksache 12/6141, 11.11.1993).

After German reunification, bias-motivated crime or hate crime became a prominent target of criminal policy. As hate violence is committed to a considerable extent by groups of juveniles and young adults, the issue of how young adults should be treated in the criminal justice system came up again. However, a proposal offered by conservative political parties to make sentencing of young adults as if they were juveniles the exceptional case received little support in the legal professions, the Parliament, or the public.

Major reforms of the sentencing process and the system of criminal penalties cannot be expected to take place in the near future (Schöch 1992, p. 11). Federal government reports in 1986 and 1992 concluded that there was no need to reconsider the system of sentencing and penalties as the current state seemed to be satisfactory. Some minor changes will be effected by recent legislation that increases maximum penalties for assault (up to five years from three years in order to lift assault sentencing ranges to those found for simple theft) and legislation that requires that restitution and offender-victim mediation be taken into account in the sentencing decision and in the decision to suspend a prison sentence. A major meeting of the legal professions in Germany (Deutscher Juristentag) discussed in 1992 an in-depth study on criminal penalties that recommended introduction of additional community sanctions, as well as the above-mentioned out-of-court restitution procedure (Schöch 1992). The debate showed that a majority in the legal professions do not favor such changes but adopt a pragmatic view not interested in complicating sentencing and the system of penalties.

Finland

Success in Finland in Reducing Prison Use
   Stan C. Proband

By means of concerted efforts to reduce the use of incarceration, Finland succeeded in lowering its incarceration rate by 40 percent between 1976 and 1992, according to a report by Patrik Törnudd (1993) recently released by Finland's National Research Institute of Legal Policy.

Table 3.9. Prison Admissions, Populations, and Median Sentences, Finland, 1976–1992

Measure	1976	1981	1986	1991	1992
Total prisoners	5,706	5,032	4,311	3,467	3,511
Prisoners/100,000 population	118	102	86	69	70
Admissions	13,457	9,840	9,216	8,874	9,851
Median sentence length (months)	5.1	4.6	3.9	3.6	NA

Source: Törnudd (1993).

Although the reduction was achieved by means of a variety of changes in sentencing laws and policies, according to Törnudd the key factor was that Finnish officials, policy analysts, and criminologists "shared an almost unanimous conviction that Finland's internationally high prisoner rate was a disgrace and that it would be possible to significantly reduce the amount and length of prison sentences without serious repercussions on the crime situation" (Törnudd 1993, p. 4).

Table 3.9 shows that the Finnish effort was successful by every relevant measure. Between 1976 and 1992, prison and jail admissions fell from 13,457 to 9,851; the prison population from 5,706 to 3,511; and the median length of sentence from 5.1 to 3.6 months (1991).

By American standards, median sentence lengths seem short. Finland, however, like most European countries, operates a unitary corrections system. Corrections data include (in American terms) both jail and prison inmates, which significantly lowers averages.

For serious and violent crimes, Finnish sentences are substantial. Average time served for nominal life imprisonment terms for homicide, for example, have declined from sixteen years in the 1970s to twelve years in the 1990s. For comparison, according to the Bureau of Justice Statistics, of persons convicted of murder who were released from prison in 1990, the median term served was six years, seven months, and the mean was eight years, seven months (Maguire, Pastore, and Flanagan 1993, table 6.108).

That the Finnish experience is not common can be seen in table 3.10 which shows prison populations at five-year intervals since 1975 in fifteen countries. The United States (and Portugal) lead the world in prison population growth. Most countries either had stable populations or increases of as much as 100 percent. Only in Finland did the prison population decline substantially.

A variety of·policy changes contributed to de-emphasis of imprisonment, including statutory changes intended to reduce penalties for theft offenses and drunk driving, a lowering in the minimum time served before eligibility for parole release, and increases in use of suspended sentences. Official crime rates, as in most European countries, were increasing during this period, and, except for a small community service program, new intermediate sanctions were not developed. For details, readers should consult Mr. Törnudd's report.

The most significant change was a shift away from confinement sentences toward suspended sentences (which will be executed if the offender reoffends within three years). In 1971, 65 percent of confinement sentences were to immedi-

Table 3.10. Combined Prison and Jail Populations in Fifteen Countries, 1975–1989

Country	1975	1979	1984	1989	% Change 1975–89
Belgium	6,150	6,137	6,637	6,761	8.5%
Denmark	2,665	2,291	3,103	3,378	26.8
Finland	5,062	5,373	4,656	3,103	(38.2)
France	27,165	34,640	40,010	45,102	66.0
Germany (F.R.)	50,140	50,395	55,806	51,729	3.2
Greece	3,173	3,221	3,557	4,564	43.8
Ireland	1,019	1,140	1,594	1,980	94.3
Italy	28,216	26,424	40,225	30,594	8.4
Norway	1,511	1,312	1,747	2,171	43.7
Portugal	2,532	5,054	6,499	8,458	234.0
Spain	14,764	10,463	13,999	31,137	111.0
Sweden	4,091	4,345	4,257	4,796	17.0
Turkey	24,397	54,671	73,488	48,413	98.4
United Kingdom	39,820	42,220	43,295	55,047	38.0
United States	365,593	459,864	677,898	1,076,460	195.4

Note: 1978 data were used to estimate 1979 jail population.

Sources: Council of Europe (1989), (1990), Bulletins 13, 14, 15; Maguire, Pastore, and Flanagan (1993), tables 6.58, 6.31; Flanagan and Macleod (1983), Table 6.16.

ate confinement and 35 percent were suspended. In 1991, 41 percent of confinement sentences were to immediate confinement and 59 percent were suspended.

The Finnish story is striking because it shows that a jurisdiction can, if it wishes, greatly diminish the use of incarceration without creating an array of intermediate sanctions and do so in the face of rising crime rates. Törnudd points out, however, that the change depended on the political climate. Finnish officials did not believe higher incarceration rates produce a safer society, and they were embarrassed that Finnish incarceration rates were higher than those in many other countries: "The decisive factor in Finland was the attitudinal readiness of the civil servants, the judiciary, and the prison authorities to use all available means in order to bring down the number of prisoners" (Törnudd 1993, p. 19).

Sentencing and Punishment in Finland
Patrik Törnudd

Finnish sentencing policies have for many years been stable and uncontroversial, and the numbers of those in prison have steadily declined. The stability results from widespread commitment to the values of proportionality and predictability in sentencing. The decline results from a concern that Finnish prisoner numbers were high compared with other countries and from a commitment over many years to reduce imprisonment.

For various historical reasons, emphasis on the rule of law and written legal norms is particularly strong in Finland. The discretions of the police, the prosecutors, and the judiciary are circumscribed in many ways. Plea bargaining is unknown in Finland—offenses of which people are convicted are expected to correspond to the offender's actual behavior. Within the criminal justice system, a high value is set on the principles of proportionality and predictability (Joutsen 1989).

Finnish Criminal Justice Ideology   The rationale of the criminal justice system is usually thought to be *general prevention*—not *general deterrence*. Outside Finland, those terms are used as synonyms by many people. In Finland they have distinctly different meanings and the preference for general prevention over general deterrence has important implications. In the Nordic countries, the concept of general prevention is strongly connected with the idea that a properly working criminal justice system has powerful indirect influences on peoples' beliefs and behavior. General deterrence is an element of general prevention, but the deterrence mechanisms are not necessarily the most important ones in maintaining respect for the law. It is, however, necessary that citizens perceive the system to be reasonably efficient and legitimate. Such a system promotes internalization and acceptance of the social norms lying behind the prohibitions of the criminal law (Anttila 1986).

This emphasis on the justice system's indirect effects has made it possible for Finns to reject both harsh punishments and highly individualized sanctions based on coercive treatment. The expectation that a properly working criminal justice system must meet certain *minimum* requirements—standards of certainty and adequacy of punishment, legitimacy of procedure, and appropriateness in the scope of the criminal law—is, in a sense, *static*. Such requirements are not directly translatable into theses about the efficacy or desirability of *changes* in the system. A strong belief in general prevention as the guiding rationale of the criminal justice system thus does not imply that changes in policy, such as increases in the severity of punishment, would be widely seen as an appropriate or cost-effective means of controlling the level of crime.

The idea of a just proportion between crime and punishment is, according to this view, an indispensable quality of a properly working criminal justice system. One important function of the proportionality principle is to set an upper limit to the penalty for any given offense. The ideas of fairness and justice within the system are also seen as absolute values. But the principal aim of the criminal justice system is not retribution or "doing justice" but the control of crime through the indirect and long-term mechanisms outlined above.

While the idea of system-based general prevention is strong in all Nordic countries, it has widest support in Finland. The rehabilitative ideal, which until recently was very popular in Sweden, never gained wide support in Finland. Nor can a system that sees the establishment of *upper* limits to punishment as a primary goal leave much room for purely incapacitative measures. Finnish law contains provisions for very dangerous, violent offenders, which authorize their continued detention in prison after the completion of their sentences, but the provisions are not used and are likely, according to recent reform proposals, to be repealed.

The Sanction System   Finnish laws provide relatively narrow ranges to govern judges' sentencing decisions (and parole release serves to reduce the durations of

Table 3.11. Sanctions Imposed in 1991 for Selected Offenses

Disposition	Theft Offenses (%)	Assault Offenses (%)	Drunk Driving (%)	All Penal Code Offenses (%)
Waiver of punishment	1.0	1.9	0.2	1.1
Fines set at trial	16.6	75.8	43.8	39.8
Summary fines	64.6			31.4
Conditional imprisonment	7.4	13.2	38.7	16.3
Unconditional imprisonment	10.4	9.1	17.3	11.3
	100.0	100.0	100.0	99.9
	(N = 32,209)	(N = 7,621)	(N = 24,054)	(N = 97,636)

Notes: The figures refer to the persons who were sentenced (N = 65,928) or absolutely discharged through waiving of punishment (N = 1,118) in general courts of first instance or who were fined in a summary procedure (N = 30,647) in 1991. The total number of offenders was almost 98,000 and the total number of separate sentences approximately 135,000. The offenders are grouped according to the most serious offense.

Source: Törnudd (1993).

prison stays). The statutes on theft provide an example. The basic statute on simple theft (sec. 28:1 of the Finnish Penal Code) states that the offender shall be sentenced to "a fine or to prison for at most one year." If, on the basis of the value of the stolen goods or certain other criteria, the offense is found to be an aggravated one (28:2), the available punishments range from four months to four years of imprisonment. Petty theft is dealt with in a separate statute allowing only fine sentences.

Unlike many other European countries, the Finnish penal system is built on only three basic sentencing alternatives: *fines* (set on the basis of the day-fine system that was introduced in 1921), *conditional (suspended) imprisonment,* and *unconditional imprisonment.* Table 3.11 indicates the use of the basic sanction options in 1991 for selected offenses.

A conditional sentence of imprisonment can be combined with an unconditional fine—this option was used in 1991 in 37 percent of all conditional prison sentences for offenses against the Penal Code. Fines can be set by the court, but for less serious offenses the police set a summary day fine, which later is approved by the public prosecutor in a routine procedure.

Waiver of punishment (the offender is found guilty but because of his or her youth or other extenuating circumstances is absolutely discharged) is also a possibility; this option was highly restricted until a 1991 amendment increased its scope.

Community service, introduced in some districts on a trial basis in 1991 and extended nationwide in 1994, allows the court to convert an unconditional sentence of imprisonment to community service. Current reform plans may create additional sanctions for juvenile offenders who were under age eighteen at the time of the offense (Lappi-Seppälä 1994, p. 203). Before 1976, the conditional sentence was reserved for first offenders and young offenders. Current law recognizes prior criminality as a criterion that in some cases may preclude suspension of sentence, but the main emphasis is on the gravity of the offense. An amendment in 1989 provides that persons under age eighteen at the time of the offense should be given an unconditional sentence only under exceptional circumstances. Prison sentences over two years cannot be suspended, but recent proposals have been made to remove this restriction.

Statutory Sentencing Standards   Apart from statutes authorizing reductions of sentences for young offenders, partially insane offenders, and certain acts of self-defense or duress, the only general sentencing rules formerly affecting the length of the prison sentence were statutes on recidivism, which stipulated extended sentences for repeat offenders. The recidivism provisions were abolished in 1976 and replaced with a set of sentencing rules (chap. 6 of the Finnish Penal Code).

The current statutes direct the courts to take into account the goal of uniformity of sentencing and all grounds for increasing or decreasing the severity of the punishment. The law specifically lists four aggravating considerations: the degree to which the criminal activity was planned; committing the offense as a member of a group organized for serious offenses; committing the offense for renumeration; and the previous criminality of the offender if the relation between it and the new offense on the basis of the similarity between the offenses or otherwise shows that the offender is particularly heedless of the prohibitions and commands of the law.

A separate statute provides three grounds for mitigating punishment: significant coercion, threat, or similar action; exceptional and sudden temptation or a similar factor that lowered the offender's ability to obey the law; and the offender's voluntary efforts to prevent or compensate for the offense or aid in clearing up the offense.

A separate provision, rarely used (only twenty times in 1991), also allows the court to reduce the sentence because of such consequences of the offense and the trial as loss of job, heavy damages, adverse publicity, or serious personal injury.

Sentencing Practices   Table 3.12 illustrates recent sentencing patterns in Finland. It shows by offense the proportions of unconditional sentences among all sentences and their mean lengths. The data in table 3.12 represent sentences per conviction offense, not aggregate sentences received by offenders for all current offenses. Summary fines are not included. The crime-specific data here do not include sentences for attempted offenses.

Long sentences are used sparingly—only 1.4 percent of all unconditionally sentenced offenders in 1991 received a prison sentence of four years or longer. A few persons each year are sentenced to life imprisonment for murder.

The parole rules assume that first-time offenders can be paroled after serving one-half of the sentence. Special rules for young offenders allow earlier release. Adult repeat offenders must serve at least two-thirds of the sentence. The paroling practices are fairly uniform and predictable.

A Trend Toward Less Imprisonment   As table 3.12 shows, there has been a trend toward fewer and shorter sentences of unconditional imprisonment. Theft offenses in particular have undergone significant changes. Theft offenses dealt with in court are more seldom than in earlier times considered to be aggravated. At the same

Table 3.12. Unconditional Prison Sentences in Finland, 1980, 1986, and 1991

Type of Offense	Unconditional Prison Sentences (in % of all sentences)			Average Length of Unconditional Sentence (in months)		
	1980	1986	1991	1980	1986	1991
Simple theft	44.2	39.5	36.2	3.4	2.7	2.0
Aggravated theft	74.0	69.1	68.7	10.8	10.3	7.7
Simple assault	16.7	11.5	9.2	3.1	2.6	2.2
Aggravated assault	50.4	58.7	54.9	15.0	13.3	13.0
Fraud	51.6	42.7	32.4	3.5	3.4	2.8
Aggravated drunk driving	25.0	31.9	33.5	3.0	3.0	2.9
Forcible rape	46.6	72.0	78.2	20.5	23.3	19.5
Intentional homicide	100.0	98.0	99.0	88.7	92.9	94.7
All Penal Code offenses	23.1	23.0	21.8	5.0	4.4	3.7

Note: Summary fines and sentences for attempted offenses are not included.

Source: Törnudd (1993).

time, the number and proportion of petty theft offenses handled by a summary day fine has dramatically increased; those data are not reproduced in table 3.12, which shows the percentages of offenders, by offense, who receive prison sentences (table 3.11 provides data on all sentences imposed, by offense). Much of the increase in petty theft results from growing numbers of shoplifting offenses.

Violent offense patterns have not changed significantly. The sentencing practices as regards intentional homicide are stable. The average age of rape offenders has increased, which is reflected in an increasing proportion of unconditional sentences, but the mean length of rape sentences has not increased.

When an offender is charged with multiple offenses, the courts before 1992 determined the proper sentence for each offense and then combined them according to certain rules. The new seventh chapter of the Penal Code, which took effect in 1992, instructs the courts to pass one single punishment for the main offense and to treat the secondary offenses as aggravating circumstances.

The prison population is declining. In 1980, Finnish prisons received 10,112 persons (including remand prisoners, fine defaulters, etc.). The average daily population was 5,088, 106 per 100,000 inhabitants. In 1992, only 9,851 persons entered Finnish prisons, the average population was 3,511, and the prisoner rate had dropped to 70 per 100,000 inhabitants.

Explaining the Trend Why has the average severity of sentences, and particularly for property offenses, declined? Crime rates have steadily increased in most offense categories and the gravity of the offenses has not declined. The average age of the offender population has slightly increased.

Finnish use of prisons has declined because Finnish policy makers decided prison use *should* decline. In the 1960s, Finnish crime specialists became aware that the Finnish prisoner rate was abnormally high compared with other Nordic and Western European countries. Sentences for property crimes were considerably harsher than in other Nordic countries. Scholars, civil servants, judges, and others involved in the system reacted strongly to these findings and called for reforms.

The decision to "normalize" the Finnish sanctioning system produced changes in crime definitions, in authorized sentences in the Penal Code, in the rules governing the choice and the severity of sanction, and in the parole system. Many of these reforms were enacted in the 1970s, including introduction of new statutes on theft in 1972, on drunk driving in 1976, and on conditional sentences in 1976 (Törnudd 1993). The effects were apparent throughout the 1980s.

Other reforms were parts of the still ongoing reform of the Finnish Criminal Code. One important change in 1991 promoted the use of very short sentences of imprisonment measured in days. The theft statutes were modernized in 1972; all authorized punishments for property offenses were reduced. The effects of this change were visible in the judicial statistics for the year 1991 (Oikeustilastollinen 1993, pp. 102–106).

The Finnish experience shows that—given the political will—the use of imprisonment can be substantially reduced without introducing new alternative sanctions. A necessary requirement has been the fairly stable and coherent criminal justice ideology of Finland and the relative lack of public controversy about it.

The Netherlands

Sentencing and Punishment in the Netherlands
Peter J.P. Tak

Dutch penal policies have become harsher in some ways since 1980 and are likely to become harsher still. Prison sentences became longer and the number of prison cells nearly tripled, rising from 3,789 in 1980 to 10,059 in 1994, with an additional 1,800 to come on line early in 1996.

During the same period, however, the use of short-term imprisonment fell, use of a new prosecutorial diversion program grew rapidly, community service orders came into use, and an experimental intensive supervision program began in 1993.

Thus, the stereotype of the Netherlands as a country with exceedingly mild penal policies is—like most stereotypes—greatly oversimplified.

This article gives an overview of changes in Dutch sentencing policy since 1980, discussing the pressures for greater penal severity, sentencing trends, the development of community penalties, and the laws of sentencing.

Background   The Dutch criminal justice system has long been noted for its mildness. Although still mild in comparison with many European countries (and more so compared with the United States), the penal climate has become harsher since the late 1970s. The incarceration rate of 24 per 100,000 population in 1980, for example, had grown to 61 per 100,000 in October 1994.

After three decades of low crime rates, a steep increase between 1976 and 1984 caused serious and widely shared concern. The number of recorded crimes had risen by around 60 percent, but public expenditure for law enforcement, the judiciary, and prison administration failed to keep pace. This resulted in a falling detection rate, a lower percentage of criminal cases dealt with by the courts, and insufficient prison capacity.

Parliament asked for a policy plan for crime prevention and for improvement of criminal law administration. In 1983, the minister of justice appointed the Roethof Committee, a seven-member panel that included a former minister of justice, a police commander, Amsterdam's chief prosecutor, the mayor of Hengelo, and several university professors. It was directed to assess the causes of the crime increase and the effectiveness of existing policies and to propose ways to improve crime prevention and control.

The Roethof Committee issued an interim report in 1984 (and a final one in 1986) and presented a set of recommendations, which were incorporated into the 1985 government policy plan *Society and Crime* (Ministry of Justice 1985). After five years' assessment, in 1990 a new plan called *Law in Motion* was published (Ministry of Justice 1990).

Both plans proposed to raise the level of criminal law enforcement and to intensify crime prevention by extending the statutory powers of the police to investigate organized crime, by improving the efficiency of the prosecution service, by increasing prison capacity, and by intensifying crime prevention programs. Numerous laws have since been enacted and measures taken to support the new policies.

Sentences, 1980–1992 As measured by the number of cases tried by courts (see table 3.13), criminality seems to have risen only slightly since 1980. However, crimes recorded by the police increased greatly. The different trends for recorded crimes and court cases result from the introduction in 1983 of a prosecutorial diversion scheme.

The 1983 Financial Penalties Act authorized prosecutors to resolve a criminal case on the basis of an arrangement in which the suspect pays a sum of money to the treasury in order to avoid further prosecution and a public trial. (This resembles German "conditional dismissals" under §153a of the German Code of Criminal Procedure.) The prosecution service can use this power for any offense carrying a potential prison term of six years or less. Only one-third of criminal cases are tried by criminal courts. The remainder are settled out of court by the prosecution service.

Table 3.13. Crimes Recorded by Police, Detected, and Tried by Courts

Offenses	1980	1985	1990	1992	Detected in 1992 (%)
Total	705,600	1,093,700	1,133,800	1,268,500	19.2
Violent crimes	26,500	37,100	49,600	58,400	48.5
Property crimes	500,900	840,600	840,400	950,100	13.7
Against public order and destruction	84,800	117,500	142,900	154,600	16.6
Detected (%)	29	24	22	19	
Cases tried by courts	79,100	77,500	88,600	87,100	

Source: Dutch Central Bureau of Statistics (unpublished data provided to author).

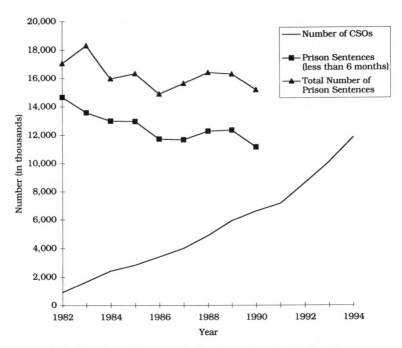

Figure 3.13. Numbers of Community Service Orders, Total Prison Sentences, and Sentences Less than Six Months, Netherlands, 1982–1994. *Source:* Dutch Ministry of Justice (unpublished data provided to author).

Criminal cases tried by courts in 1993 differed significantly from those tried in 1980. Between those years, the number of convictions for sexual offenses and capital offenses each increased by 250, and the numbers of violent threats, drug offenses, and robberies increased by 400, 800, and 1,000, respectively. As figure 3.13 shows, the number of prison sentences remained stable. This is partly due to the increased use of the prosecutorial diversion scheme, partly to a sizable increase in the use of community service orders, and partly to a 2,000 fall in the number of prison sentences for drunk driving.

Because courts are trying more severe offenses, the average prison term has increased considerably—by one month for drug offenses, by sixteen months for murder and homicide, by thirteen months for rape, and by eight months for violent theft.

Although the number of prison sentences imposed in 1991 does not differ much from that of 1980 (see table 3.14), the number of years of detention ordered differs considerably: 2,848 in 1980, 4,887 in 1985, 5,958 in 1990, and 6,442 in 1991.

Prison Reduction Policy   Penal policy in the 80s was characterized by strong tendencies to reduce the use of short-term imprisonment and to expand the use of noncustodial sanctions.

Table 3.14. Prison Sentences, 1980–1991

Years	Total	One Month	One to Twelve Months	One to Three Years	Three to Six Years	Six Years or More
1980	15,309	8,944	5,602	646	175	—
1985	16,348	6,719	8,665	1,281	215	72
1990	15,182	4,946	8,772	1,437	394	144
1991	15,683	5,052	9,019	1,614	519	148

Source: Dutch Central Bureau of Statistics (unpublished data provided to author).

The 1983 Financial Penalties Act. The aim was to improve the enforcement of fines so that fines could better function as an alternative to short-term prison sentences.

Since 1983, the fine is legally presumed to be the appropriate penalty. All offenses, including ones subject to life imprisonment, may be sentenced with a fine. The minimum fine for all offenses is 5 Dutch guilders (roughly $3.00). The maximum depends on the fine category into which a crime or infraction is placed. The 1983 act created six categories with maximums of 500 guilders, 5,000 guilders, 10,000 guilders, 25,000 guilders, 100,000 guilders, and 1,000,000 guilders (category VI).

Category VI fines can be imposed on corporate bodies and on individuals under a few special criminal laws—such as the Economic Offenses Act and the Narcotic Drug Offenses Act.

Community Sanctions. In 1989, following eight years of experimentation, a new principal penalty, "the performance of unpaid work for the general good," also known as the community service order (CSO), was introduced in the Penal Code.

CSOs function as a substitute for unconditional prison sentences of six months or less and can be imposed only with the consent of the offender and subject to a maximum of 240 hours. There is no minimum. A CSO of less than 120 hours must be carried out within six months, otherwise within twelve months.

The prosecution service supervises compliance with CSOs, with assistance from the probation service. If the work is not carried out properly (roughly 10 percent of all CSOs), the judge may, at the request of the prosecutor, replace the CSO with a prison sentence to be served in full or in part.

The number of CSOs imposed increased from 213 in 1981—when the experiment started—to 8,585 in 1992. The percentage of cases disposed of by means of a community service order, as figure 3.14 shows, increased from 1 percent in 1982 to nearly 10 percent in 1990. The 1990 policy plan *Law in Motion* contemplates a further annual increase of 10 percent. Recent research indicates that CSOs in many cases have not substituted for short-term prison terms but for suspended sentences. However, it is unknown to what extent this was the case.

In 1993, new community-based sanctions have been introduced by way of experiments on a restricted scale including an intensive day program and intensive supervision.

Earlier Parole Release. Another target of penal policy in the 1980s was to reduce

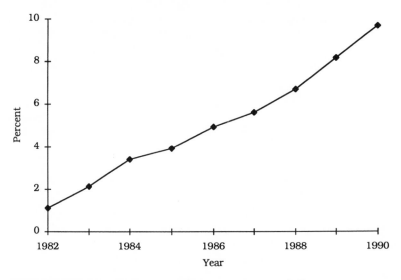

Figure 3.14. Community Service Orders as a Percentage of All Sentences, Netherlands, 1982–1990. *Source*: Dutch Ministry of Justice (unpublished data provided to author).

the effective term of imprisonment. Under a 1987 change in the parole legislation, prisoners serving a sentence up to one year must be released after serving six months plus one-third of the remaining term. Prisoners sentenced to more than one year must be released after serving two-thirds of the sentence. The prisoner's right to be released early can be restricted only in a few exceptional cases.

Sentencing Process   The Dutch judiciary is vested with wide discretionary power in choosing the type and severity of sanctions. The few statutory rules are general and do not limit judges' choices in individual cases. The court can fully individualize the sentence, giving full consideration to the crime, the situation, and the offender.

The Penal Code lists four principal penalties in order of severity—imprisonment, detention, community service, and fine. Maximum terms of imprisonment are specified and reflect the gravity of the worst possible case—for murder, either a life sentence or imprisonment up to twenty years; twelve years for rape; six years for domestic burglary; and four years for theft. Few crimes are subject to life imprisonment, and for these there is an alternative of a fixed term up to twenty years.

The code does not prescribe life imprisonment or a maximum prison term in any circumstances or for any crime. The statutory minimum term of imprisonment is one day and is the same for all crimes, regardless of the generic seriousness of the offense.

The choice of sanction lies with the court but is subject to procedural requirements concerning the judge's reasoning. Section 359(6) of the Code of Criminal

Procedure (CCP) requires the court to give special reasons whenever a custodial sentence is ordered instead of a fine, which is by law preferred as the chief sentence.

Section 359(7) CCP requires a statement of reasons when the court imposes a more severe sentence than the prosecution service has requested. Furthermore, section 359(8) CCP requires reasons when the court denies the defendant's offer to perform a community service order.

The choice of sentence is in principle determined by the seriousness of the offense and the aims of sentencing. In theory, factors connected with the suspect's personality and criminal history are to be considered only in the case of recidivists. When the defendant has displayed indifference to sentences imposed after earlier convictions, this also may justify a sentence that would not be indicated on the basis of the seriousness of the offense.

The statutory aims of sentencing are retribution, special or general deterrence, reformation, protection of society, and reparation, and the court may choose among them in each individual case.

Although the choice of sanction lies with the court, subject only to statutory requirements concerning reasons, Supreme Court case law offers guidance on matters to be considered in determining the severity of the sentence.

Various personal or isolated factors may be reasons to adjust the sentence upward or downward. An upward adjustment may be justified by the criminal past of the accused (Hoge Raad [Supreme Court] 19 June 1986, Nederlandse Jurisprudentie [Dutch Case Law] 1987, 61), by the negative attitude of the accused during the examination in court, for example, an accused who consequently denies having committed the crime (Hoge Raad 10 December 1984, Nederlandse Jurisprudentie 1985, 358) or an accused who wishes to evade a sentence by making several false statements (Hoge Raad 27 January 1987, Nederlandse Jurisprudentie 1987, 711), by the motives that compelled him or her to commit the offense, for instance, jealousy and hate (Hoge Raad 15 April 1986, Nederlandse Jurisprudentie 1987, 25), by the circumstance that the accused did not want to cooperate in a psychiatric evaluation (Court Amsterdam 19 December 1991, Nederlandse Jurisprudentie 1992, 142), or by the fact that the accused fails to understand that his or her behavior was wrong (Hoge Raad 12 November 1985, Nederlandse Jurisprudentie 1986, 409).

A downward adjustment may be indicated by serious delay between the time of committing the crime and the trial, by the accused voluntarily offering compensation for damages inflicted, by expression of regret by the accused, by lack of previous convictions, or by positive probation prospects.

The absence of mandatory rules for sentencing and sentencing guidelines may contribute to the present mild penal climate but may also result in great disparity in sentencing, as recent research has shown. A number of proposals have been developed to reduce disparity in sentencing. Some of them will likely lead to considerably higher sentences and a harsher penal climate.

Policy from 1993   Despite all efforts to reduce the number and the length of prison sentences, it has been clear since 1985 that a serious prison construction program had to be set to meet the need for prison capacity. The actual need for capacity has repeatedly proven higher than expected mainly due to developments like the internationalization of criminality, the rise of organized crime, and the improvement of law enforcement policies.

Due to the prison construction program, prison capacity has nearly tripled from 3,789 prison cells in 1980 to 10,059 cells in 1994, and decisions were made to increase the number of prison cells to 11,818 to be available early in 1996. Nevertheless, a prison capacity shortage of 1,600 cells in 1998 is expected. To reduce the expected shortage, proposals will be developed to detain drug-addicted or disturbed prisoners in other than penal establishments, to improve the use of alternatives to imprisonment, to implement conditional release after having served half of the sentence due to good behavior, and to begin experiments with use of electronic monitoring in the last phase of a long-term sentence.

Prisoners today are detained for more serious offenses than in the 1970s and 1980s. Furthermore, a large number of prisoners are foreigners, addicted to drugs or suffering from psychiatric disturbances. Those prisoners are difficult to handle. This has had a negative impact on the prison regime.

Recently, a new prison memorandum concerning the prison regime has been submitted to Parliament. The high level of aspiration, a full resocialization, which formed the backbone of previous prison regime memoranda, has been reduced to a basic target: a safe and effective but humane detention. The central issue in the regime is labor plus the statutorily guaranteed fresh air, visits, recreation, and sports. An additional vocational training is possible. A regime of restricted contacts with inmates is applied to a prisoner who is not participating in these activities.

Sentencing became harsher in the Netherlands between 1980 and 1993; in the near future, as well, implementation of sentences will become harsher. No one can deny that Dutch penal policy has changed considerably.

Netherlands Successfully Implements Community Service Orders
Peter J.P. Tak

Community service orders (CSOs) have been in use in the Netherlands since 1982. Established as part of an effort to reduce use of short-term imprisonment, CSOs are by statute to be imposed only on offenders who otherwise would receive a prison sentence of up to six months. Today they are commonly imposed and offenders must work until their obligations are satisfactorily completed. There is considerable evidence, however, that judges often (no one knows how often) use CSOs for some offenders who otherwise would receive nonincarcerative sentences.

Background Dutch sentencing policy has long been characterized by efforts to reduce use of short-term imprisonment. In 1974 the Committee on Alternative Penal Sanctions was set up to advise the government on the development of new sentencing options. An earlier Committee on Financial Penalties had restricted its proposals to use of fines instead of short-term imprisonment. The rationale was that large fines have the same punitive value as short-term imprisonment. The proposals, generated in a period of increasing affluence, were implemented in a period of economic crisis. The need for a different technique to reduce prison use became evident.

Resolution (76) 10 of the Committee of Ministers of the Council of Europe and positive experiences in England and Scotland suggested CSOs as a sentencing option. In 1979, the Committee on Alternative Penal Sanctions proposed a CSO experiment. A working party then proposed a legal framework for the experiments and how CSOs could be implemented. Experiments were initiated on February 1, 1981, in a few jurisdictions. In 1982 the experiments were extended to all jurisdictions.

Ministerial guidelines directed that the experiments take place within the existing statutory framework. During the experiments, a CSO could be imposed by the prosecution service as a condition to a decision to waive prosecution or to settle a case without a court hearing. A CSO could also be imposed by the trial judge as a condition attached to a decision to suspend pretrial detention, to suspend a sentence, to postpone the sentence, or to grant pardon.

The experiment's first two years' operation were evaluated by the Research and Documentation Center (RDC) of the Ministry of Justice (Bol and Overwater 1984). The generally positive results are summarized below. On the basis of the evaluation, a bill was formulated in 1986. Statutory provisions governing the CSO were introduced in the Penal Code on December 1, 1987.

Statutory Framework Under Penal Code (PC) section 9, the CSO is a distinct sentence option and considered to be a restriction of a person's liberty that is less severe than the custodial sentence and more severe than a fine. A CSO may not exceed 240 hours. If less than 120 hours, the work must be completed within six months; otherwise, within twelve months (sec. 22d PC). The trial judge may impose a CSO only if he or she would otherwise impose an unconditional prison sentence of six months or less or a part-suspended/part-unconditional prison sentence of which the unconditional part is six months or less (sec. 22b PC). Community service may not be used as an alternative to a suspended prison sentence, a fine, or a fine-default detention.

A judge may impose community service only when there has been a proposal from the accused that he or she is willing to carry out nonremunerated work of a type described in the proposal (sec. 22c PC). The accused's "compulsory consent" is a statutory requirement to avoid contravening international conventions prohibiting forced labor. The judge has to state in the sentence the prison sentence for which the community service is a substitute and specify the number of hours' work to be carried out, the period within which it must be completed, and the nature of the work (sec. 22d PC).

The prosecution service is responsible for overseeing CSOs, and it may request information from individuals and organizations involved in probation work for this purpose (sec. 22e PC). When the prosecution service is satisfied that the work has been carried out properly, it must notify the convicted person as soon as possible.

If the convicted person has not carried out the work properly, the prosecution service may request revocation of the CSO and imposition of the prison term mentioned in the sentence. The judge must take into account work that has been properly carried out (sec. 22g PC). The prosecution service must make its revocation request within three months after the end of the completion period for the community service (sec. 22i PC).

The probation service is responsible for administering CSOs, and coordinators have been appointed for each of the nineteen jurisdictions. The coordinator's job is to canvass for projects, maintain a project bank, maintain contacts with the project institutions, and write final reports.

The coordinator decides on the nature of the work to be carried out, taking into account the offender's skills, education, and vocational training. If the work requires team effort, the coordinator decides whether the offender fits. If the place or nature of work needs to be changed, the coordinator contacts the prosecution service, which has authority to make these changes. The convicted person is informed of the changes and may object within eight days to the sentencing judge (sec. 22f PC).

Community service work must benefit the community. It can be with public bodies like the government or private organizations involved in health care, the environment and the protection of nature, and social and cultural work. To discourage unfair competition with paid workers, regional review committees check that no regular workplaces are being used for community service.

No offenses are statutorily excluded from punishment with CSOs. Given the boundary of the six-month prison sentence, however, community service operates mainly for mid-level crimes and is seldom ordered for more serious offenses unless there are mitigating circumstances.

Evaluation Results Use of community service orders has been twice evaluated by the Research and Documentation Center of the Ministry of Justice. The first study (Bol and Overwater 1984) covered the period from February 1981 to May 1983. The second (Spaans 1994) covered the year 1987. Results from the first study are described below, with results from the second study shown in parentheses.

Over 90 percent (87 percent) of the CSOs were completed successfully. The failures can be ascribed partly to circumstances beyond the person's control and partly to uncooperativeness or prosecution for other offenses.

The average length of community service was 100 hours (116 hours). In 7.5 percent (16 percent) of cases, community service was imposed exceeding the recommended upper limit of 150 hours.

Community service was imposed for property offenses in 48.6 percent (66.4 percent) of cases, and for traffic offenses in 23.8 percent (9.8 percent) of cases. Almost 8.6 percent (12.4 percent) concerned aggressive offenses and 3.0 percent (5.3 percent), sex or drug offenses.

Of persons sentenced to perform community service, 93.4 percent (91.9 percent) were male and 4.6 percent (8.1 percent) female; 25.3 percent (22.2 percent) were between eighteen and twenty years old; 22.4 percent (29.6 percent) between twenty-one and twenty-four years old, and 17.5 percent (18.4 percent) between twenty-five and twenty-nine years old.

Of persons sentenced with a community service order, 63 percent (63.4 percent) had previous convictions, while 37 percent were first offenders.

Almost 60 percent (54.9 percent) of the persons sentenced with community service were unemployed and dependent on social security benefits.

The work in over 50 percent (28.6 percent) of cases involved maintenance, repair, and painting, usually for institutions in the welfare sector. Housework and forestry and gardening made up a further 14 percent (27.6 percent) and 12 percent (4.3 percent), respectively.

Recidivism Research   Both evaluations examined recidivism by CSO offenders. Reoffending by persons sentenced to community service was compared with the experience of persons sentenced to unconditional short-term imprisonment. The two groups were matched for the nature of the offense, age, where the sentence was imposed, criminal records, sex, and suspicion of drug use.

Reoffending was examined in terms of both the severity of the new offense and how soon it occurred within a three-year period after the first conviction. A distinction was made between reoffending in general and committing another offense of the type for which the person was originally convicted.

Significant differences were found only in relation to recidivism in general. Of those who had performed community service, 42 percent reoffended (Spaans 1994, 62 percent), whereas 54 percent of those who had served short-term prison sentences reoffended (Spaans 1994, 76.4 percent).

The second evaluation (Spaans 1994), however, demonstrated that those sentenced to CSOs had significantly fewer and less severe previous convictions than those sentenced to short-term prison sentences and therefore that comparisons with respect to recidivism are suspect.

Net Widening   No nationwide research has investigated whether CSOs are indeed used only as substitutes for short-term imprisonment (a statutory requirement). Strong indications exist that CSOs in some jurisdictions are imposed in cases previously sentenced with fines and suspended sentences (Hanewinkel and Lolkema 1990). This is in line with the results of interviews with members of the judiciary and the prosecution service (Bol and Overwater 1984, 1986; Kockelkorn, van der Laan, and Meulenberg 1991). The majority of the interviewed judges and prosecutors admitted that they had no objections to use of a CSO instead of a suspended sentence or a heavy fine. Spaan's (1994) recent research suggests a considerable net-widening effect as well. Community service orders were imposed on offenders in 1987 who might well have received a suspended sentence. There is, without further research, no reliable evidence about net widening.

Other European Countries

Sentencing in Switzerland
   Martin Killias, André Kuhn, and Simone Rônez

Swiss sentencing laws and practices have long been influenced by several legal traditions, including the French Code Civil and German legislation. In criminal law, Switzerland has a typical continental system, with one unified criminal code at the federal level and twenty-six autonomous procedural systems at the cantonal level. Despite that formal diversity, the cantonal systems have many common features, such as the inquisitorial system, the limited discretion of police and prosecutors, and the protection of fundamental procedural rights of defendants as guaranteed by the European Convention of Human Rights.

The definitions of offenses are contained in the Swiss Criminal Code and a few major federal laws, such as the Narcotics Act and the Road Traffic Act. Sentences are imposed, as elsewhere in Europe, by the same bench of judges who found the defendant guilty. The panels consist usually of three to five judges, with a senior judge in the chair. They by simple majority decide guilt and the sentence to be imposed. Typically, the sentence is meted out immediately after the verdict and at the same hearing.

For most offenses, judges have wide sentencing discretion. In most cases, imprisonment of several years, or a few days, or even a fine are available options. Custodial sentences can be suspended if they do not exceed eighteen months and the defendant's record does not warrant an immediate custodial sentence. In practice, sentences are often set either at eighteen months in order to make suspension possible, or just beyond in order to evade the question of whether the offender's record would permit suspension (Kuhn 1993, pp. 113–117).

There are no sentencing guidelines; there are a few general rules in the criminal code concerning aggravating and mitigating factors and a few more general principles. In practice, however, judges are required to give the criteria on which they choose a certain type of punishment and to explain why they reached a particular sentence. Over the last few years, these requirements have been substantially increased by the Federal Supreme Court of Switzerland (Nay 1994). Although the Supreme Court does not review sentences as such, it watches over the conformity of sentences imposed by lower courts (and the reasons given) in light of criminal code criteria and what the Court considers appropriate interpretation of these principles. In the lower courts and for mass offenses—such as minor thefts, drunken driving, and drug offenses—sentencing conventions have developed on an informal basis. They are seldom quoted in opinions.

Inmates are eligible for parole after having served two-thirds of their sentence, but not less than three months (Criminal Code sec. 38). Given the overcrowding of correctional institutions, about 80 percent of inmates are granted parole nowadays (La Liberation Conditionnelle 1994, p. 83).

Prevailing Sentencing Patterns Short custodial sentences are widely used in Switzerland. In 1993 over 78 percent of persons released had served less than three months (Office fédérale de statistique [OFS] 1994a, table 4d). Switzerland resembles the Netherlands and Scandinavian countries in this regard, but a few qualifications are necessary.

First, the high prevalence of short sentences may be due not, as in the Netherlands, to a general pattern in sentencing but to frequent use of short sentences in road traffic offenses (such as drunken driving), which make up 46 percent of all immediate custodial sentences (OFS 1994c, table 2). For more serious crimes, sentences tend to be long in a European perspective.

Second, for minor offenses (assaults, thefts, and most drug offenses) the probability of being sentenced to an unsuspended custodial sentence is in line with the

Table 3.15. Unsuspended Custodial Sentences, Seven Countries, 1990

Country	Persons Sentenced to an Unsuspended Sentence (per 100 convicted)			Length of Unsuspended Sentences; Percentage under 1 Year and Average in Years (in parentheses)					
	Assaults	Theft	Drugs	Assaults		Theft		Drugs	
England/Wales	14	13	13	66%	(1.2)	78%	(0.8)	43%	(2.4)
France	17	27	45	85	—	88	—	51	—
Germany	7	8	18	67	—	69	—	36	—
Netherlands	13	35	66	96	(0.3)	94	(0.3)	68	(1.1)
Sweden	33	11	30	86	(0.4)	92	(0.4)	72	(1.0)
Switzerland	22	27	41	89	(0.6)	91	(0.4)	71	(1.0)
United States	72	64	73	—	(2.9)*	—	(1.5)*	—	(2.1)*

* = estimated time to be served in prison.

Sources: Council of Europe (1995), tables 2d (II, Va, VIa) and 2e (II, Va, VIa); Langan, Perkins, and Chaiken (1994), tables 3, 4, 5, and 11.

general European trend (see table 3.15). This is also true for the proportion of sentences of less than one year and for the average length of sentences.

Although in Europe the probability of incarceration is considerably lower than in the United States, American sentences in 1990 were roughly twice as long as those in Switzerland for many offenses. It should be kept in mind, moreover, that table 3.15 shows for the United States net time to be served, whereas the data for Switzerland and the other European countries pertain to announced sentence length. In Switzerland, time actually served might approximate 75 percent of sentence length on average. Thus the data understate the differences between the United States and Europe. Finally, however, the American data pertain to aggravated assault, whereas the European data include common assaults, which means that differences for that offense may be overstated.

Compared with other European countries, sentences tend to be long in Switzerland for homicide, robbery, and rape, and short for drug trafficking (see table 3.16, pt. C). The proportion of sentences under one year is also low (table 3.16, pt. B). However, the probability of receiving an immediate custodial sentence is lower than in most other countries (table 3.16, pt. A).

The likeliest explanation for these patterns is that those receiving unsuspended custodial sentences, on average, have committed more serious offenses, or have a longer record, or both. The relatively mild average sentences for drug trafficking may result from the widespread practice among drug addicts to earn their living through small-scale deals, which means that those convicted are often low-scale user-dealers. If more serious drug trafficking (involving larger quantities) is considered, the average sentence length (2.9 years) is comparable to that in England.

Compared with American practice in 1990, the probability of unsuspended custodial sentences is low in Switzerland, and the estimated time to be served is about half as long. In light of the more complete data available here, it is doubtful whether American and European sentences are as similar as suggested by Lynch (1993).

Table 3.16. Unsuspended Custodial Sentences, Seven Countries, 1990

Country	A. Persons Sentenced to an Unsuspended Sentence (per 100 convicted)			
	Homicide	Robbery	Rape	Drug Trafficking
England/Wales	84	68	94	47
France	97	67	95	51
Germany	87	45	59	—
Netherlands	80	73	82	—
Sweden	95	70	95	—
Switzerland	78	64	68	41
United States	95	90	86	77
	B. Persons Sentenced to an Unsuspended Term of Less than 1 Year (per 100 receiving an unsuspended sentence)			
England/Wales	4	20	4	34
France	1	49	2	25
Germany	1	13	4	—
Netherlands	36	72	56	—
Sweden	1	27	19	—
Switzerland	3	13	0	53
United States	—	—	—	—
	C. Average Length of Unsuspended Sentences			
England/Wales	5.3	3.1	5.8	2.8
France	—	—	—	—
Germany	—	—	—	—
Netherlands	2.6	1.0	1.4	—
Sweden	5.5	1.6	2.3	—
Switzerland	9.6	3.1	3.1	1.6
United States	8.5*	4.2*	6.2*	2.2*

* = estimated time to be served in prison.

Sources: Council of Europe (1995), tables 2d (Ia, III, IV, VIb) and 2e (Ia, III, IV, VIb); Langan, Perkins, and Chaiken (1994), tables 4, 6, and 11.

Wide use of short prison sentences has led to a comparatively high prevalence of imprisonment in the general population. According to a survey on the cohort born in 1955, 6.5 percent of male Swiss who live in Switzerland had experienced incarceration at least once by age thirty-three, and 24 percent had been convicted at least once between age eighteen and thirty-three (Killias and Aeschbacher 1988). This is a similar prevalence rate of incarceration as in England and Wales (Harvey and Pease 1987), although convictions (among males) are considerably more prevalent there (30 percent by age twenty-eight).

Changes in Sentencing Patterns during the Last Decade   Sentencing patterns may change over time. Figure 3.15 shows average custodial sentences between

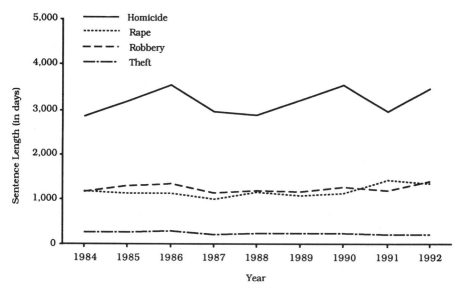

Figure 3.15. Average Length (in days) of Unsuspended Custodial Sentences, Switzerland, 1984–1992. *Source:* Office fédéral de la statistique (unpublished data provided to authors).

1984 and 1992 for intentional homicide (including attempts), rape, robbery, and theft.

Overall, sentencing patterns remained remarkably stable. The exceptions are rape and theft. Sentences for rape increased by 24.3 percent since 1989, and by 12.0 percent since 1984. For theft, sentence length decreased from 1984 to 1992 by 16.4 percent. For homicide and robbery, sentences fluctuate within a 10 to 15 percent range.

As Figure 3.16 shows, the same overall stability characterizes sentencing in drug cases. This stability illustrates the power of unwritten sentencing standards. Even in the absence of sentencing guidelines, judges look at their and their colleagues' previous sentences in similar cases. The prevailing practice of entrusting a bench rather than a single judge with sentencing decisions may facilitate the flow of relevant information concerning the "usual" sentence length in similar cases.

Two qualifications to the preceding comments about stability in sentencing are warranted. First, substantial numbers of offenders convicted of any of the offenses considered here have been simultaneously convicted (and thus sentenced) for other offenses. It is not likely, however, that offense combinations have changed in a way that would invalidate the generalizations offered here. Second, a stable trend in length of custodial sentences might mask shifts in the frequency of such sentences compared with other sentencing options. As the following section demonstrates, custodial sentences have indeed decreased slightly since 1989, but the prevailing trend has been remarkably stable.

Trends in Prison Populations The incarceration rate was comparatively low for many years, even by Western European standards. From 42 per 100,000 in 1972, it

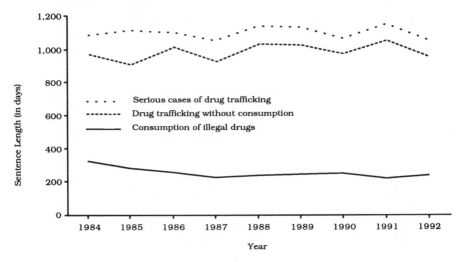

Figure 3.16. Average Length (in days) of Unsuspended Custodial Sentences, Switzerland, 1984–1992. *Source:* Office fédéral de la statistique (unpublished data provided to authors).

rose to 73 in 1988 (Killias 1991, p. 383) and to 90 in 1993. According to the Council of Europe *Prison Information Bulletin,* Switzerland in 1983 ranked seventh among eighteen countries in incarceration rate and eighth among twenty-five countries in 1992. Thus, Switzerland's prison population (in absolute numbers and per 100,000 population) has increased at about the same pace as in the rest of Europe.

Not unlike the Netherlands, France, and the Scandinavian countries, Switzerland once had much higher prison populations—around 150 per 100,000 population in the early 1940s, and over 120 per 100,000 in the 1920s (Killias 1991, pp. 369, 378). It is not obvious why the use of imprisonment declined so greatly after World War II. Perhaps the Swiss Criminal Code, which took effect in 1942 and replaced the codes of the then twenty-five cantons, reshaped sentencing policies and, especially, reduced the length of custodial sentences. But given the absence of data on prison populations between 1942 and the 1970s, this explanation is speculative at best. Unemployment, which was a major problem in those times and which, after fifty years of virtually no unemployment, has reached about 5 percent of the labor force, can almost certainly be ruled out as an explanatory variable. For one thing, the recent increase in prison populations preceded the new employment crisis. For a second, labor market fluctuations have always correlated more with pretrial detention than with imprisonment as a penal sanction (Killias and Grandjean 1986), a pattern that has been noted in other European countries (Melossi 1995). This time, it is imprisonment rather than pretrial detention that has increased dramatically.

Increasing prison populations have led to overcrowding in many institutions. Despite signs of overcrowding in the early 1980s (OFS 1985), major building programs did not start until the late 1980s. More recently, crowding has reduced the space available for pretrial confinement and constrained police from making arrests

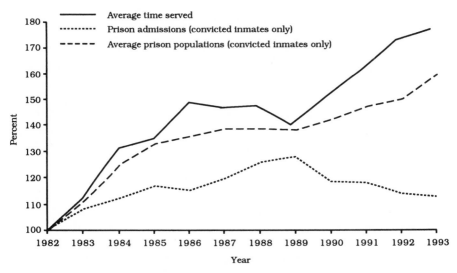

Figure 3.17. Percentage Changes in Prison Populations, Prison Admissions, and Time Served, Switzerland, 1982–1993 (1982 = 100). *Source:* Office fédéral de la statistique (unpublished data provided to authors).

even in more serious cases. This phenomenon has been most pronounced in Zurich in connection with the internationally known needle park and the high prevalence of nonresident drug dealers.

Figure 3.17 shows overall trends in prison admissions, prison populations, and average time served from 1982 to 1993, in each case using data from 1982 as a base (1982 = 100) and showing changes in proportion to that base.

Time served and overall population numbers have increased much more than has the number of custodial sentences imposed (and served). For the latter, there was first an increase between 1982 and 1989, followed by an almost identical decrease until 1993. Time served, however, steadily increased, from an average of 74 days in 1982 to 131 in 1993.

One might suppose that judges are today sentencing offenders to longer prison terms. The data presented in figures 3.15 and 3.16, however, do not suggest such an explanation. Average sentence lengths have been remarkably stable, except for rape, where, given the low numbers involved, the longer sentences may have increased the absolute number of inmates in 1993 only by about forty over their number in 1984.

The likeliest explanation is an increase in serious offenses that typically receive long custodial sentences. When sentencing patterns in 1984–1985 are compared with those in 1991–1992, moderate shifts become apparent for homicide (up 15 percent), robbery (down 19 percent), rape (up 6 percent), and theft (up 5 percent); this sums up to a moderate overall increase in convictions for serious offenses. The large shifts involve drug crimes. For all forms of drug trafficking, the increase is 38 percent, and serious forms of drug trafficking (involving usually large quantities) rose by 65 percent.

Thus, offenses involving large numbers of offenders and lengthy sentences increased particularly strongly. The average sentence for serious drug trafficking was three years in 1991–1992, and 788 offenders were sentenced to immediate custody in 1992, approximately 7.5 percent of all prison admissions. Not surprisingly, a large proportion of prison inmates have been convicted for drug offenses (34 percent in 1984 and 33 in 1991, OFS 1994b, table 33). The apparent stability may mask a major shift from minor to large-scale drug dealers.

Current Policy Initiatives   Mainstream criminal law teachers and policy makers consider short prison sentences harmful. Consonant with that belief, a committee of criminal law experts, on behalf of the federal government, proposed a draft of a reshaped criminal code in 1993. Major proposals include a minimum prison sentence duration of six months (instead of three days under current legislation), authority to judges to suspend sentences up to thirty-six (instead of eighteen) months, and introduction of new sanctions such as day fines and community service orders.

This reform agenda inspired by German reforms of the early 1970s is being debated and may or may not survive the political process. However, community service orders have been introduced as an alternative to execution of custodial sentences in about half of Switzerland's twenty-six cantons. Offenders sentenced to relatively short sentences may, in the cantons that adopted this system on an experimental basis, perform eight hours of community work per day they would otherwise serve in prison. Technically speaking, these offenders are sentenced to unsuspended custody. Since these programs are part of an experiment, they are being evaluated by researchers. In one canton, it has been possible to randomize community service orders with traditional prison sentences (Fichter 1994).

So far, only preliminary evaluation findings have been published (Fichter 1994; von Witzleben 1994), and the scale of use of community service orders is too small for them to be significant sentencing options in quantitative terms. Other sentencing reforms have not been introduced since 1980.

Sanctions and prisons in Switzerland have changed dramatically since 1980, but the changes "just happened"—unplanned, uncoordinated, and widely unanticipated. Besides plans to build new and enlarge existing prisons, no policy has yet been designed to address problems of prison overcrowding. The only option seriously debated has been the abolition of short prison sentences, but more from traditional hostility among German and Swiss criminal law teachers toward these sentences than from concern to find responses to the current crisis.

Any serious effort to reduce the prison population must address the issue of shortening long sentences, particularly in the area of drug offenses. Amendments to the Narcotics Act in 1975, which introduced harsher penalties for drug trafficking, were a major cause of the current crisis (Kuhn 1987). Unfortunately, there are few indications that government agencies or policy makers will turn their attention to these aspects of the problem. The most likely option, therefore, is that problems will get worse until they become really acute. This could happen if the current prison construction programs fail or, more likely, do not bring the relief most policy makers expect.

Sentencing Reform in Sweden
Andrew von Hirsch

Swedish sentencing laws, overhauled in 1988, offer an alternative model to American guidelines for jurisdictions that want to make sentencing fairer and more proportionate. Although responding to the same concerns that underlay American reforms—skepticism about rehabilitative programs, concern about disparities, a wish to make judges more accountable—the Swedish reforms are based on statutory rules, principles, and presumptions rather than on numerical grids. The 1988 changes have affected sentencing processes but have not increased prison populations. This article describes the recent Swedish developments.

The Swedish Court System   Sweden has a three-tier court system. Ordinarily, criminal cases are tried in the district courts (Tingsrätt), where cases are heard by panels consisting of a professional judge acting as chair and a specified number of lay magistrates. For minor crimes, the professional judge hears the case alone. The judge or panel decides all aspects of the case—facts and law, including the sentence. There is no jury.

Either party may appeal to one of six courts of appeals (Hovrätt). Appeals ordinarily are heard by panels of three professional judges and two lay magistrates.

The Supreme Court has discretion to consider further appeals.

Sentencing Practice before the 1988 Reforms   Before enactment of the 1988 sentencing reforms, Sweden had little explicit regulation of sentencing. The Penal Code provided in a brief paragraph that the choice of sanction should depend on sanctions' general preventive effects and on the offender's rehabilitation. The reference to rehabilitation was generally interpreted to apply only to the choice among sanctions—for example, between probation and imprisonment. The mention of general prevention was read as a referring to the gravity of the criminal offense. The effect was a two-track system: usually, sentences were to be imposed according to an informal "tariff," based mainly on the courts' perceptions of the seriousness of the crime. Where the offender's apparent treatment needs so indicated, however, a rehabilitative disposition could be substituted.

Sweden's incarceration rates have been considerably lower than those of England and the United States. Between 1988 and 1993, the number of convicted imprisoned offenders varied between 55 and 65 per 100,000 population (*Statistisk Årsbok* 1995, table 400). By comparison, the English rate was half again higher and the American rate was six to eight times higher.

One reason for this lower rate is that Sweden makes much more extensive use of fines. In 1993, 74 percent of those convicted were fined—largely day fines, based on a proportion of earnings. Probation and conditional sentences (essentially, suspended sentences) made up 12 percent of cases. Prison was the disposition in about 10 percent of cases (*Statistisk Årsbok*, table 395).

Where imprisonment is ordered, moreover, terms tend to be short. In 1993, 44 percent of those admitted to prison received sentences of two months or less, and 62 percent had sentences of five months or less. Sentences of a year or more constituted only 18.3 percent of admissions. Sentences of four years or more were rare— only 2.6 percent (*Statistisk Årsbok*, table 398).

While noncustodial penalties are thus widely relied on, the number of types of such penalties has been kept restricted. The main sanctions have been fines, conditional sentences, and probation, with the latter two imposed much less frequently than fines. Swedes have resisted the proliferation of penalty-types that has been in evidence in the United States and England.

Incentives for Change   Sweden was not suffering from any major prison overcrowding crisis during the period when momentum was developing for the changes eventually embodied in the 1988 legislation. The legislation was not aimed at altering overall incarceration rates.

The most important impetus for change was a widely shared sense that the courts had inadequate legal guidance in making sentencing decisions. Judges understood that in ordinary cases, the seriousness of the crime should be an important determinant. However, the reasoning steps for deciding a sentence were not adequately specified. The law did not spell out, for example, how offense gravity should be assessed, what aggravating or mitigating factors are relevant, or what weight should be given to prior offending.

Another impetus was growing doubts about rehabilitation as the basis for sentencing decisions, as documented in an influential 1977 report published by the National Council on Crime Prevention. Research on treatment indicated low rates of success. Moreover, doubts were raised about the fairness of basing sentencing decisions on treatment, particularly where the result would be a sanction more severe than that suggested by the gravity of the crime.

Penologists thus began to gravitate toward reform legislation that would make it clear that sentences should be proportionate to the gravity of the offense and that would provide clearer guidance to judges. This thinking received momentum from changes in sentencing law in Finland; that country, in 1976, adopted a sentencing statute emphasizing proportionality (Jareborg 1995).

In 1979, the Swedish minister of justice appointed the Committee on Imprisonment to review sentencing policy and recommend statutory changes. The committee, aided by a panel of experts consisting of several of the country's leading penologists, issued its report, *Sanctions for Crimes*, seven years later in 1986 (Fängelsestraffkommittén 1986). As is customary in Sweden, the committee circulated its proposals to courts, prosecutorial and correctional agencies, law faculties, and other interested parties. The proposed provisions received generally favorable comment, and early in 1987, the government decided to support them. They were enacted by Parliament in 1988 as new chapters 29 and 30 of the Penal Code.

Structural Features of the 1988 Reforms   Proportionality is the central principle of the new legislation: punishments are to be proportionate in severity to the gravity of offenses. However, there was consensus that the reform should not establish a pre-scribed list of penalties—that was to remain the province of the courts. The law thus would provide general statutory criteria, not an American-style sentencing grid. The challenge was to set forth statutory criteria that provided useful guidance.

For that purpose, a statutory declaration calling in general terms for proportion-ate sanctions was not enough. The law needed to spell out the conceptual steps involved in setting proportionate sentences.

The key concept is an offense's "penal value"—the seriousness of the criminal conduct (Swedish Penal Code, chaps. 29, 30; Jareborg 1995). Punishments are to be imposed "according to the penal value" of the offense, defined in terms of the harmfulness of the conduct, and the degree of culpability of the actor. The court is first to gauge the "abstract" penal value, that is, the characteristic degree of serious-ness of that type of offense. Next, the court is to make a judgment of the "concrete" penal value—the seriousness of the crime in the particular circumstances. The court should take into account the offense's manner of commission—for example, the extent of threatened violence in a robbery. Mitigating and aggravating factors are to be considered, and the statute provides a list of such factors. Once the con-crete penal value is ascertained, a limited adjustment is permitted for the offender's previous criminal record and a penalty reduction is authorized in certain special circumstances where the deserved sanction seems uncharacteristically onerous—such as ill health or advanced age (chap. 29, sec. 5(6)).

Juvenile offenders are subject to the law but with substantially reduced sanc-tions; offenders under age eighteen are to be imprisoned "only if there are extraor-dinary reasons." What these provisions do is to put some flesh on the bones of the general idea of a proportionate sentence.

The statute also provides guidance on the choice of type of sanction. Imprison-ment is made the appropriate sanction where the crime (robbery, for example) has a high penal value, irrespective of whether the offender has a record. The crime will be deemed to have a high penal value if the applicable statutory minimum is one year or more in prison. However, the court is authorized to go below the applicable minimum on grounds of mitigating circumstances. Imprisonment is also indicated for crimes having a penal value in the intermediate range (say, burglary or the more substantial thefts) if the offender also has a significant criminal record. If the offender has a modest record or none, a conditional sentence (supplemented by a fine) or probation is called for instead. For crimes in the lower seriousness range—routine thefts, for example—day fines are ordinarily to be employed, even where the offender has a significant criminal record.

The statute has, however, one important provision (Swedish Penal Code, chap. 30, sec. 4) that reduces its clarity. Offenders who have been convicted of upper-intermediate offenses—but who have no criminal records or only modest ones—may be sent to prison (instead of receiving a conditional sentence or probation, as the statute's provisions normally suggest) if imprisonment is called for by the "nature of the crime." This provision is designed to permit the imposition of deter-rent penalties for certain special categories of offenses, such as drinking and dri-ving, illegal possession of weapons, and tax fraud. The more-than-deserved penalty in such cases is rationalized on the basis that the conduct in aggregate has unusually severe consequences and seems more than typically amenable to deterrence.

Impact on Judicial Reasoning What is the sentence reform law's impact on the reasoning process of courts? Without a full law-in-action study of sentencing in Swedish courts, which is not available, no definitive answer is possible. However, my conversations with Swedish judges and academic colleagues suggest that there has been a change in the direction of more reasoned decision making. Sentencers seem to be engaging more frequently in the kind of explicit reasoning the statute is designed to promote—that is, discussion of such issues as how great the offense's penal value is, what mitigating or aggravating factors (if any) are present, whether the prior criminal record is sufficient to adjust the penalty, and so forth.

An indication that sentencing is developing more structure concerns the addition of a new noncustodial penalty. In 1989, the Swedish Parliament authorized for the first time the use of community service. Without the sentencing reform law, this would have become simply another sentencing option for courts to use—one that elsewhere is often employed for lesser infractions, in lieu of fines. With the 1988 sentencing law, the new penalty could be better targeted. Community service has been classified as a somewhat severer variant of probation—a sanction that is available only for upper-intermediate-level offenses. The new measure was aimed at reducing imprisonment somewhat, by drawing its recruits from cases on the borderline between ordinary probation and imprisonment. However, this sanction is not invoked sufficiently often to be likely to have a traceable impact on aggregate imprisonment levels.

Impact on Imprisonment What has been the impact of the legislation on the use of imprisonment? It is difficult to say with any confidence, as a formal evaluation has not been conducted. However, a recent essay by a Swedish criminal-law scholar, Nils Jareborg (1995), furnishes some aggregate data. These suggest that the legislation has been neutral in its impact on the aggregate use of imprisonment.

Jareborg compares the use of imprisonment in 1987 (a year before enactment of the law) with that in 1991 (two years after its effective date). In the two periods, the number of convicted persons was about the same—about 150,000 persons. Of these, 16,500 (or about 11.5 percent) were imprisoned in 1988 and about 14,300 (or about 9.5 percent) in 1991. These figures are the persons sent to prison each year. The average prison population is below one-third this amount (because of the short duration of most prison sentences).

The modest decrease appears largely to be accounted for by a 2,700 reduction in the number of persons imprisoned yearly for aggravated drinking and driving. The reduction was occasioned by a bill enacted in 1990 that reduced the level of permissible blood-alcohol content for drivers (from .05 to .02). A note in the preparatory materials for that bill suggested that the normally applicable rules regarding penal value should apply to such convicted drunken drivers, not the "nature of crime" exception authorizing deterrent penalties. The courts took up this suggestion, nearly halving the number of drinking drivers sent to prison.

Parliament, however, objected to this change and later changed the definition of aggravated drinking and driving rules so as to create a stronger presumption favoring imprisonment (at .10 blood alcohol). Judges anticipated this change before actual enactment, and as a result, drinking drivers are being imprisoned at a rate approximating the pre-1990 rate.

Sentencing Changes, 1991–1994   The 1988 sentencing law was enacted under a Social Democratic government that did not have toughness-on-crime as one of its political priorities. In 1991, the Social Democrats lost to a more conservative coalition that did emphasize the issue: the main campaign slogan of the largest coalition party (the Conservatives) was: "They should sit inside so you can go outside."

This newly elected government made it one of its aims to raise the number of incarcerated persons substantially. (Jareborg notes that the aim was to produce a 20 percent increase.) Two of the steps envisioned to achieve this were making parole release rules more restrictive and increasing prison sentences for recidivists (Jareborg 1995, pp. 119–120).

The first change—alteration of the parole rules—was enacted in 1993. For prison sentences of between two months' and two years' duration, automatic release from prison to parole supervision was postponed from one-half to two-thirds of the sentence. The Committee on Imprisonment, which drafted the 1988 reforms, had also proposed such a step, but with a corresponding reduction of the statutory minimums. The new government opposed any reduction in minimums. Jareborg has estimated that this measure would increase prison populations by about 10 to 12 percent, and recent figures show approximately this increase between 1991 and 1995.

The second change, tougher sanctions for recidivists, would have been the most destructive to the structure of the 1988 reforms—as a proportionalist sentencing scheme is designed to emphasize the gravity of the offense, rather than the criminal record. However, the coalition government was defeated in 1994, before being able to put forward legislation. The Social Democrats, who took office again, are not inclined to support such a measure.

The events just described have raised the specter of law-and-order politics— something previously absent in Sweden. The 1988 reforms were passed with widespread support in all parties. The 1991 election was the first in which harsher punishments were a major campaign theme. The justice minister from 1991–1994, Gun Hellsvik, spoke throughout her tenure in ways evocative of some American and English rightwing politicians: themes of toughness-on-crime employed as appeals to popular resentment. It remains to be seen whether Sweden will resist such appeals in the future.

Implications for Other Jurisdictions   Sweden was not facing a major overcrowding crisis when the 1988 reforms were enacted. How useful, then, are its reforms in a place where serious crowding exists?

The Swedish reform relies (as that country's legislation customarily does) on the accompanying preparatory materials: much of the change can be understood only through consultation of those materials. England and the United States permit less reliance on the legislative history, and hence would require the rules to be spelled out more fully in the statute itself.

General principles for sentencing, such as those in the Swedish statute, are not self-executing and require active cooperation and implementation by the courts. In Sweden, cooperation could be expected because of the courts' tradition of giving legislative declarations of purpose considerable weight. The less potentially responsive any given jurisdiction's courts are to such outside policy guidance, the greater will be the difficulty of this route to reform.

FOUR

Race and Sentencing

The cruelest irony of the modern American sentencing reform movement is that diminution of racial discrimination in sentencing was a primary aim and exacerbation of racial disparities is a major result. The aim was to make it less likely that officials would exercise broad unreviewable discretions in ways harmful to minority defendants and offenders. The result has been the establishment of rigid rules and laws that narrow officials' discretion but that also punish minority offenders disproportionately harshly. Racial disparities in the justice system that are unprecedented in American history, and steadily getting worse, are the result.

Many people have forgotten that the civil rights and prisoners' rights movements were major precipitants of the earliest sentencing initiatives. One major objection to indeterminate sentencing was that its broad discretions, justified in the name of individualization of punishment, permitted biased officials to discriminate against minority offenders under circumstances in which no appeals of such decisions were possible. The solution proposed was enactment of revised sentencing laws that limited officials' discretion and narrowed scope for biased decisions (e.g., American Friends Service Committee 1971).

The reformers' solution might have worked had the sentencing policies and politics of the early 1970s continued to the 1990s. In Minnesota, the first state to adopt presumptive sentencing guidelines, the early solution did work. Racial differences in sentences declined and overall severity of punishments did not increase (Knapp 1984).

In the longer term, the use of crime as a central issue in partisan politics and ideological debate has fundamentally changed the legal framework within which sentences are imposed and vastly increased the severity of punishments imposed. Two stark effects of these changes are the American incarceration rate on June 30, 1995, of 403 per 100,000 people, a rate six to ten times higher than that of any other developed Western country, and a combined prison and jail population on June 30, 1995, of 1,600,000, four times higher than in 1972.

Among the harsh penal policies of recent years that have disproportionately affected black Americans, four stand out. First, the "war on drugs" launched in the 1980s gave priority to arrests of street-level drug dealers. These arrests are much easier to make in disadvantaged inner-city areas than elsewhere, and hundreds of thousands of disadvantaged young people were arrested. Arrests of whites for drug crimes in the 1980s fluctuated around 300 per 100,000 throughout the 1980s. The arrest rate for blacks tripled and approached 1,600 per 100,000 by the late 1980s (Blumstein 1993), five times the white rate. Laws mandating lengthy prison terms for drug dealers compounded these developments and are the major reason why racial disparities in prison worsened throughout the 1990s (Tonry, 1995, chap. 3; Mauer and Huling 1995).

Second, particularly in the federal system and to a lesser extent in some states, legislatures enacted laws punishing offenses involving crack cocaine, a substance more commonly used and sold by blacks, much more harshly than offenses involving powder cocaine, a pharmacologically indistinguishable substance more commonly used and sold by whites. The extreme case occurs under the U.S. Sentencing Commission's sentencing guidelines that require that 1 gram of crack cocaine be treated in sentencing as equivalent to 100 grams of powder cocaine. The federal sentencing commission proposed elimination of that bias in 1995, but the U.S. Congress rejected the proposal (U.S. Sentencing Commission 1995).

Third, a general trend toward harsher penalties and more mandatory penalties has also disproportionately affected black offenders. From 1975 through 1994, between 43 and 46 percent of the people arrested each year for murder, robbery, rape, and aggravated assault were black. That stable pattern means that increases in the percentages and absolute numbers of blacks in prison did not occur because of a relative increase in violent crimes by blacks. The overrepresentation of blacks among violent offenders, however, does mean that harsher penalties for violent crimes disproportionately affect black offenders.

Fourth, many rigid modern sentencing systems forbid judges to mitigate sentences because of individuals' personal characteristics, including their education and employment records, their residence patterns and family status, and the effects of a sentence on their dependents. The extreme case is again the federal sentencing guidelines, which provide that age, education and vocational skills, mental and emotional conditions, employment records, family ties and responsibilities, and community ties are "not ordinarily relevant in determining whether a sentence should be outside the applicable [narrow] guideline range."

Such provisions are sometimes justified as an effort to prevent preferment of white, middle-class offenders who are likelier to be well-educated, employed, and in stable families than are other offenders. This was a plausible claim in the late 1970s when the first presumptive guidelines systems were being developed in Minnesota and Pennsylvania, but it was based on a mistaken premise that significant numbers of white middle-class offenders are sentenced in felony cases. The premise is false. Most felony defendants come from disadvantaged backgrounds, lack strong educational and employment records, and a majority are black and Hispanic. In practice, such provisions have harmed minority defendants by preventing judges from mitigating punishments of people who have to some degree overcome disadvantage and whose characteristics give them a greater likelihood than others of putting their lives back together and becoming law-abiding, self-supporting citizens.

The articles in this chapter document the assertions in this introduction. Those by Marc Mauer show that in 1990, one in four black men aged twenty to twenty-nine were in jail or prison or on probation or parole; by 1995, a third of such young men were under criminal justice system control. Articles by Jerome Miller show that in recent years 42 and 56 percent of young black men in Washington and Baltimore were under justice system control. Carl Pope shows that racial disparities in the juvenile system are worse than those in the adult system and less explicable in terms other than discrimination. Articles by Michael Tonry document the increasing racial disproportions in American prisons since 1980 and the role of drug policies in causing that trend. Douglas McDonald and Kenneth Carlson summarize research showing that the federal 100-to-1 crack cocaine policy is the primary cause of sharp racial disparities in federal sentencing.

Young Black Men and the Criminal Justice System
   Marc Mauer

The Sentencing Project's February 1990 report on "Young Black Men and the Criminal Justice System" shocked the nation. The report showed that, on any day, one in four black males aged twenty to twenty-nine is in prison or jail or on probation or parole. The 609,690 young black males under the control of the criminal justice system on one day far exceeded the 436,000 black males of all ages enrolled in higher education on the same day.

The report was featured in more than 700 newspapers and magazines, and it attracted radio and television coverage including the network evening news, the "Geraldo" show, and National Public Radio's "All Things Considered." More significantly, the report has led to a broad range of efforts to reduce the disproportionate impact of the criminal justice system on African-American males.

The American Bar Association's Committee on Minorities in the Criminal Justice System is developing a report on policies and programs that can reduce the criminal justice system's disproportionate impact on minorities.

National organizations, including the National Association of Pretrial Services Agencies and the American Society of Criminology, held annual meeting sessions on strategies to respond to this problem. State agencies and criminal justice reform groups in Connecticut, Michigan, New York, and Virginia have convened public discussions.

The Suffolk County (Long Island), New York Department of Probation convened a working group of county criminal justice officials and community organizations to develop a mentoring program. The program will serve both as a preventive measure for black youth and as a diversion program for young offenders in the criminal justice system.

A prison warden in Missouri, who realized that state furlough screening criteria result in far more white offenders qualifying for furloughs than blacks, is attempting to analyze the reasons for this disparity and to see if any bias exists in the screening device.

State and national policy makers have looked carefully at these problems. Not surprisingly, African-American officials have taken the lead.

The Congressional Black Caucus sponsored sessions at its annual legislative conference to analyze racial disparities in the criminal justice system and to examine links between educational failures and entry into the criminal justice system.

The New York State Black and Puerto Rican Legislative Caucus initiated forums to solicit community suggestions for responding to the large-scale incarceration of black males. The first forum in Harlem attracted 700 people and was broadcast live on radio.

Professionals and organizations not primarily involved in criminal justice have also responded. The *Boston Globe* reported that "campus discussions of black male enrollment have been stimulated by a report released by The Sentencing Project." Educators concerned with school dropouts and declining enrollment in higher education have been discussing a variety of approaches to design curricula and structure schools to meet the needs of black youth more effectively. Other groups, such as "100 Black Men" in Memphis, have begun mentoring programs to provide positive role models for young people in their community.

Criminal justice professionals should be pleased with the widespread interest the report has revived in problems that have long been acute in the criminal justice system. The nature of this interest offers some important lessons.

First, although the reasons for racial disproportion in the system are complex, this has not prevented criminal justice officials from facing these issues. Too often, we hear that the criminal justice system is the "end of the line," the institution that steps in when all else has failed. While there is truth in this, recent actions of criminal justice personnel indicate that many want to try to address these problems.

Second, there has been increased recognition that criminal justice problems cannot be solved in isolation from the larger community. Whether developing mentoring programs or working with the religious community, criminal justice personnel have a potentially deep source of support for working with offenders.

Reaction to the report challenges the myth that the public is uniformly "tough on crime" and has no sympathy for examining the underlying causes of our high national crime rates. Editorials across the country, both conservative and liberal, echoed similar themes. As the Charleston, South Carolina, *Post/Courier* stated, "If the report does nothing else, its horrifying statistics should ignite a national debate on a subject that has become too critical to ignore any longer."

Racial Disparities Getting Worse in U.S. Prisons and Jails
   Michael Tonry

Racial disparities in U.S. prisons and jails have gotten much worse since 1980, according to data compiled by the Bureau of Justice Statistics. In 1991, blacks constituted 54 percent of state prison admissions (up from 42 percent in 1980), 52 percent of persons confined in state prisons (up from 44 percent in 1980), and 48 percent of jail inmates (up from 40 percent in 1980).

Table 4.1. Black, White, and Other Incarceration Rates—1990

Ethnicity	General Population	Prison Population	Jail Population	Prison + Jail Population	Rate per 100,000
White	199,686,000	369,485	206,713	576,198	289
Black	29,986,000	367,122	190,500	557,622	1,860
Other	19,038,000	37,768	8,106	45,874	241
Total	248,710,000	774,375	405,319	1,179,694	474

Sources: Bureau of Justice Statistics, *Correctional Populations in the United States, 1990*, tables 2.1, 2.3, 5.6; U.S. Department of Commerce (1992), *Statistical Abstract of the United States*, table 6.

These stark increases are not associated with increases in commission by black Americans of the violent crimes—murder, rape, robbery, and aggravated assault—that have long resulted in prison sentences.

Among those arrested for serious violent crimes, the percentage of blacks has been stable for nearly twenty years. Blacks constituted 47.5 percent of such arrests in 1976 and 44.1 percent in 1979; the percentage has since fluctuated within those limits. In both 1991 and 1992, blacks constituted 44.8 percent of such arrests.

There is substantial disagreement over the reasons why blacks, who constitute 13 percent of the national population, make up so large a percentage of arrestees. Few deny that bias remains a troubling problem. However, most scholars and officials believe that arrest patterns closely parallel offending patterns, and that disproportionate arrests of blacks for violent crimes are largely the result of disproportionate commission of violent crimes by blacks. Many political activists, however, to the contrary believe that racial bias plays a large role. Wherever the truth lies, the arrest pattern has not changed significantly for twenty years.

This article presents data documenting racial patterns in incarceration. A second article in the June 1994 issue of *Overcrowded Times* demonstrates that the increasing black percentages in prisons and jails are largely a product of the recent "war on drugs" which, for a variety of reasons, has had especially devastating effects on young black males.

Disproportionate Black Incarceration. In 1990, black incarceration rates were six and a half times higher than white rates. Table 4.1 shows the racial compositions of prisons and jails on counting dates in 1990. For every 100,000 black Americans, 1,860 were in confinement. For every 100,000 whites, 289 were in prison or jail. Thus, the chances that a black American was confined were 6.44 times higher (1,860 divided by 289 = 6.44) than the chances that a white was confined.

Racial Arrest Patterns. As table 4.2 shows, arrest patterns by race have changed very little since 1976. The data, drawn from the Federal Bureau of Investigation's annual reports, *Crime in the United States,* summarize arrests for the serious offenses known as "Index Offenses," divided into violent, property, and total crimes. Rates are shown at three-year intervals (except for 1992). In every year, 50 to 54 percent of violent crime arrests are of whites, and 44 to 47 percent are of blacks. In the

Table 4.2. Percentage of Black and White Arrests, Violent and Property Crime, 1976–1991 (three-year intervals) and 1992[a]

Type of Crime	1976		1979		1982		1985		1988		1991		1992	
	White	Black	White	Black	White	Black	White	Black	White	Black	White	Black	White	Black
Violent crime[b]	50.4	47.5	53.7	44.1	51.9	46.7	51.5	47.1	51.7	46.8	53.6	44.8	53.6	44.8
Property crime[c]	67.0	30.9	68.2	29.4	65.5	32.7	67.7	30.3	65.3	32.6	66.4	31.3	65.8	31.8
Total crime index	64.1	33.8	65.3	32.4	62.7	35.6	64.5	33.7	62.4	35.7	63.2	34.6	62.7	35.2

a. Percentages do not equal 100 because "other" arrests of persons recorded as neither black nor white are excluded.
b. Violent crimes include murder, forcible rape, robbery, and aggravated assault.
c. Property crimes include burglary, larceny-theft, motor vehicle theft, and arson.

Sources: Sourcebook of Criminal Justice Statistics (various years); Federal Bureau of Investigation (1993), Crime in the United States 1992, table 43.

most recent years, the black percentage is lower than in earlier years. The patterns for property and total offenses also are essentially stable.

If arrests for serious violent crimes were the principal determinant of incarceration patterns, the percentages of blacks and whites in prison and jail should have fluctuated little since 1976, and in 1991 and 1992 should have declined. As figures 4.1 through 4.3 demonstrate, however, black percentages increased rapidly from the mid-1980s onward.

Racial Confinement Patterns. Figure 4.1 shows racial percentages among people admitted to state and federal prisons from 1960 to 1991. From 1970 to 1984, with minor fluctuations, black and white percentages held stable at 60 to 40. The racial pattern flip-flopped after 1985. The percentages rapidly narrowed, and by 1992, blacks made up 54 percent of prison admissions and whites only 46 percent.

Figures 4.2 and 4.3 show similar racial patterns for prison and jail populations. Relatively stable from 1970 to 1985, the patterns radically changed after 1985. By 1991, more than half of prison inmates and nearly half of jail inmates were black.

Anyone thinking about the data presented in this article must be struck by the enormous differences by race in incarceration rates, by the abrupt increases in black percentages in incarceration populations, and by the absence of relative increases in violent crimes by blacks.

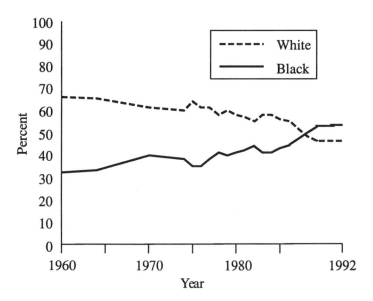

Figure 4.1. Admissions to Federal and State Prisons by Race, 1960–1992. *Note:* Hispanics are included in black and white populations. *Sources:* Bureau of Justice Statistics, various years.

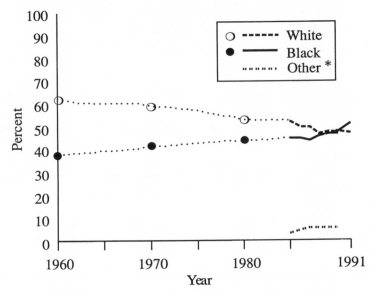

Figure 4.2. Prisoners in State and Federal Prisons on Census Date, by Race, 1960–1991. * Hispanics in many states, Asians, Native Americans. *Sources:* Bureau of Justice Statistics, various years.

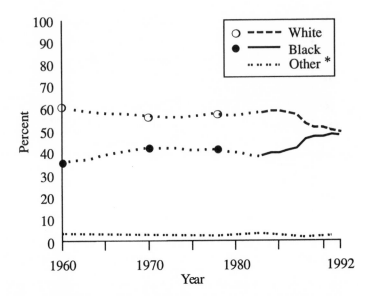

Figure 4.3. Jail Inmates at Mid-Year, by Race, 1960–1992. * White and Black figures for 1988, 1991, and 1992 are estimated: white non-Hispanic, black non-Hispanic, and Hispanic reported; Hispanic racial breakdown assumed to be the same as in 1990 for which racial data were reported. *Sources:* Bureau of Justice Statistics (various years).

Forty-Two Percent of Black D.C. Males, Ages Eighteen to
Thirty-Five, under Criminal Justice System Control
Jerome Miller

On an average day in 1991, 42 percent of black males aged eighteen to thirty-five
in Washington, D.C., were in jail or prison, on probation or parole, out on bond
awaiting disposition of criminal charges, or being sought on an arrest warrant.

The disproportionate numbers of blacks ensnared in the criminal justice system
have been accelerating. Each year a larger percentage of a diminishing pool of
black males is swept into D.C.'s criminal justice system. While the number of black
eighteen- to thirty-five-year-old males in the District *decreased* by 17 percent
between 1980 and 1990, the number of adult arrests *increased* by 60 percent.

In 1980, black males aged eighteen to thirty-five living in the District of Columbia numbered 65,117; by 1990 that number had fallen to 54,305. During the same
period, the number of arrests in the District, of which nearly two-thirds are of black
males aged eighteen to thirty-five, increased from 30,000 to 48,000.

In Jails and Prisons   On an average day in 1991, approximately 12,500 District residents were in District of Columbia or federal correctional facilities. Approximately
7,800 of these were black males aged eighteen to thirty-five. Figure 4.4 shows their

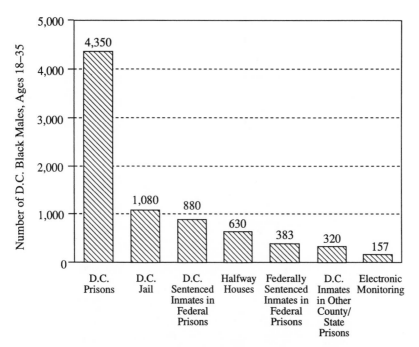

Figure 4.4. Eighteen- to Thirty-Five-Year-Old Black Males from Washington,
D.C., Residing in Correctional Facilities, 1991. *Source:* Miller (1992a).

distribution among correctional facilities. *This represented 15 percent of all black males aged eighteen to thirty-five living in the District of Columbia.*

On Probation and Parole   Approximately 6,000 black males aged eighteen to thirty-five were on probation in D.C. on an average day in 1991. An additional 3,700 were on local parole, and 1,300 were on federal probation or parole. Thus, approximately 11,000 black males aged eighteen to thirty-five were on probation or parole in D.C. on an average day. *This represented 21 percent of all black males aged eighteen to thirty-five living in the District of Columbia.*

Criminally Charged/Arrest Warrants   About 3,500 black males aged eighteen to thirty-five were out on bond awaiting disposition of criminal charges on any given day in 1991. There were also approximately 1,500 outstanding felony warrants and 2,000 outstanding misdemeanor warrants for the arrest of eighteen- to thirty-five-year-old black men. Allowing for duplication and repeat offenders, at least 3,000 of these 7,000 black males were either being sought on warrants for their arrest or were out on bond awaiting disposition of criminal charges on an average day in

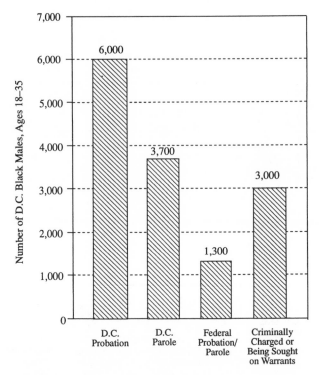

Figure 4.5. Eighteen- to Thirty-Five-Year-Old Black Males on Probation or Parole, Criminally Charged, or Being Sought on Arrest Warrants, 1991. *Source:* Miller (1992a).

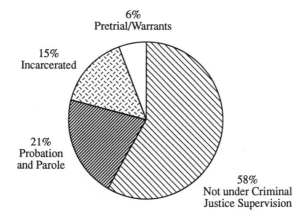

Figure 4.6. Percentage of All Black Males Aged Eighteen to Thirty-Five Living in D.C. under Various Types of Criminal Justice Supervision on Any Given Day, 1991. *Source:* Miller (1992a).

1991. *This represented 6 percent of all black males aged eighteen to thirty-five living in the District of Columbia.*

Figure 4.5 shows numbers of young black males on probation or parole, out on bond, or subject to outstanding warrants.

Figure 4.6 summarizes the percentages of eighteen- to thirty-five-year-old black males living in D.C. under each type of correctional supervision.

The implications are grim. If present policies continue, approximately 75 percent of black males in the District of Columbia will be arrested and jailed or imprisoned at least once during their years between ages eighteen and thirty-five. Given current practices, the lifetime risk of arrest of a black D.C. male approaches 90 percent. One must question any public policy that results in this type of outcome.

Fifty-Six Percent of Young Black Males in Baltimore
under Justice System Control
Jerome Miller

Fifty-six percent of young black males in Baltimore were under criminal justice system supervision on any given day in 1991.

Of Baltimore's 60,715 black males aged eighteen through thirty-five, as shown in figure 4.7, 34,025 were in jail or prison, on probation or parole, out on bond awaiting disposition of criminal charges, or being sought on an arrest warrant.

This extraordinary finding is not unique. A variety of studies in various jurisdictions have shown that black males are grossly disproportionately entangled in the criminal justice system. An earlier study of Washington, D.C., for example, showed that 42 percent of young black males were subject to justice system control.

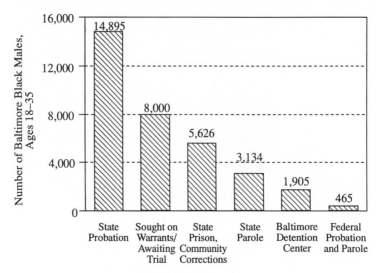

Figure 4.7. Eighteen- to Thirty-Five-Year-Old Baltimore Black Males in the Criminal Justice System. *Source:* Miller (1992b).

These studies reveal a sad truth—that serious unanticipated consequences have befallen black men as a result of America's reliance on law enforcement to deal with a wide range of personal problems and social pathologies. The best example is the "war on drugs."

The War on Drugs Since its inception, the war on drugs has focused mostly on arrest and imprisonment as the strategy of choice. It was foreseeable, and foreseen, that many of the "enemy casualties" would be young, disadvantaged, minority males. Prison admission statistics in many states show that recent drug offenders are more disproportionately black than are other offenders.

The war on drugs has created a rationale for harsher police action, tougher sentences, and, in Baltimore, even suggestions for martial law. The public has been told the drug epidemic has created unprecedented violence in the inner city. However, despite the violence created by and associated with the drug war, Baltimore's criminal justice resources are not being used predominantly for violent offenders but for nonviolent offenders. In 1991, drug arrests constituted one-fourth of all arrests (the largest single arrest category). Eight percent of arrests were for violent crimes, and the balance, 68 percent, were for nondrug, nonviolent offenses.

While blacks represent 60 percent of Baltimore's population, they are significantly overrepresented in arrest statistics. In 1991, of 52,000 arrests, 78 percent were of blacks.

This disparity is most alarming, as figure 4.8 illustrates, when drug arrests are isolated. In 1991, five times as many blacks as whites were arrested for drug offenses.

Available data on juveniles in Baltimore are even bleaker. As figure 4.9 illustrates, black youth accounted for 82 percent of all juvenile arrests while white

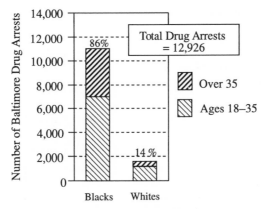

Figure 4.8. Baltimore Drug Arrestees, 1991. *Source:*
Baltimore Police, Commissioner's Office (1992).

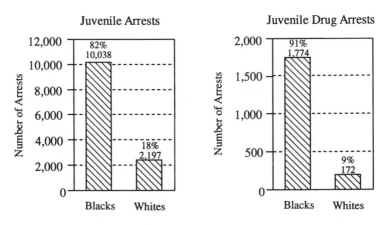

Figure 4.9. Baltimore Juvenile and Drug Arrests, 1991. *Source:* Miller (1992b).

youth accounted for 18 percent. However, when juvenile drug arrests were isolated, white youths accounted for only 9 percent while black youths made up 91 percent.

Figure 4.10 shows that three times more black juveniles were charged with possession and one hundred times more were charged with the sale of drugs in Baltimore than were white youths.

The decision to charge possession is highly subjective. Being labeled a seller rather than a user often determines whether the youth is held in detention, dealt with by juvenile justice agencies, prosecuted as an adult, or released to his or her family. Only fifteen white juveniles were arrested for sale of drugs in 1981. In 1991, only thirteen white juveniles were arrested. By contrast, eighty-six black juveniles

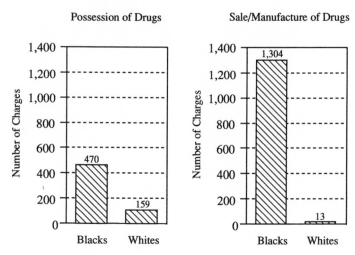

Figure 4.10. Baltimore Juvenile Drug Charges, 1991. *Source:* Baltimore Police, Commissioner's Office (unpublished data provided to author).

were arrested for drug sales in 1981, in contrast to 1991 when 1,304 black youths were arrested on that charge.

Blacks are routinely arrested and imprisoned at rates that far exceed their involvement with drugs or their representation in the general population.

While some might attribute these disparities to alleged greater drug use by blacks, national studies indicate otherwise. The National Institute on Drug Abuse estimates that whites make up 77 percent of all drug users, blacks 15 percent, and Latinos 8 percent. These estimates suggest that drug usage is roughly proportionate to the population of each group in the United States.

Conclusions Current urban policies on crime control need fundamental reexamination. These startling racial disparities are not due to any peculiar incorrigibility, moral deficiency, or inherent violence among black males. They are the predictable consequence of having replaced the social safety net with a dragnet. The political rhetoric of both major political parties, calling for the death penalty, removing "weeds," and demanding ever-longer prison sentences, is a poor substitute for thoughtful policy. Policy makers can no longer conceal the racial implications of public policies that ignore the disintegration of the inner cities while a generation of black men is sent off to the nation's reform schools, jails, prisons, detention centers, and camps.

Drug Policies Increasing Racial Disparities in U.S. Prisons
 Michael Tonry

Law enforcement and sentencing policies associated with the Reagan and Bush administrations' war on drugs are the primary cause of recent worsening in black/white disparities in American jails and prisons. Although black/white proportions among persons arrested for serious violent crimes have held steady for nearly twenty years, by 1991 the percentage of blacks among people admitted to state prisons reached 54 percent (up from 42 percent in 1980). Black percentages among those confined in prisons and jails followed a similar pattern.

The increase in incarceration of blacks results primarily from an interaction between vast increases in arrests of blacks for drug crimes and substantial increases in the severity of sentences imposed for drug crimes. In many jurisdictions, and especially in the federal system, recent laws mandate lengthy prison sentences for drug offenses. Because black arrest rates for drug crimes are up, and drug offenders are likelier to receive long sentences, the effect is to send more blacks to prison and to keep them there longer.

This article presents the arrest and confinement data on which the preceding observations are based. It then considers whether the trend toward increased arrests of blacks for drug crimes results from increases in drug dealing by blacks or from selective enforcement aimed at minority neighborhoods. The evidence is close to compelling that increased black drug arrests are the product of selective enforcement.

Confinement for Drug Crimes   Recent trends in racial patterns of incarceration might be thought of as a mystery. The percentages of blacks among persons arrested for serious violent crimes are stable or declining slightly. In 1976, for example, 47.5 percent of those arrested in the United States for murder, rape, robbery, or aggravated assault were black; 44.8 percent were black in 1991. In between, the black percentage fluctuated between 44 and 47 percent. Why then are incarceration rates for blacks increasing faster than those for whites?

The solution might begin with figure 4.11, which shows the percentages of

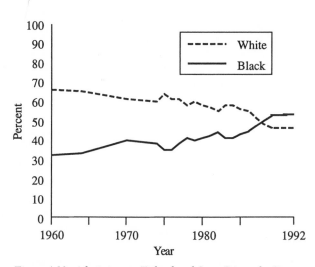

Figure 4.11. Admissions to Federal and State Prisons by Race, 1960–1992. *Note:* Hispanics are included in black and white populations. *Sources:* Bureau of Justice Statistics (various years).

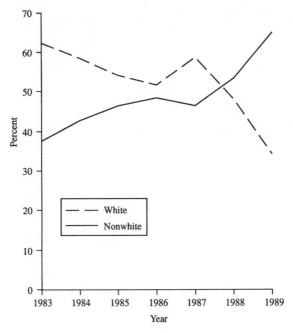

Figure 4.12. Percentage of New Drug Commitments by Race, Virginia, Fiscal Years 1983–1989. *Source*: Austin and McVey (1989).

blacks and whites among persons admitted to state and federal prisons from 1960 to 1992. From 1970 to 1985, albeit with year-to-year fluctuations, whites made up about 60 percent of prison admissions and blacks about 40 percent. In the mid 1980s, however, the period of stability ended and the black percentage began to climb. In the early 1990s, the pattern stabilized again with blacks making up 52 percent of state and federal admissions and whites 47 percent.

Drug crimes made the difference. According to the *1991 Survey of State Prison Inmates* compiled by the U.S. Bureau of Justice Statistics, 25 percent of black inmates in that year had been convicted of drug crimes, compared with 12 percent of white inmates. By contrast, in 1986 only 7 percent of black inmates had been convicted of drug crimes, compared with 8 percent of whites. In other words, the black percentage increased by three and a half times, the white percentage by half.

The racial effects of drug sentencing can be seen even more clearly by looking at experiences in individual states. Figure 4.12 shows white and nonwhite prison commitments for drug crimes in Virginia from 1983 to 1989. Sixty-two percent were of whites in 1983; 38 percent were of nonwhites. By 1989, the percentages had flipped: 65 percent of drug commitments were of nonwhites, 35 percent were of whites.

Figure 4.13 shows prison commitment patterns by race, sex, and offense in 1980 and 1990 in Pennsylvania. The number of black males committed for drug crimes

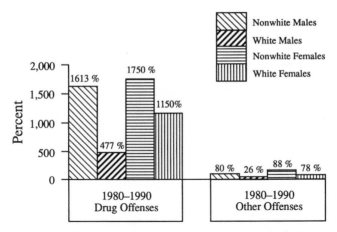

Figure 4.13. Percentage Growth in Prison Commitments in Pennsylvania, by Race, Sex, and Offense, 1980–1990. *Source:* Clark (1992).

grew 1,613 percent, the number of white males by 477 percent. The percentage increase for black females was even larger.

These data demonstrate that sentencing for drug crimes is a major reason why the number of blacks in prison has grown so much faster than the number of whites. That, however, is not the solution to the black incarceration mystery. The next step is to examine the reasons why relatively so many more blacks than whites are imprisoned for drug crimes.

Arrests for Drug Crimes   The principal reason black incarceration for drug crimes is so high is that black arrests for drug crimes are high. Table 4.3 shows FBI drug arrest data for the United States from the Uniform Crime Reports for the years 1976 to 1992. Less than a fourth of those arrested for drug crimes in the 1970s were black. That percentage crept up during the 1980s and by 1989 reached 42 percent.

The data in table 4.3 may not fully depict the changes in drug arrest patterns. Because seven times as many Americans are white as are black, any time that a black proportion increases relative to whites, the absolute increase for blacks is much larger. Thus when racial drug arrest patterns for 1976 and 1989 are compared, the percentage of blacks among arrestees grew from 22 to 42 percent and many would say that the black percentage grew by 20 percent. However, in absolute numbers, between 1976 and 1989, the number of white arrests grew by 70 percent and the number of black arrests grew six times faster, by nearly 450 percent.

This can be seen in figure 4.14, which shows white and nonwhite arrest rates per 100,000 for the same race population from 1965 to 1991. The white rate is basically flat from 1980 onward, fluctuating around 300 per 100,000 with only a modest increase at the height of the drug wars. The black rate, initially somewhat higher, nearly triples after 1980 and by 1989 is five times the white rate.

Table 4.3. U.S. Drug Abuse Violations by Race, 1976–1992

Year	Total Violations	White	White %	Black	Black %
1976	475,209	366,081	77	103,615	22
1977	565,371	434,471	77	122,594	22
1978	592,168	462,728	78	127,277	21
1979	516,142	396,065	77	112,748	22
1980	531,953	401,979	76	125,607	24
1981	584,776	432,556	74	146,858	25
1982	562,390	400,683	71	156,369	28
1983	615,081	423,151	69	185,601	30
1984	560,729	392,904	70	162,979	29
1985	700,009	482,486	69	210,298	30
1986	688,815	463,457	67	219,159	32
1987	809,157	511,278	63	291,177	36
1988	844,300	503,125	60	334,015	40
1989	1,074,345	613,800	57	452,574	42
1990	860,016	503,315	59	349,965	41
1991	763,340	443,596	58	312,997	41
1992	919,561	546,430	59	364,546	40

Sources: Federal Bureau of Investigation (1993), *Sourcebooks of Criminal Justice Statistics—1978–1992*, various tables.

The next step is to ask why drug arrests of blacks increased so much more than arrests of whites. The most obvious possible answers are that police arrest practices are biased against blacks or that blacks are many times likelier than whites to use and sell drugs. Neither "obvious" answer appears to be true. Although I know of no studies of police bias in making drug arrests per se, the leading reviews of the literature on bias in police arrest practices generally conclude that little systematic bias can be demonstrated. Research reviews both by white conservatives (e.g., Wilbanks 1987) and by black liberals (e.g., Mann 1993) reach that conclusion. That many major urban departments are headed by black police executives and that many more police are black than in earlier times makes that conclusion not implausible.

However, there is also little evidence that blacks are much likelier than whites to be drug dealers. We have no representative surveys of drug traffickers, but the next closest thing, surveys of drug users, shows that blacks are no likelier than whites to be drug users. Table 4.4 shows findings for 1990 from the National Institute on

Table 4.4. U.S. Percentage of Drug "Ever Used," by Race, 1990

	Alcohol	Marijuana	Cocaine	Hallucinogens	Heroin
White	85.2	34.2	11.7	8.7	0.7
Black	76.6	31.7	10.0	3.0	1.7
Hispanic	78.6	29.6	11.5	5.2	1.2

Source: Flanagan and Maguire (1992).

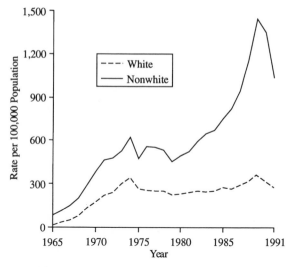

Figure 4.14. Arrest Rates for Drug Offenses, by Race, 1965–1991. *Source:* Blumstein (1993).

Drug Abuse's national household survey on self-reported drug use. Larger percentages of whites than blacks report that they have ever used every illicit substance except heroin, and heroin is no longer a major drug of abuse. If—a not unreasonable assumption—most drug dealers are also drug users, then the proportion of drug dealers among blacks should be little higher than that among whites.

The explanation for racial differences in arrest patterns apparently lies neither in police bias nor in vastly greater participation by blacks in drug use (and, by implication, in drug trafficking). The likeliest explanation lies in the politics of crime control and in the conscious drug control policies of federal and state officials. Arrests are a dramatic demonstration that "something is being done," and policy makers encouraged police to make more arrests. Arrests are much easier to make in deteriorated, often minority, inner-city neighborhoods than in stable working- or middle-class neighborhoods. Thus policy makers' insistence on more arrests resulted in vastly more arrests of poor inner-city blacks.

Penalties for Drug Crimes   Police arrest policies, however, are only part of the solution to the black incarceration mystery. Crime control politics have also resulted in penalty increases that have exacerbated the effects of drug arrest policies on blacks. Mandatory penalties for drug crimes, and increases in the lengths of minimum sentences for drug crimes, are likely to have disproportionate effect on blacks if blacks are disproportionately likely to be arrested for drug crimes, and that is what has happened.

The extreme instance is the federal "100-to-1" rule, under which drug penalties vary directly with the quantity involved, and quantities of crack cocaine are multiplied by 100 to determine applicable penalties. Every gram of crack cocaine, a substance commonly used and sold by members of minority groups, is equivalent to 100 grams of powder cocaine, a pharmacologically indistinguishable substance commonly used and sold by whites. Although a number of federal district courts have declared the 100-to-1 rule unconstitutional because of its disparate impact on blacks, and the Minnesota Supreme Court declared an equivalent state law unconstitutional, every federal court of appeals that has considered the law has upheld it.

The effect of harsh drug penalties on minority drug offenders is illustrated by a report on racial disparities in federal sentencing recently released by the U.S. Bureau of Justice Statistics (McDonald and Carlson 1993). There were three major findings. First, prior to full operation of the current federal sentencing guidelines (which track the mandatory penalty laws), "white, black, and Hispanic offenders received similar sentences, on average, in federal district courts." Second, during a subsequent eighteen-month period, whites (72 percent) were much less likely than blacks (78 percent) or Hispanics (85 percent) to be sentenced to prison. Third, *"on average, black offenders sentenced to prison during this period had imposed sentences that were 41 percent longer than for whites (21 months longer)"* (McDonald and Carlson 1993, p. 4). The second and third findings, the report's authors concluded, were primarily the result of the operation of mandatory sentencing laws for drug crimes.

The mystery of disproportionate and increasing black incarceration turns out not to have a tidy solution. Increases in serious violent crimes by blacks are not the answer. Nor is disproportionate black participation in drug use. Nor apparently is police bias. The villains, if there be any, are either the elected officials who adopted policies calling for greatly increased arrests and harsher sentencing policies or the electorate whose will elected officials were implementing. Whether politicians' law-and-order electoral appeals heightened voters' fears and anger about crime, which in turn produced support for harsh policies, or voters' outrage about crime led politicians to focus on the issue and to adopt harsh policies, is a subject beyond the scope of this article. What is clear, however, is that black Americans have uniquely borne the burdens of the harsh drug control policies of recent years.

Drug Policies Causing Racial and Ethnic Differences
in Federal Sentencing
  Douglas McDonald and Kenneth E. Carlson

Believing that unwanted disparities were rampant in the federal criminal justice system a decade ago, Congress passed the Sentencing Reform Act (SRA) of 1984 and ushered in dramatic changes in sentencing and paroling procedures. But an odd thing happened: racial and ethnic differences in imprisonment sentences grew wider *after* implementation of guidelines. Sentences imposed on blacks grew increasingly more severe, on average, relative to whites and Hispanics. The average maximum prison sentence for a black federal offender was 41 percent longer than the white average. Very little of that difference appears to result from bias by judges. The black/white differences result mostly from sentencing for cocaine trafficking. Most (two-thirds) of the difference results from a congressional decision to impose much harsher penalties for crack cocaine offenses than for powder cocaine offenses. One-third of the difference results from the U.S. Sentencing Commission's decision to raise sentences for many cocaine offenses even higher than the minimum levels mandated by Congress.

At the request of the U.S. Department of Justice's Bureau of Justice Statistics, we undertook an analysis of all sentencing decisions in all federal district courts between January 1, 1986 and June 30, 1990 (McDonald and Carlson 1993). This period spanned the months immediately before and after full implementation of the guidelines.

Offenders convicted of crimes committed after November 1, 1987, were subject to radically different sentencing procedures. Judges were required to follow guidelines that specify a relatively narrow range of permitted imprisonment sentences. The possibility of parole release was abolished. And offenders would serve at least 85 percent of the declared sentences, assuming maximum time off for good behavior.

Our study compared decisions in all "old-law" cases brought to sentencing during 1986–1988 and all "new-law" (or guidelines) cases sentenced between January 20, 1989, and June 30, 1990.

We analyzed information in offenders' presentence reports and court records information in the Federal Sentencing and Supervision Information System. These files provided data on charges and offense characteristics on offenders and on case processing, such as the judicial district in which the case was prosecuted and the method of conviction (trial or plea).

Racial and Ethnic Differences in Sentences before and after Implementation of Guidelines During 1986–1988, in old-law cases, white, black, and Hispanic offenders in district courts received similar sentences, on average. Prison sentences were imposed on 54 percent of both white and black offenders, and the lengths of these sentences were similar: a maximum of fifty months, on average, for whites, and fifty-three months for blacks. Hispanics were more frequently imprisoned (69 percent), but their sentences were of similar length: fifty-two months.

After full implementation of the guidelines, the differences grew larger. Between January 20, 1989, and June 30, 1990, 78 percent of all blacks and 85 percent of all Hispanics convicted of federal crimes in guidelines cases were sentenced to prison, compared with 72 percent of whites. Blacks' maximum sentences were 41 percent longer than whites': seventy-one months, on average, compared with fifty months for whites and forty-eight months for Hispanics. This pattern of sentencing Hispanics occurred because large numbers were convicted of immigration law offenses, for which incarceration rates were high but for which sentences were relatively short. Because the large white/black difference is less readily explained, we focused attention on it.

Why Did These Differences Emerge? We explored several possible explanations of this growing gap in sentences imposed on whites and blacks. First, judges may have begun to discriminate against black offenders in guidelines cases more frequently or more systematically than before. Second, the guidelines themselves might have produced this gap by giving greater weight to characteristics for which blacks score worse than whites. Third, white and black offenders appearing in federal courts after implementation of the guidelines might have become more dissimilar, compared with the prereform period. This could have resulted from changes in charging practices or the frequency with which state and local prosecutors referred white and black defendants to federal prosecutors. Finally, sentencing differences might have stemmed from changes in federal law not related to the sentencing guidelines—such as Congress's enactment of mandatory minimum sentences for some crimes. To test these various hypotheses, we conducted a number of analyses, using a variety of statistical techniques.

What We Found   White/black differences were concentrated in a few types of crime, especially drug trafficking, bank robbery, and weapons offenses. Between January 1989 and June 1990, the average imprisonment sentences for those convicted of drug trafficking under the guidelines were 36 percent longer for blacks than whites (or 96 versus 70 months). Among bank robbers, blacks' sentences were 16 percent longer, on average (or 104 months versus 90 months). For those convicted of federal weapons offenses, blacks' sentences averaged 55 percent longer (56 months versus 36 months). There were also differences in the proportions of blacks and whites sentenced to prison for drug trafficking (96 percent of blacks, 92 percent of whites) and weapons offenses (91 percent of blacks, 78 percent of whites). For other crimes, differences were either much smaller, or average sentences for blacks were *less* severe than for whites.[1]

Drugs. Cocaine trafficking is by far the most prevalent federal drug crime. Among those convicted of this crime, there were substantial black/white differences in sentencing. Blacks were sent to prison more often (96.7 percent; 94.1 percent for whites), and their average sentences were substantially longer: 102 months versus 74 months—a 37 percent difference.

This does not indicate invidious discrimination by judges, however. Rather, we discovered a significant difference in the frequency with which whites and blacks were convicted of trafficking in crack as opposed to powdered cocaine, which accounts for much of the apparent racial difference in sentencing.

In the Anti-Drug Abuse Act of 1986, Congress revised federal law to distinguish between crack and powdered cocaine. Mandatory minimum sentences were prescribed for both types of cocaine trafficking, but the thresholds for crack were much lower than for powdered cocaine. Persons convicted of selling fifty grams or more of crack—or possessing that much with the intent to sell—must be given no less than ten years in prison, or no less than twenty if previously convicted of a drug crime. Those convicted of selling five or more grams, but less than fifty, face minimum sentences of five years, or ten for second offenders. These punishments are

---

1. For example, white offenders imprisoned for fraudulent property offenses were sentenced to an average of 15.2 months, compared with 12.5 months for black offenders and 9.1 months for Hispanics. White offenders imprisoned for larceny were sentenced to an average of 19.9 months, compared with 15.8 months for black offenders and 14.1 months for Hispanics. Specific characteristics of property offenses—particularly the economic value of the loss—explain most of these differences. The information presented here does not necessarily represent policies or positions of the U.S. Department of Justice or the Bureau of Justice Statistics, which supported the research summarized here.

the same as those prescribed for persons convicted of selling one hundred times these amounts of powdered cocaine.

To accommodate the principle of proportionality, the U.S. Sentencing Commission extended this differentiation. It defined break points between the quantity thresholds specified by Congress and created guidelines ranges above the thresholds for both types of cocaine. At each level, the guidelines sentence for crack is the same as for one hundred times the weight of cocaine powder.

Blacks and whites convicted of trafficking in powdered cocaine were sentenced approximately the same. Nearly all were sent to prison (95 percent of whites, 96 percent of blacks), and for about the same lengths of time (73 and 71 months, respectively). Furthermore, the proportions of whites and blacks going to prison following a conviction for crack trafficking was identical (99 percent), and blacks received only slightly longer sentences, on average, than whites: 140 months, compared with 130 months. Statistical analyses showed that these differences resulted from characteristics other than the offender's race and did not, consequently, indicate racial bias on the part of judges. *However, the majority (83 percent) of those convicted of this harshly punished crime were black.* Only 4 percent of all white cocaine traffickers were convicted of crack offenses, compared with 27 percent of all black cocaine traffickers. This largely explains the overall difference in sentences imposed on black and white offenders convicted of drug offenses in the federal courts.

We did not attempt to find out why blacks were more frequently convicted of crack trafficking than whites, but surveys have found that crack is more likely to be used by blacks than whites. This may be not because blacks strongly prefer crack but because crack is marketed more heavily in neighborhoods with minority populations (Lillie-Blanton, Anthony, and Schuster 1993). Another more worrisome possibility is that black offenders are likelier than whites to be referred to federal prosecutors by state and local authorities (Berk and Campbell 1993).

We conducted a simulation to learn how much of the racial disparity resulted from the federal sentencing commission's decision to set sentences for some drug offenses higher than the mandatory minimums. We assumed that judges conformed strictly to the mandatory minimum sentencing requirements but did not increase punishments to comply with the guidelines. Had this occurred, the difference in white and black sentences would have narrowed. For selling any kind of cocaine, blacks' average sentences would have been 11 percent longer than whites—a difference about one-third of that actually observed.

Bank Robbery and Weapons Offenses. We also found white/black sentencing differences for bank robbery, federal weapons offenses, and drug offenses other than cocaine trafficking. Among those convicted of bank robbery, blacks' sentences averaged 16 percent longer (104 months versus 90 months). For weapons offenses, blacks' sentences averaged 55 percent longer (56 months versus 36 months). The proportions sentenced to prison for weapons offenses also differed: 91 percent of blacks convicted of weapons offenses, compared with 78 percent of whites.

To explore whether these differences indicated racial discrimination by judges, we constructed statistical models of sentencing decisions that included a large number of characteristics of the offenders, their crimes, and the processing of their cases that might affect sentences. This gave us a way to evaluate whether the observed racial differences resulted from consideration of characteristics that were correlated with the offenders' race but legitimately considered, such as criminal records, or the offenders' role in the offense. Most—but not all—of these observed racial differences could be attributed to considerations that legitimately influence the severity of sentences.

The residual black and white differences that were not accounted for in our statistical analyses resulted from some factor for which data were not available. It is possible that they signal discrimination against blacks, but we think it improbable. The unexplained differences concern specific types of bank robberies and weapons offenses. Were federal judges acting with discriminatory intent, a more generalized gap in sentencing would be evident across a wider spectrum of cases.

Implications for Policy   Our study did not propose policy changes. However, the analysis does suggest a number of questions for policy makers. First, should the fact that imprisonment sentences fall more heavily on blacks because they are more frequently convicted of trafficking in crack cocaine be considered a "disparity"? Second, even if one concludes that this is not a disparity, does the fact of heavier sentences falling on blacks impose a special burden on justifying the difference drawn in law between crack and powdered cocaine? Should laws that have racial effects in their applications, even if unintended, be subject to some special scrutiny? Third, even if trafficking in crack should be sentenced more severely than powdered cocaine, is the 100-to-1 ratio justifiable? Fourth, if crack is indeed the more harmful form of cocaine, it might be instructive to learn whether the 100-fold increase in sanctions produces any appreciable increase in deterring the sale or possession of crack.

Racial Disparities in the Juvenile Justice System
Carl Pope

Racial disparities adverse to minorities, that cannot be explained by differences in offending patterns, are far more evident in the juvenile justice system than in the adult system. The U.S. Congress directed all states to examine racial disproportions in their juvenile justice systems. Of forty-two states that have released reports on secure detention facilities, forty-one found minority overrepresentation. Of thirteen states releasing reports on other phases of the juvenile system, all thirteen found evidence of minority overrepresentation. This article summarizes findings from those reports and from a major federally funded survey of the scholarly literature on juvenile disparities. It also reports on subsequent research findings. As part of a larger project funded by the federal Office of Juvenile Justice and Delinquency Prevention (OJJDP), W. Feyerherm and I (Pope and Feyerherm 1993) examined two decades of research on race and the juvenile justice system. Approximately two-thirds of that literature showed that minority youth were disparately treated within selected juvenile justice systems. These studies found disparately severe treatment of minorities even after relevant variables were taken into account by means of statistical controls.

Among the more pronounced findings:

1. The preponderance of studies revealed both direct and indirect race effects or a mixed pattern (racial effects present at some stages but not others).
2. Studies finding evidence of racial disparities were generally no less sophisticated methodologically than studies that found no such evidence, nor were the data of lesser quality.
3. Racial disparities can occur at any stage of juvenile justice processing. In some studies, disparate treatment by race was noted throughout the system; in others it occurred at selected points such as intake or detention.
4. Small initial racial differences can accumulate and become more pronounced as minority youth are processed deeper into the juvenile justice system. In particular, statewide data from California and Florida illustrated this pattern.

Findings indicated that racial disparities were not limited to any one stage but instead concerned all stages (e.g., intake, detention, adjudication, disposition), leading us to conclude, "there is substantial support for the statement that there are race effects in operation within the juvenile justice system, both direct and indirect in nature" (Pope and Feyerherm 1990, p. 335).

Recent Research Findings   Since that review was completed, additional research has substantiated its conclusions. For example, the May 1994 issue of the *Journal of Research in Crime and Delinquency*, a special issue on race and punishment, contains six articles, four of which examine race effects in the adult and juvenile justice systems. Conley (1994) presents a qualitative analysis of minority overrepresentation in the juvenile justice system of a western state. Focusing on police encounters with youth of color, she used observations, interviews, and focus groups to document variations across minority groups and community settings. Minority youth were often found to be more at risk depending on how police assessed gang affiliations within specific communities.

In the same issue, Wordes, Bynum, and Corley (1994) examine detention differences in processing minority and white youth in five Michigan counties. Using a stratified random sample, they extracted and coded information from agency case records. Their findings suggest: "Across all analyses, youth who were African American or Latino were consistently more likely to be placed in secure detention. This was observed in the detention practices of both the police and the courts" (Wordes, Bynum, and Corley 1994, p. 162). Minority youth were more at risk of being detained even when statistical controls to take account of relevant legal variables were used.

A forthcoming edited book, *Minorities in the Juvenile Justice System* (Kempf-Leonard, Pope, and Feyerherm [1995]), contains seven original studies on this subject. These studies use quantitative and qualitative techniques to examine disparities among black, Hispanic, and American Indian youth, compared with whites, and find that minority status affects processing decisions in various ways. For example, in one of the few studies examining processing differences for white and American Indian youth, Poupart (1995) reports greater disparities in one northern Wisconsin county than in others. In that county, American Indian youth, compared with white counterparts, are substantially more likely to have their cases referred to intake (61 percent versus 37 percent). Similarly, while the absolute differences are small, American Indian youth experienced disparate outcomes at detention, filing of a petition, and disposition. Poupart concludes (1995, p. 28), "[T]he overall trend of this analysis suggests that American Indians are at a disadvantage in the juvenile justice system in the county examined."

Wordes and Bynum (1995) employ quantitative and qualitative methods to examine differential treatment of youth of color by law enforcement agencies in selected Michigan counties. Their analysis suggests that law enforcement referral and custody decisions for felony, misdemeanor, and status cases varied by race and gender.

In sum, this volume reports findings documenting the disproportionate processing of minority youth in jurisdictions in California, Florida, Michigan, Minnesota, Pennsylvania, Wisconsin, and Washington.

Available evidence strongly suggests that minority youth are at greater risk of being held in secure facilities and receiving more severe outcomes than white youth, and that action should be taken to further address and alleviate race disparities.

The National Agenda   In January 1989, the National Coalition of Juvenile Justice Advisory Groups produced a report, A *Delicate Balance*, which was presented to the president, the Congress, and the administrator of OJJDP. This report identified the problems faced by minority youth within the juvenile justice system, as well as their overrepresentation in secure facilities. It urged that Congress identify this problem as a priority issue.

Congress responded by amending the Delinquency Prevention Act to deal with the overrepresentation of minority youth in the juvenile justice system. In Phase I, states are asked to demonstrate whether minority youth are overrepresented in secure facilities with regard to their population base. If such overrepresentation is found, the states are asked to take steps to account for it. Phase II contemplates examinations of other stages in juvenile processing (e.g., intake, detention, adjudication, disposition). This often involved additional data collection and development of policies and programs designed to reduce minority overrepresentation.

As of 1992, forty-two states and territories had complied with the Phase I requirement. Only Vermont reported no apparent overrepresentation of minority youth. In general, the Phase I data indicated that overrepresentation of minority youth in secure facilities is a national problem but that the degree of overrepresentation and its location varies by jurisdiction.

Moreover, the degree of overrepresentation varies by minority group. Black youth generally have the highest index of overrepresentation in secure facilities.

The degree of overrepresentation typically increases as youth penetrate further into the juvenile justice system. Overrepresentation tends to be lowest at arrest, with increasing levels of overrepresentation progressing toward secure confinement or transfer to adult court.

With regard to the Phase II mandate, three of the earliest state studies of juvenile processing found racial differences to be prominent. Bishop and Frazier (1990) used statewide data over a three-year period to examine case processing through Florida's juvenile justice system. Being nonwhite made a difference with regard to outcome decisions. Disparities existed for filing of a petition, secure detention, commitment to an institution, and transfers to adult court.

Lockhart et al. (1990) examined racial disparity within Georgia's juvenile justice system through the analysis of case records from 150 counties. The severity of the current charge and the extent of prior contact with the juvenile system were major determinants of outcome. Black youths, compared with white youths, tended to have more prior contacts and to be arrested for more severe offenses. These results pointed to the possibility that offense and prior record may not be legally neutral factors as some have argued. If bias influences these decisions, then race differences will be accentuated throughout the system.

Kempf, Decker, and Bing (1990) examined the processing of minority youth through Missouri's juvenile justice system for both urban and rural courts. In the urban courts, black youth were more likely than their white counterparts to be held in detention and to be referred for felony offenses. Parental factors, such as whether the youth resided in an intact or nonintact home and whether parents provided support for the youth, also affected outcomes. For rural courts, however, black youth received more severe outcomes at the disposition stage; they were more likely than white youth to be removed from their homes.

The Missouri study found glaring differences between urban and rural courts. Two different types of juvenile court appear to be operating—a legalistic court located in urban areas and a more traditional pre-Gault court in rural areas, each providing disparate treatment to black youths.

Approximately thirteen states have completed Phase II research efforts and all have found either direct or indirect race effects. Although there are variations between jurisdictions, all have found that minority youth are at risk and that race makes a difference in juvenile justice processing. One of the most recent Phase II studies from Michigan used multiple levels of data collection (case files, interviews, and observations) to examine race and gender differences in juvenile processing across counties (Bynum, Wordes, and Corley 1993). Serious overrepresentation was found for youth of color and females (as status offenders). Moreover, being black or female placed these youth at serious risk: "[R]ace and gender had an impact on law enforcement and juvenile court decisions. Consistent findings were observed in the decision to refer to court and the decision to secure detain juvenile offenders. In some cases the effect was direct while in other situations the influence was indirect through factors such as the family and social situation of the youth" (Bynum, Wordes, and Corley 1993, p. 135).

Conclusion   Concerns regarding fairness within juvenile justice systems are not unwarranted. Minority youth, especially black youth, are often treated more severely when compared with comparable white youths. Depending upon the jurisdiction, it is not uncommon to find that minority youth are more likely to be held at intake, be detained prior to adjudication, have petitions filed, be adjudicated delinquent, be placed in secure confinement facilities, and be transferred to adult court.

It is not surprising that local, state, and national efforts are increasingly focusing on the problem of minority overrepresentation within juvenile justice systems and secure facilities. As the National Coalition of Juvenile Justice Advisory Groups stated in its most recent annual report to Congress: "The 1992 reauthorization of the Act elevated the focus on minority over-representation to one of the four Act mandates, and this report represents our commitment to address the serious problem of the over-representation of children of color in America's juvenile justice system as a major priority."

One in Three Young Black Men Is Ensnared in the Justice System
Marc Mauer and Tracy Huling

One in three young black men aged twenty to twenty-nine is under criminal justice supervision on any given day—in prison or jail, on probation or parole. The cost to taxpayers for control of these 827,440 young black men is about $6 billion a year. And although the absolute numbers of women under control are much smaller, black women have experienced the greatest increase in criminal justice supervision of all demographic groups—up by 78 percent between 1989 and 1994.

In 1990, the Sentencing Project, a nonpartisan public interest organization in Washington, D.C., released a report showing that 23 percent of black men in the age group twenty to twenty-nine were in prison or jail, on probation or parole. That report received extensive national attention and helped generate concern, discussion, and activity on the part of policy makers, community organizations, and criminal justice professionals.

Many of the factors contributing to high rates of criminal justice control for black men remain unchanged or have worsened in the last five years, and the extraordinary one-in-three social statistic (32.2 percent) is the result. Public policies ostensibly designed to control crime and drug abuse have exacerbated racial disparities while having little impact on the problems they were supposed to address.

Drug policies constitute the single most significant factor contributing to the rise in criminal justice populations in recent years, with the number of incarcerated drug offenders having risen by 510 percent from 1983 to 1993. A 1991 census of state prisoners provides data on racial patterns of confinement for drug crimes. As table 4.5 shows, nearly three times as many black non-Hispanic men were in prison for drug crimes as white non-Hispanic men. The ratio of black-to-white female drug prisoners was 2 to 1.

Between 1986 and 1991, the number of minority prisoners incarcerated for drug offenses increased much faster than the number of white prisoners. The percentage increases for black women (828 percent) and men (429 percent) far outpaced those for white women (241 percent) and men (106 percent).

All of the racial contrasts would be worse today than in 1991 because the numbers both of prisoners and of drug offenders in prison have continued to grow rapidly.

While arrests of blacks for violent crimes—45 percent of such arrests nationally—are disproportionate to their share of the population, this proportion has not changed significantly for twenty years. For drug offenses, though, the black proportion increased from 24 percent in 1980 to 39 percent in 1993, well above the black proportion of drug users nationally.

Table 4.5. State Prisoners Incarcerated for Drug Offenses by Race, Ethnicity, and Sex, 1986 and 1991

	1986		1991		% Increase	
	Male	Female	Male	Female	Male	Female
White (non-Hispanic)	12,868	969	26,452	3,300	106%	241%
Black (non-Hispanic)	13,974	667	73,932	6,193	429%	828%
Hispanic	8,484	664	35,965	2,843	324%	328%
Other	604	70	1,323	297	119%	324%
Total	35,930	2,370	137,672	12,633	283%	433%

Source: Bureau of Justice Statistics. State prisoner data for 1986 and 1991.

Racial disproportions are worst for drug possession offenses. Figure 4.15 shows that blacks make up 12 to 13 percent of the national population and of monthly drug users but 35 percent of drug possession arrestees, 55 percent of convictions, and 74 percent of prisoners sentenced for drug possession.

Overrepresentation of Black Men   While debate will continue on the degree to which the criminal justice system overall contributes to racial disparities, there is increasing evidence that the policies and practices of the "war on drugs" have been an unmitigated disaster for young blacks and members of other minorities. Whether these policies were consciously or unconsciously designed to incarcerate

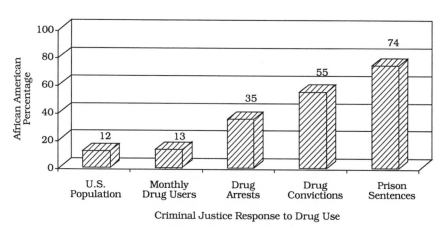

Figure 4.15. African Americans and Drug Possession. *Note:* Data are for 1992 or 1993 depending on the most recent available figures. *Sources:* National Institute on Drug Abuse; Uniform Crime Reports; Bureau of Justice Statistics Reports on Felony Sentencing and Corrections for 1992.

more members of minority groups is a question that may be debated. Those policy choices have, however, not only failed to reduce drug trafficking or use but have seriously eroded the life prospects of disadvantaged minority Americans.

*Increased Arrests.*Arrest policies since the early 1980s have disproportionately affected black and other minority citizens through greatly increased numbers of drug arrests and through an increased proportion of minority drug arrests. Drug arrests increased from 471,000 in 1980 to 1,247,000 by 1989. As the number of arrests grew, so did the proportion of blacks among arrestees, from 24 percent of all drug arrests in 1980 to 39 percent by 1993.

Some people might contend that blacks are arrested in larger numbers because of their higher rates of drug use and sales, but there is no reliable evidence that this is true. Data are not available on the overall racial composition of drug sellers, but reasonably good data are available on drug use and possession through the annual household surveys of the U.S. Department of Health and Human Services. The most recent survey reveals that blacks comprise 13 percent of monthly drug users, compared with their 1993 arrest proportion of 39 percent (U.S. Department of Health and Human Services 1994). Even if only arrests for drug possession, which should be reflective of drug use, are considered, blacks still constitute 34.7 percent of such arrests. Although the drug use surveys, like all surveys, have limitations, the disparity between drug use patterns and drug possession arrests is so large that it clearly points to disproportionate arrest practices.

Prosecution Policies. Prosecutors' decisions can worsen the disproportionate impact of drug policies on minorities. A recent survey of prosecutions for crack cocaine offenses conducted by the *Los Angeles Times* revealed that not a single white offender had been convicted of a crack cocaine offense in the federal courts serving the Los Angeles metropolitan area since 1986, even though a majority of crack users are white (Weikel 1995). During the same period, though, hundreds of white crack traffickers were prosecuted in state courts. While federal prosecutors contend that they target high-level traffickers, the *Times* analysis found that many blacks charged in federal court were low-level dealers or accomplices.

The consequences of these prosecutorial practices are quite serious, since federal mandatory sentencing laws require five- and ten-year minimums even for first offenders. The *Times* study found that whites charged with crack offenses and prosecuted in California state courts receive sentences as much as eight years less than do blacks in federal courts.

Sentencing Policies. Recent changes in sentencing policies have also disproportionately affected blacks. The passage of a new generation of mandatory minimum sentencing statutes, now in place in all states and the federal system, has led to dramatic increases in the number of incarcerated drug offenders.

The effects of these policies can be seen in several ways. First, the chance that a person arrested for a drug offense will be incarcerated was more than five times higher in 1992 (104 per 100,000) than in 1980 (19 per 100,000), an increase far greater than for any other offense (Beck and Brien 1995). This has led to a 510 percent increase in the number of incarcerated drug offenders between 1983 and 1993, with one in four inmates now serving time or awaiting trial for a drug offense.

The full effects of these policies have yet to be seen, since many of the mandatory sentences began to be applied in large numbers only in the late 1980s. In state prison systems, therefore, average time served can be expected to rise in the years ahead due to mandatory sentencing and other harsh policies. In the federal system, the effects of these changes are already being felt.

Young Men and the Drug Trade Even though increasingly larger numbers of black men have come under justice system supervision as a result of drug offenses, the nature of the drug distribution process and the recruitment of individuals involved in it remain poorly understood. There are, however, solid indications that many drug sellers are not incorrigible, antisocial outlaws, but disadvantaged young people trying to lead ordinary lives.

Several recent studies provide insight into the lives of young men who become involved in the drug trade. A 1990 study by Peter Reuter and colleagues at RAND examined the criminal histories and demographic characteristics of groups of young black men arrested for drug distribution in Washington, D.C. (Reuter, Mac-Coun, and Murphy 1990). The researchers documented the vast extent to which drug dealing has become a source of income for this group, with fully one-sixth of the black men born in 1967 having an arrest for drug distribution by the age of twenty, and projections of one-quarter having an arrest by the age of twenty-nine. The study found that about two-thirds of offenders were employed at the time of arrest, primarily at low-wage jobs with a median income of $800 a month. Thus, drug dealing was a type of "moonlighting" for some of these young men.

Research by John Hagedorn on black and Latino gangs and drug dealing in Milwaukee found great variation both in the extent to which gang members were involved in drug dealing and in their orientation toward conventional lifestyles. While a small proportion were committed to drug dealing as a career, the majority "were not firmly committed to the drug economy." The main characteristics they shared were working regularly at legitimate jobs, with occasional drug dealing; conventional aspirations toward economic security; and conventional ethical beliefs about the immorality of drug dealing, even while justifying their drug sales as necessary for survival (Hagedorn 1994).

The RAND researchers found that despite the actual and perceived risks of drug dealing in Washington—the chances of arrest or physical harm being significant— "such risks failed to deter substantial numbers of young males from participating in the trade." They speculate that the prospects of immediate rewards combined with adolescents' lesser concern for physical harm and their future prospects make drug selling very appealing. They conclude, "The prospects for raising actual and perceived risks enough to make for markedly more deterrence through heavier enforcement against sellers do not appear promising."

Family and Community The compounding adverse effects of parental arrest and incarceration on children, particularly if that parent is the primary caretaker, are well documented, and include traumatic stress, loss of self-confidence, aggression, withdrawal, depression, gang activity, and interpersonal violence (Johnston 1993).

As more and more inner-city children lose not only their fathers but their mothers, most often the primary caretakers, to the justice system, their own risks for future involvement in crime and incarceration increase dramatically.

The high rate of incarceration of black men raises concerns about its effects not only on the individuals who are incarcerated but on their families and communities. As increasing numbers of young black men are arrested and incarcerated, their life prospects are seriously diminished. Their possibilities for gainful employment are reduced, thereby making them less attractive as marriage partners and unable to provide for children they father. This in turn contributes to the deepening of poverty in low-income communities.

Large-scale rates of incarceration may contribute to the destruction of the community fabric in other ways. As prison becomes a common experience for young men, its stigmatizing effect is diminished. Gang or crime group affiliations on the outside may be reinforced within the prison only to emerge stronger as individuals are released back to the community. With so few men in underclass communities having stable ties to the labor market, ubiquitous ex-offenders and gang members may become the community's role models.

While we should not ignore that most people under justice system control have committed crimes, current crime control policies may actually be increasing pressures for future offending, particularly when other options for responding to crime exist.

Public Opinion and Sentencing

T he severity of criminal penalties in the United States in the 1990s is unprecedented in American history and unequaled in any Western country in this century. That severity is, however, recent. The incarceration rate per 100,000 was stable from the 1920s through the early 1970s. A scholarly literature emerged to explain that stability and hypothesized that each country had a natural incarceration rate that was little affected by short-term changes in crime patterns or changes in sentencing laws (e.g., Blumstein and Cohen 1973). That the theory was wrong is shown by the uninterrupted increase in incarceration rates since 1971 to a level in 1995 four times higher than the 1972 rate.

Historians looking back at our time will try to explain what happened. Changes in crime rates are clearly not the answer. Changes in officially recorded crime rates, including sharp declines in the early 1980s and again in the early 1990s, had no dampening effect on incarceration rates. Whether crime rates fell or rose, incarceration rates steadily increased.

One explanation that will be considered is that popular attitudes toward crime became steadily harsher, and public officials responded to those changes by enacting steadily harsher policies. The key question is whether public attitudes became harsher, and public officials followed, or whether conservative politicians cynically used crime as an issue to frighten Americans and then offered to respond to those fears by enacting and carrying out harsh policies, in effect heightening fears and then promising to assuage them.

Although such questions can be convincingly answered only from historical hindsight, there is considerable evidence that the answer will be cynical politics. Few people who have lived through the past quarter century can have failed to recognize that conservative politicians have used crime control policies, welfare, and recently immigration as "wedge issues" to appeal to the basest instincts and the genuine fears of working-class whites in order to break their traditional links to the Democratic Party. There is a growing historical literature that documents the partisan use of these wedge issues, often as a way to appeal to racial animosities while appearing to focus on neutral issues like crime and welfare (Edsall and Edsall 1991; Carter 1995). More important, a substantial body of public opinion research shows that Americans are not as single-minded in their punitiveness as many politicians claim.

Professional pollsters distinguish between public opinion and public judgment. The first is what is elicited when people are called at home at dinnertime and asked out of context to give yes or no answers to complicated questions. It is true that most people when asked, cold, "Are sentences too harsh, too lenient, or about right?" will answer "too lenient." Public opinion researchers have established, however, that such answers are generally premised on mistaken beliefs, nurtured by newspaper accounts of sensational crimes and anomalous cases, that sentences are much less harsh that they really are.

Public judgments are the views that people express when they know enough to base their views on solid information. Substantial literatures in the United States and in most Western countries show that public judgments about punishment are much subtler and more complicated than public opinion surveys typically show or than conservatives commonly claim. Ordinary people want wrongdoers punished, but they also want them rehabilitated. Ordinary people believe that offenders are morally responsible for their actions, but they also believe that offenders' disadvantaged backgrounds or circumstances are a major cause of their wrongdoing. Ordinary people want people who commit serious violent crimes to be sent to prison, but they want many other offenders to be sentenced to meaningful intermediate punishments, especially if they include credible treatment elements. Most frontline practitioners share the same set of conflicted and complicated views.

This chapter contains a selection of articles on public opinion and crime. The first, by Julian Roberts, summarizes the massive international and American literatures on public opinion about punishment. The others summarize the findings of major surveys in Alabama, Delaware, Pennsylvania, and Oregon of public judgments about punishment. They all show the pattern of conflicted views described above and a much greater public support for intermediate punishments than anyone who listens to politicians' speeches would believe possible.

American Attitudes about Punishment: Myth and Reality
Julian V. Roberts

Much of the conventional wisdom about Americans' attitudes toward criminal punishment is wrong. Most Americans believe in rehabilitation and are willing to spend tax dollars to rehabilitate offenders. Most Americans want to see criminals punished but for many offenders would rather see community-based punishments imposed than prison sentences.

These findings, which contravene the assumption that Americans want ever more offenders put in prison, is based on a recent review of public opinion research conducted in the United States (and elsewhere) over the past thirty years.

The myth that the public is harshly and single-mindedly punitive results largely from the limits of public opinion surveys. The questions asked are often too simple.

Table 5.1. Summary of Principal Findings: Public Opinion and Sentencing

- Public opposition to alternatives to incarceration has been overstated.
- The public knows little about alternatives.
- The public's first response to crime is in terms of imprisonment.
- Informing the public about alternatives reduces support for prison. This is true for a range of crimes, including some violent crimes.
- The American public believes in rehabilitation and favors rehabilitative programs.
- Sentencing stories in the media usually involve violent crimes and sentences of imprisonment.
- Many politicians have misread public views of crime and punishment.
- The public is not more punitive than are judges.

The answers are often premised on worst-case stereotypes of offenders and on beliefs that punishments are much less harsh than they really are.

Table 5.1 summarizes principal recurring findings from public opinion research concerning sentencing. These findings are drawn from research using different methods and different sample populations in the United States and elsewhere.

Simplistic Questions Generate Simplistic Answers   Consider the poll that asks a single question: "Are sentences too harsh, about right, or not harsh enough?" For many years, this kind of question was used to investigate public attitudes toward sentencing. Most people respond that sentences are "not harsh enough." And from that response politicians have inferred that sentencing policies in America, Great Britain, Canada, and elsewhere should not be modified to promote greater use of alternatives to imprisonment, for to do so would alienate the public. But is this a correct reading of public opinion? I think not.

The weakness with the interpretation is that it fails to consider three important additional findings from the survey literature. First, when answering a question like this, most people are thinking of violent offenders who have criminal histories. The worst-case scenario, in short. Second, the public has little idea of the actual severity of sentencing practices. Most people *underestimate* the severity of sentencing. Third, and perhaps the most glaring weakness associated with addressing a complex issue like sentencing by means of a single question, is that it fails to take into account that the public knows little about sentencing alternatives such as house arrest, intensive probation, restitution, and community service.

The public tends to think about sentencing exclusively in terms of imprisonment. To many people, the question is not so much whether the offender should be imprisoned, but for how long.

Recent research, using new techniques, shows that public opinions about sentencing are more complicated than is generally acknowledged.

Given Alternatives, the Public Will Use Them   What happens when the public is asked questions about sentencing *after* receiving information about sentencing alternatives? Support for imprisonment declines significantly. This important finding emerges from a number of recent studies.

The Public Agenda Foundation carried out research in Alabama and Delaware to determine what would happen to public attitudes if people were given information about sentencing alternatives. The results were remarkably consistent across the two states, and they conform to a pattern emerging from surveys in other countries including Canada and Australia. The methodology of the foundation's research is straightforward. Groups of people are given brief descriptions of crimes. They are then asked to choose between incarceration or probation for the offenders in these descriptions. At this juncture, members of these groups heavily favor imprisonment for most offenders. For example, participants in Delaware were asked to sentence an armed robbery committed by a juvenile with no previous convictions. Four out of five favored incarceration. This is the "before" phase of the study.

Participants are then shown a videotape that describes five alternative sentences: intensively supervised probation; restitution; community service; house arrest; boot camp. A discussion period follows, after which participants are asked to "resentence" the offenders they sentenced in the first phase. This second sentencing is the "after" phase of the study. Support for incarceration declined substantially for a number of offenses. For example, in the case of the juvenile first offender who was convicted of armed robbery, only 38 percent of the participants favored incarcerating the offender after learning about alternative punishments.

Of course, the substantial decline in support for incarceration did not occur for every offense. For rape, for example, the 98 percent of respondents who favored imprisonment in the "before" phase declined only to 86 percent. Nevertheless, these offenses were in the minority. In the "before" phase in Delaware, participants wanted to incarcerate seventeen out of twenty-three offenders. This declined to only five out of twenty-three in the "after" phase.

Similar results emerged in Alabama: incarceration was favored in eighteen out of twenty-three cases in the first phase and only four cases afterward. Illustrative data for the two studies, showing remarkable consistency in the before and after findings, are shown in table 5.2.

Table 5.2. Support for Incarceration before and after Receiving Information about Sentencing Alternatives

Crime	Alabama		Delaware	
	Before	After	Before	After
Theft (5th offense)	90%	46%	83%	47%
Armed robbery	78	47	72	47
Shoplifting	74	22	71	19
Burglary	68	19	70	22
Drunk driving	13	2	16	4
Embezzlement	71	30	71	31

Sources: Doble and Klein (1989); Doble, Immerwahr, and Richardson (1991).

Different Methodologies Find Similar Results   The Public Agenda studies used a focus-group approach, in which relatively small groups of participants focus on a particular problem and discuss the issues involved. The principal alternative approach is to conduct surveys of large, representative samples of the public, in which less information is given, and respondents have less time in which to respond. This is the standard way to measure public opinion in areas such as politics and advertising. What is striking, however, is that similar findings emerge from conventional surveys. For example, Galaway (1984) conducted an experiment that used a representative survey of the public in New Zealand. Half the respondents were simply asked to sentence an offender. The other half were asked to sentence the same offender but were explicitly given the option of court-ordered restitution. Results indicated that the public was willing to accept a reduction in the use of incarceration if offenders were required to pay restitution. Comparable results were found in a Canadian survey by Doob and Roberts (1983), indicating that support for imprisonment is based in part on ignorance of the alternative sanctions available.

Further evidence of public support for alternatives comes from other questions examined in the Public Agenda Foundation studies. In Delaware, for example, the following question was posed: Some states are experimenting with alternatives to prison and probation such as restitution, community service, and house arrest. How do you feel about using alternative sentences?

More than nine out of ten respondents expressed support for these alternatives. Even more strikingly, support for alternatives was equally substantial among victims of crime. Support for alternatives appears to be particularly strong in the case of juvenile offenders. This is apparent from a number of surveys in the United States. One recent poll in Ohio by Knowles (1987) found that over 90 percent of respondents approved of use of alternatives for juveniles.

The Public versus the Courts   Conventional wisdom suggests that "sentences" preferred by the public would be much harsher than the sentences actually imposed by judges. This is not true. Mande and English (1989), for example, show that the public surveyed in Colorado was more, not less, supportive than judges of community-based sanctions. This was true even for some serious crimes. Aggravated robbery provides a good illustration. The Colorado researchers found that approximately three-fourths of offenders convicted of robbery were imprisoned. However, when the views of the public were examined, only 30 percent of respondents favored incarceration; almost half preferred to impose a community-based alternative. Table 5.3 presents an example of a scenario used in a Colorado study, in which public views were contrasted with those held by criminal justice officials. For some crimes the public was close to the public officials in terms of sentencing preferences. The overall result is striking, however: sentence lengths recommended by this representative sample of Colorado residents were on average twelve months shorter than sentences for the same crimes recommended by criminal justice officials.

This finding—of a public that is not harsher in its sentencing preferences than are judges—has now been replicated in a series of studies in the United States, Canada, Great Britain, and Australia. Taken together, these studies do not support the conclusion that the public is considerably more punitive than the courts.

Table 5.3. Sentencing Preferences, Public versus Criminal Justice Officials

Case: Offender pled guilty to manslaughter, was drinking at the time of the offense, is employed, is twenty-five years old, married with a three-year-old child. He has a prior conviction for assault.

Sentence	Public	Officials
Probation	1%	2%
Probation/Jail	19	5
Intensively Supervised Probation	11	12
Community Corrections	20	21
Prison	49	60
	100	100
Median Prison Term	36 months	48 months

Source: English, Crouch, and Pullen (1989).

These results suggest that increased use of alternatives to incarceration can proceed without fear of widespread public opposition, so long as the public is provided with enough information.

Politicians Misread the Public   Public education is only part of the problem, however. It is also necessary to educate criminal justice professionals about the true mood of the public toward sentencing. Criminal justice policy makers fail to appreciate the support among the public for important reforms. For example, Immarigeon (1986) found that a sample of policy makers in Michigan believed that fewer than one resident in four would support an increased use of alternatives to incarceration. In fact, two-thirds of the Michigan public surveyed supported this policy.

Public Support for Rehabilitation   Another misperception concerns rehabilitation. It is commonly suggested that the public favors punishment over all other sentencing objectives and that support for rehabilitation has evaporated. This is simply not true. A national survey recently posed the following question. In dealing with those who are in prison, do you think it is more important to punish them for their crimes, or more important to get them "started on the right road" (i.e., rehabilitate them)?

There was more support for rehabilitation than for punishment (Flanagan and Maguire 1990). Support for rehabilitation in the correctional sphere is also manifested in substantial support for parole. A national survey of Americans conducted a few years ago found that eight of ten respondents endorsed use of parole in the criminal justice systems of America.

Consequences for the Criminal Justice System   The results summarized here have important implications. The first step toward reducing America's alarming prison populations is to reduce public opposition to such a change. The American public is not implacably opposed to alternatives that will lead to a reduction in the number of persons committed to prison. This is true for a wide range of property offenders and also for some offenses involving violence.

Public education is clearly a priority. A substantial degree of public opposition to alternatives is founded on ignorance of these dispositions. It is imperative that the system educate the public in this regard. Information about "intermediate sanctions" that lie between probation and imprisonment needs to be communicated to the public.

At present, the public tends to respond to the problem of sentencing offenders by considering prison as the primary criminal justice response. For this the media bear some responsibility. Most crimes reported in the news are crimes of violence; most sentences reported are terms of imprisonment. But the public's reflex imprisonment response to crime does not spring from a deep-seated desire to be punitive. Rather, it springs, in part at least, from a lack of awareness of the alternatives available. When the public becomes more aware of alternatives, its enthusiasm for prison declines significantly. There is an important lesson to be learned here.

Survey Shows Alabamians Support Alternatives
John Doble

Alabamians favor the use of alternatives to imprisonment for nonviolent offenders, according to a recent public opinion survey conducted by the Public Agenda Foundation, a nonpartisan, not-for-profit research organization, with the assistance of Dr. Philip Coulter of the University of Alabama.

The survey shows broad and deep support for alternative sentencing in Alabama, one of the most conservative states in the country. Once people have had a chance to learn about prison overcrowding and sentencing alternatives, they become much more supportive of use of alternatives.

Nearly three dozen states are under court order to reduce prison overcrowding. Alabama's prison system is not now subject to court orders, but its prisons operate at or near 100 percent of capacity and over 1,000 state prison inmates are backed up in county jails.

The main options facing state officials, building more prisons or releasing large numbers of inmates early, are politically unacceptable. Numerous surveys show that Americans oppose higher taxes for prison construction and that they want more emphasis on basic law and order; indeed, according to the conventional wisdom, most people want to lock offenders up and "throw away the key."

Caught between the rock of overcrowded prisons and the hard place of limited resources, states are considering the use of intermediate sanctions or alternatives to incarceration such as house arrest, community service, restitution, intensive supervision probation (ISP), boot camp, and day-reporting centers as a way to relieve prison overcrowding. Given the intensity of public sentiment concerning crime and punishment, it is prudent to ask, "Under what circumstances, if any, will the public approve of the use of alternative sentences? Would people accept alternatives if they knew more about them and why they are being considered?"

To shed light on these questions, the Public Agenda Foundation, with support from the Edna McConnell Clark Foundation, explored the views of a cross section of 422 people in the state of Alabama. Using a newly developed research technique, the study was designed not only to gauge people's initial views about crime, sentencing, and prison overcrowding but also to determine their considered judg-

Table 5.4. The Sentencing Alternatives Given to the Alabama Respondents

1. Regular Probation
   Offender visits the probation officer once a month.
   Length of Sentence: Up to 2 years.
   Cost: $1,000 per year.
2. Strict Probation
   Offender sees the probation officer up to five times a week.
   Length of Sentence: Up to 2 years.
   Cost: $3,000 per year.
3. Strict Probation Plus Restitution
   Offender must pay back the victim.
   Length of Sentence: Up to 2 years.
   Cost: $3,500 per year.
4. Strict Probation Plus Community Service
   Offender must perform community service to pay back the community.
   Length of Sentence: Up to 6 months.
   Cost: $5,500 per year.
5. House Arrest
   Offender must stay home except to go to work, church, or a doctor.
   Length of Sentence: Up to 1 year.
   Cost: $4,500 per year.
6. Boot Camp
   Offender must complete a basic training–style program in a building near the prison but
   separate from the regular prisoners.
   Length of Sentence: 3 to 6 months.
   Cost: $8,500 per year.
7. Prison
   Cost: $10,000 to $30,000 a year, depending on the state.

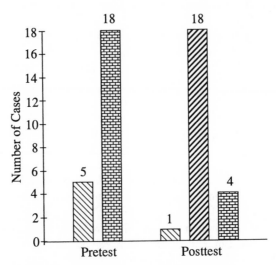

Figure 5.1. Comparison of Sentences in Twenty-Three Cases, Pre- and Posttests. *Source:* Doble and Klein (1989).

ments once they understood more about the issues. The results show far more public support for alternatives than conventional wisdom would suggest.

The Alabamians were asked to sentence twenty-three different offenders whose crimes ranged from petty theft and joyriding to rape and armed robbery. Respondents were given information on each offender's crime, his or her role in the crime, and his or her prior criminal record. When initially asked in the "pretest" whether each offender should be put in prison or on probation, majorities chose prison for eighteen of the twenty-three and probation for five. But when asked a second time, in the "posttest," after seeing a video that detailed the problem of overcrowding and described five "generic" alternative sentences plus probation and prison (see table 5.4), and after discussing the issue at some length, majorities chose prison for only four: three violent offenders and a drug dealer convicted for the fifth time. The other fourteen originally slotted for prison were sentenced to one of the alternatives. See figure 5.1. Once Alabamians learned about alternatives they wanted to make widespread use of them for a wide variety of offenders, including, among others, a burglar and a drunk driver with previous convictions, a man who embezzled $250,000, the accomplices to an armed robbery, and petty thieves with as many as five convictions. Table 5.5 shows the percentages of respondents preferring prison sentences for each of the twenty-three offenders during the pretests and posttests.

But other results may be less pleasing to corrections officials with limited resources. After learning about alternatives, respondents also wanted to use them for most of the offenders they had originally placed on probation. With more choices, respondents chose probation for only one of the twenty-three offenders, a first-time shoplifter guilty of stealing about $250 worth of merchandise. While less expensive than prison, the alternatives are substantially more expensive than "regular" probation as practiced in most states. And when given more choices, respondents sometimes preferred more expensive options.

Beyond revealing judgments about particular categories of offenders, the study suggests that the public approaches criminal justice issues from a different perspective than the one used by professionals. Alabamians believe the top priority for the criminal justice system is not to relieve overcrowding, or reduce costs, or protect the rights of the accused, or even to promote justice; rather, they think the top priority is to protect people from becoming victims of crime. While concern about overcrowding necessarily preoccupies criminal justice professionals, the general public is most concerned about public safety. To the extent that people believe safety will not be compromised, they express considerable support for using alternatives.

Alternatives were popular for several reasons. First, they were felt to have greater potential to rehabilitate. Alabamians believe in rehabilitation and think that a stint in prison, by itself, does little to rehabilitate. They believe that hard work and paying back society are the ways by which offenders can gain the work experience and self-respect they need to turn their lives around. As a result, alternatives featuring work and repayment, such as restitution and community service, were far more popular than those like ISP and house arrest that focus on controlling an offender's activities. More generally, 75 percent agreed that "alternatives improve the chance that an offender will be rehabilitated." Numerous people talked about the importance of stirring an offender's conscience by making the offender come face-to-face

Table 5.5. Comparison of Sentencing Pretest and Posttest

Circumstances of Offender	Pretest Prison %	Posttest Prison %	Alternatives %
In Cases of Petty Theft			
Petty theft, first offense	13	2	47*
Petty theft, fifth offense, works 2 jobs, has wife and 4 kids	59	27	66
Petty theft, third offense	74	22	76
Petty theft, fifth offense	90	46	52
Petty theft, fifth offense, woman working 2 jobs, has 4 kids	48	18	78
Petty theft, fifth offense, woman	71	23	74
In Cases of Drug Selling			
Drug selling, first offense	54	21	62
Drug selling, third offense, addict seeking treatment	62	30	67
Drug selling, fifth offense, addict seeking treatment	74	55	41
In Cases of Burglary/Embezzlement			
Burglary, first offense, 15-year-old	20	4	58
Burglary, first offense	31	7	72
Burglary, second offense	68	19	78
Armed burglary, second offense	83	46	51
Embezzling $250,000, first offense	71	30	67
In Cases of Drunk Driving/Joyriding			
Joyriding, first offense	9	2	53
Drunk driving, first offense	13	2	54
Drunk driving, second offense	51	9	89
In Cases Involving Armed Robbery, Force, or the Threat of Force			
(A case involving three men—second offense for each)			
Offender #1 who goes into a liquor store, shoots a clerk in the arm, and steals money	97	79	21
Offender #2 who drives the getaway car	78	46	49
Offender #3 who fell asleep in the car after agreeing to robbery	50	29	60
Rape, first offense (A man mugged and raped a young woman in the park.)	94	76	21
Armed robbery, third offense (A man pointed a loaded gun at a woman and took her purse.)	93	60	38
A 15-year-old armed robber, first offense (He slashed a woman on her arm with a knife then took money.)	78	33	63

*In the posttest, 51 percent would have sentenced this offender to regular probation.

Source: Doble and Klein (1989).

with his or her victim or crime. A man from the Florence area said, "When I was ten, my friends and I shot out some schoolhouse windows. The windows cost a dime then, and we had to pay to replace them. I've respected glass ever since."

People also thought that alternatives would result in more appropriate sentencing. Ninety percent agreed that alternatives "give judges the flexibility to make the punishment fit the crime." A woman from Mobile said, "You need all these options, as many as you can get, because the crimes are so varied and the offenders so different." Others made the same point indirectly. "If it's a crime against society, not one person, like damaging property, I'd favor [community service]," a man from the Birmingham area said. Others suggested that middle-class offenders should have to make restitution, but that poor offenders who committed the same offense should be sentenced to community service.

A third reason why Alabamians favored using alternatives was to save money. Though they underestimated the extent of the problem, a large majority knew the state's prisons are overcrowded, with 69 percent (in the first questionnaire or pretest) saying the state needs more prisons. A woman from the Huntsville area said, "Our jails are certainly overcrowded. They put pictures of it on TV all the time." In the discussions, numerous respondents said one reason to use alternatives was because of lower cost. However, people were not persuaded by cost considerations alone.

People also liked alternatives because they were felt to be hard. Many Alabamians think that most prison inmates sit idle all day, relaxing, watching television, and playing cards instead of working productively. But especially with the alternatives that feature a mandatory work component, respondents felt that offenders would have to work hard to pay for their crime, an outcome that promoted both justice and rehabilitation.

The final and most important reason why alternatives were popular is that people thought they accomplished all these goals without unduly endangering public safety. Respondents thought that alternatives should be reserved for offenders who pose little risk of violence to the community. In addition, more than two-thirds wanted any alternative sentence to be strictly enforced. Even minor violations should not be tolerated, people felt, because they saw alternatives as a second chance that should be taken very seriously. A man from Montgomery put it this way: "Hey, [an offender who violates the terms of his alternative sentence] had his chance. Now if he violates it, he should be put away."

In sum, support for using alternatives, though widespread and strong, was conditional. It was not open-ended and should not be misread. The Alabamians favored using alternatives for a wide variety of *nonviolent* criminals. But if people think that violent offenders are being allowed in their communities or that alternatives are loosely administered "revolving doors" for career criminals, public support for using them will evaporate.

Delawareans Favor Prison Alternatives
John Doble and Stephen Immerwahr

People in Delaware are much more supportive of use of intermediate sanctions than conventional wisdom suggests, according to a recent study conducted by the

Public Agenda Foundation. Prison alternatives are supported because they are seen as fairer, more flexible, and more conducive to offenders' rehabilitation.

Support for alternatives is, however, conditional. Delawareans want to restrict the use of alternatives to nonviolent offenders, and they want alternatives to be strictly enforced.

How society punishes criminals is an important and hotly debated issue. Over the last decade, the prison population in the United States has increased dramatically. This country has the highest incarceration rate in the world. Costs of prison construction and operation have increased accordingly.

But what is the public's view of criminal punishment? What do people think about sentencing options other than prison or probation? What do they want to do in the face of increasing costs and overcrowded prisons? In an attempt to discover what public opinion analyst Daniel Yankelovich calls the "boundaries of political permission," or the limits set by public opinion on policy makers' choices, the Public Agenda Foundation, a nonpartisan, not-for-profit research organization, studied the public's views about crime, corrections, and the use of intermediate sanctions in the state of Delaware. This article describes the study and discusses some of its findings.

The Delaware Context In many ways, Delaware is a microcosm of the United States. Wilmington, the state's largest city, is a scaled-down version of most cities in the Northeast and Midwest. The state's hub and traditional business center, Wilmington has a rich, ethnic diversity coupled with the social problems associated with large numbers of people, disproportionately black and Hispanic, trapped in chronic poverty. Most of the surrounding New Castle County is made up of suburban, middle-class housing developments interspersed with small towns such as historic New Castle, settled by the Swedes in the seventeenth century, and Newark, home of the University of Delaware. Dover, the state capital, represents the state's midpoint in terms of commerce and culture, as well as geography. The state's two southernmost counties, Kent and Sussex, resemble much of the southern United States: more sparsely populated, more dependent on agriculture, and more traditional in terms of values and lifestyle.

Delaware parallels the rest of the United States in terms of crime and criminal justice. As in the country as a whole, crime, particularly crime related to illegal drug abuse, has recently been a subject of heightened public concern. Like most states, Delaware faces the problem of increasingly crowded prisons and jails. Delaware has also suffered from the economic recession. The government faces large budget deficits and revenue shortfalls that require cutbacks in services, higher taxes, or both.

The key to the dilemma of overcrowded prisons and scarce resources may lie with the public. For more than a decade, Americans have made it clear they do not want higher taxes. At the same time, numerous polls show that citizens believe crime is on the increase, that illegal drug use is the cause, and that most offenders, especially drug dealers and those convicted of violent crime, should be put behind bars. These opinions create an impasse for officials in state after state: people want to incarcerate ever-increasing numbers of offenders, yet they oppose building the requisite prison space if that means higher taxes or cutbacks in services.

A *Study of Public Opinion*   In 1991, the Public Agenda Foundation, a research organization founded and headed by former Secretary of State Cyrus Vance and public opinion analyst Daniel Yankelovich, set out to determine how the people of Delaware would resolve this dilemma if they had more information. How would they feel about using intermediate sanctions or alternatives to prison and probation when they understood more about prison overcrowding, the cost of corrections, and how alternatives work? Under what circumstances, if any, would they support using alternatives?

In February 1991, a total of 432 Delawareans reflecting a cross section of the state's population in terms of geography and key demographic characteristics were brought together for three-hour sessions. Their opinions were assessed before and after an educational discussion to determine whether their thinking changed when they learned more about prison overcrowding, the cost of corrections, and a variety of alternative sentences or intermediate sanctions. The process included four major steps.

First, each participant filled out a questionnaire (called the pretest) sentencing twenty-three hypothetical offenders to either prison or probation, the only options available to judges in most of the United States. Second, they watched a twenty-two-minute video about the problem of prison overcrowding and five alternative sentences—strict probation (or intensively supervised probation), restitution, community service, house arrest, and boot camp—along with the main arguments for and against using alternatives.

Third, they met in small groups of about fifteen people to discuss the issues for about ninety minutes under the leadership of a neutral Public Agenda moderator.

Fourth, they filled out a second questionnaire (the posttest) sentencing the same twenty-three offenders, this time with the possibility of using the five alternatives, as well as the two original options, prison and probation.

Table 5.6 offers brief descriptions of the twenty-three cases and shows Delawareans' preferences concerning prison sentences in the pretest and the posttest.

While the primary goal was to learn when, if ever, the Delawareans would sentence offenders to alternatives, the group was also asked a variety of other questions including their views about the performance of various state agencies, the extent of crime and illegal drug use in Delaware, the causes of crime, what constitutes a "violent" crime, the use of mandatory sentences, and paying higher taxes to build more prisons or to provide more treatment space for drug addicts.

Given a Choice of Prison or Probation, Delawareans Favor Incarceration   The twenty-three hypothetical cases ranged in severity from joyriding to armed robbery and rape. Many of the cases were designed to fall into gray areas in which a sentencing decision would be difficult. Examples include an addict seeking help in an overcrowded treatment center who sells a small amount of cocaine to an undercover police officer for the third time; an offender convicted of nonviolent, petty theft for the fifth time; burglars and embezzlers who neither use force nor threaten its use; teenage offenders or offenders with dependent children; and those who were accomplices to a crime.

In the pretest questionnaire, in which they were given only two sentencing options, the Delawareans wanted to incarcerate seventeen of twenty-three offend-

Table 5.6. Comparison of Sentencing Pretest and Posttest

Circumstances of Offender	Pretest Prison %	Posttest Prison %	Posttest Alternatives %
In Cases of Petty Theft			
Petty theft, first offense	13	2	54
Petty theft, fifth offense, works 2 jobs, has wife and 4 kids	52	25	70
Petty theft, third offense	71	19	77
Petty theft, fifth offense	83	47	50
Petty theft, fifth offense, woman working 2 jobs, has 4 kids	43	16	77
Petty theft, fifth offense, woman	69	23	73
In Cases of Drug Selling			
Drug selling, first offense	56	25	62
Drug selling, third offense, addict seeking treatment	64	33	61
Drug selling, fifth offense, addict seeking treatment	75	60	33
In Cases of Burglary/Embezzlement			
Burglary, first offense, 15-year-old	17	3	68
Burglary, first offense	32	6	81
Burglary, second offense	70	22	74
Armed burglary, second offense	88	71	28
Embezzling $250,000, first offense	71	31	66
In Cases of Drunk Driving/Joyriding			
Joyriding, first offense	8	2	58
Drunk driving, first offense	16	4	68
Drunk driving, second offense	69	22	75
In Cases Involving Armed Robbery, Force, or the Threat of Force			
(A case involving three men—second offense for each)			
Offender #1 who goes into a liquor store, shoots a clerk in the arm, and steals money	98	93	7
Offender #2 who drives the getaway car	72	47	49
Offender #3 who fell asleep in the car after agreeing to robbery	41	25	59
Rape, first offense (A man mugged and raped a young woman in the park.)	97	86	13
Armed robbery, third offense (A man pointed a loaded gun at a woman and took her purse.)	98	71	27
A 15-year-old armed robber, first offense (He slashed a woman on her arm with a knife then took money.)	79	38	58

Source: Doble, Immerwahr, and Richardson (1991).

ers, sentencing the remaining six to probation in which the offender would see a probation officer about once a month. When they only had a choice of prison or probation, the Delawareans generally favored incarceration. A large number of Delawareans also said they believed the crime rate was rising, that judges in Delaware and across the country are "too soft" on convicted criminals, that most prison sentences should be longer, and that the state should make greater use of capital punishment. To oversimplify, many of the pretest results suggest that Delawareans favor a hard-line, law-and-order approach to the problem of crime and corrections. But a closer analysis shows that their thinking is more complex.

Delawareans Favor Alternatives for a Wide Variety of Nonviolent Crimes   As with many national problems, public opinion about crime and criminal justice is multi-faceted, not easy to pigeonhole. The Delawareans' opinions were not ideological, neither liberal nor conservative; instead, they were a rich mixture of views stemming from a variety of beliefs and convictions. In the posttest, after learning about alternative sentences, the Delawareans wanted to incarcerate only five of the twenty-three offenders—four violent offenders, and a drug dealer convicted for the fifth time. In the other eighteen cases, the group wanted to use one of the five alternative sentences. Figure 5.2 depicts the shift in sentencing preferences.

Support for alternatives was broad and deep. Large majorities of Delawareans from all parts of the state and in all key demographic groups—including whites and

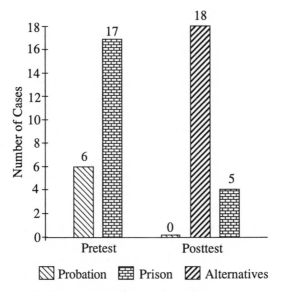

Figure 5.2. Comparison of Sentences in Twenty-Three Cases, Pre- and Posttests. *Source:* Doble, Immerwahr, and Richardson (1991).

nonwhites, old and young, the well- and the not-so-well-educated—favored using alternatives in a wide variety of criminal cases. They liked the idea that alternatives give judges the flexibility to make the punishment fit the crime, they thought that alternatives were appropriate punishment, and they believed that alternatives improve the chances of rehabilitation, a principle the Delawareans believed in deeply.

It is important to stress that the group liked alternatives on their merits, not because they were less expensive; the idea of cutting the costs of corrections, though considered a plus, was a secondary consideration. People thought that alternatives make sense. Indeed, much of the support for alternatives stemmed from unhappiness with the current systems of prison and probation. Most of the Delawareans felt that prisons do not keep offenders productively occupied, let alone provide job training or other rehabilitative measures. Most also felt that seeing a probation officer once a month is an inadequate punishment that fails to provide protection to the public and does little to rehabilitate. In the posttest, the group chose alternative sentences for six offenders whom they originally put on probation, in effect "widening the net."

The Importance of a Work Component   Three of the alternatives—restitution, community service, and boot camp—were more popular among the Delawareans surveyed than house arrest or intensively supervised probation. The three preferred alternatives not only keep track of an offender's whereabouts but also feature a work component that literally makes offenders pay for their crimes, by reimbursing either the victim (restitution) or society as a whole (community service and boot camp).

The Delawareans liked the idea of a work component for five reasons. First, they saw hard work as an appropriate punishment for lawbreakers. Second, they thought that, in principle, offenders should be kept busy and productive. Third, they liked the idea of making offenders give something back. Fourth, they thought working might help offenders acquire job skills. Fifth, they thought working would help offenders internalize the work ethic.

But the willingness to use alternatives was conditional. With the exception of a fifteen-year-old first offender, the Delawareans wanted to limit alternatives to nonviolent offenders. And they wanted strict enforcement. Nearly two-thirds endorsed a more severe sentence for an offender on strict probation who overslept and missed a meeting with his probation officer. The group was even tougher with offenders on strict probation caught shoplifting or using illegal drugs.

Prison Overcrowding   This study was designed to assess public support for alternative sentences as a way to alleviate prison overcrowding. But while the Delawareans support using alternatives, in general they were unconcerned about the problem of prison overcrowding. The Delawareans did not think that putting offenders in overcrowded prisons was, in and of itself, such a terrible thing or a violation of the constitutional prohibition against cruel and unusual punishment. Rather than supporting alternatives as a solution to the problem of prison overcrowding, the Delawareans supported alternatives for their own merits.

The Causes of Crime   The study also explored Delawareans' beliefs about a number of other issues. Asked to identify the causes of crime from a list of sixteen possibilities, the Delawareans named illegal drug use as the principal cause, followed by an array of social problems including "a breakdown of the family structure, especially among poor families," lack of education or poor schools, poverty/economic hardship, low self-esteem, and the availability of handguns. Not enough police, "judges who are too lenient," and "not enough emphasis on basic law and order" were near the very bottom of their list of possible causes of crime.

Summing Up   The study found that a cross section of people from the state of Delaware were willing to use alternative sentences or intermediate sanctions in a wide variety of criminal cases. The "boundaries of political permission" about this issue are wider than conventional wisdom would suggest. The Delawareans favored using alternatives for a number of reasons: they believed alternatives are more just, more flexible, and offer a better chance for rehabilitation. They were also dissatisfied with the existing systems of prison and probation, as they understand them, feeling they do a poor job of rehabilitating and punishing offenders, or protecting public safety. Saving money or controlling the costs of corrections was a secondary consideration of those in the survey. But support for using alternatives is conditional. The Delawareans want to restrict the use of alternatives to nonviolent offenders, and they want alternatives to be very strictly enforced.

Pennsylvanians Prefer Alternatives to Prison
Steve Farkas

The Pennsylvanian public strongly supports the use of alternative sanctions for nonviolent offenders, according to a recent study conducted by the Public Agenda Foundation. Pennsylvanians did not perceive alternative sentences to be a lenient, "slap-on-the-wrist" response to crime. While policy makers like alternative sentences principally for the budgetary savings they promise, citizens saw alternatives as inherently attractive options. Citizens liked alternatives because they allow the criminal justice system to react in a calibrated way to make the punishment fit the crime, because they hope alternatives will change the future behavior of offenders, and because they think alternatives will save money.

Description of the Study   In October 1992, Public Agenda—a nonprofit, nonpartisan research and public education organization—conducted a survey for the Edna McConnell Clark Foundation with a demographically and geographically representative sample of about 400 citizens in six areas across Pennsylvania. Participants' attitudes about criminal justice issues were gauged through two questionnaires and through focus-group discussions. Each participant:

- filled out a questionnaire (the pretest) that gauged attitudes on criminal justice issues and asked the respondent to sentence twenty-four hypothetical offenders to prison or probation;
- watched a twenty-two-minute video produced by Public Agenda about prison over-

crowding and five alternative sentences—strict probation, strict probation plus restitution, strict probation plus community service, house arrest, and boot camp—
along with the main arguments for and against using the alternatives;

- met in a small group of about fifteen people to discuss the issues for about ninety
minutes under the guidance of a neutral moderator. The alternatives were summarized on a sheet of paper handed out to each participant during the discussion; and
- filled out a second questionnaire (the posttest) sentencing the same twenty-four
offenders, but with the five alternative sentences added to the sentencing options.

Initial Results Initially, Pennsylvanians expressed typical "get-tough" attitudes
toward the crime problem. Most wanted longer prison sentences for convicted
offenders, most wanted convicted offenders to serve at least some time in prison,
and most thought judges were "too soft." Although the vast majority (85 percent)
knew their state's prisons were overcrowded, strong majorities rejected building
more prisons if this meant increasing taxes or cutting public services. By a two-to-
one margin, Pennsylvanians were not concerned whether prison overcrowding
amounted to cruel and unusual punishment. In the focus-group discussions, many
respondents said overcrowding added to the deterrent value of prisons and
expressed little sympathy for incarcerated offenders. A woman from the Wilkes-
Barre area expressed this sentiment when she said, "It's [prison] not supposed to be
nice for you. They're in there to be punished."

Predisposition for Alternatives These pretest attitudes seemingly illustrate the
dynamic that created the present budgetary and overcrowding pressures. But the
public's approach to the crime problem is actually more complex and flexible than
these initial "lock-'em-up" attitudes indicate. Even in the pretest—before viewing
the video and participating in discussion—an overwhelming 85 percent of Pennsylvanians favored finding "new ways to punish offenders that are less expensive than
prison but harsher than probation." In the pretest, three-fourths (76 percent)
favored the use of alternative sentences such as restitution and community service.
Pennsylvanians support alternative sentencing not only in the abstract. When asked
to sentence a variety of twenty-four hypothetical offenders, majorities preferred
applying alternatives over prison and probation to most cases.

Sentencing of Twenty-Four Hypothetical Offenders Table 5.7 briefly describes the
twenty-four hypothetical cases and Pennsylvanians' sentencing preferences for
those offenders in the pretest and posttest. The cases range in severity from joyriding to rape, but the bulk of the offenses were of moderate severity (e.g., property or
drug crimes)—the types of crimes criminal justice experts often cite as most appropriate for alternative sentences. The cases also present scenarios that introduce gender, juvenile offenders, and mitigating circumstances such as being the household
provider.

As figure 5.3 shows, in the pretest—before the educational intervention and
given only the options of prison and probation—majorities sentenced fifteen
offenders to prison and nine to probation. In the posttest—after the intervention

Table 5.7. Comparison of Sentencing Twenty-Four Hypothetical Cases

Description of Offense	Pretest Prison %	Posttest Prison %	Posttest Alternatives %
Burglary/Embezzlement			
Burglary, first offense, 15-year-old, unarmed	9	2	69
Burglary, first offense, unarmed, $5,000 stereo from a store	37	5	83
Embezzlement, first offense, $250,000 in forged checks	64	24	71
Burglary, second offense, armed, $5,000 stereo	87	47	51
Force or Threat of Force			
Armed robbery, first offense, pointed a loaded gun at a victim	76	39	56
Rape, first offense, forced rape in a park	96	77	22
Petty Theft			
Shoplifting, first offense, male, $150 radio from a store	13	3	60
Purse snatching, second offense, 15-year-old male	37	4	91
Shoplifting, fifth offense, female, $150 dress from a store	71	22	73
Same offense as above, but head of household	45	13	81
Shoplifting, third offense, male, $150 radio from a store	82	20	78
Shoplifting, fifth offense, male, $150 radio from a store	92	47	51
Same offense as above, but head of household	56	27	69
Drunk Driving/Joyriding			
Joyriding, first offense	13	2	61
Drunk driving, first offense	18	4	64
Drunk driving, second offense, crashed into fire hydrant	62	18	80
Bar Brawls			
Bar brawl, second offense, no injury	17	2	78
Bar brawl, second offense, injured victim	62	18	79
Drug Crimes			
Drug possession, first offense, two grams of cocaine — a probable user	23	9	71
Drug possession, first offense, ten grams of cocaine — a probable drug dealer	48	17	76
Drug dealer, first offense	52	18	67
Drug dealer/addict, third offense, sought treatment	63	28	69
Drug dealer/addict, fifth offense, sought treatment	75	53	41
Others			
Statutory rape, first offense, 21-year-old male with a 15-year-old female, no force involved	24	15	57

Source: Farkas (1993).

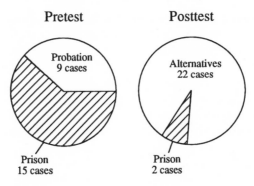

Figure 5.3. Comparison of Sentencing in Twenty-Four Hypothetical Cases. *Source:* Farkas (1993).

and with five alternative sentences added to their sentencing options—majorities sentenced twenty-two of the twenty-four offenders to alternative sentences. Only two offenders were sentenced to prison in the posttest, and no offender was sentenced to probation. These results are very similar to comparable studies that Public Agenda conducted in Alabama in 1989 and in Delaware in 1991. A woman from Exton explained her support for alternative sentences by saying, "I like the idea of progressive punishment. In every school discipline code you don't go from one suspension to expulsion—there are steps all the way up. If they do it properly, you're going to stop them somewhere down the line."

A brief review of the public's sentencing patterns illustrates the preference for alternative sentences, especially in cases involving nonviolent offenses. Majorities sentenced four of the five drug offenders to prison in the pretest, but in the posttest four of the five were instead sentenced to alternatives. For example, while a 48 percent plurality of pretest respondents favored prison for a first-time offender convicted of selling ten grams of cocaine, three-fourths (76 percent) in the posttest sentenced him to alternatives. Policy makers heartened by the potential savings engendered by moving an offender from prison "down" to alternatives should note that the public also moved offenders from probation "up" to alternatives. For example, while two-thirds (66 percent) of pretest respondents had opted to sentence to probation an offender convicted of possession of two grams of cocaine, 71 percent sentenced him to alternatives in the posttest. A Philadelphia man said of probation, "It's a joke! They feel 'Okay, I commit a crime, you put me on probation. Probation is a joke so I'll go do something else.' They should put them in there, maybe boot camp, on the first offense. Maybe it'll make them think twice."

The public drew the line, however, when it came to violent criminals or the most persistent offenders. More than three-fourths (77 percent) in the posttest opted to imprison an offender convicted of mugging and raping a woman (96 percent had wanted to imprison him in the pretest). A 53 percent posttest majority wanted to imprison a drug dealer convicted of his fifth offense (75 percent had wanted to imprison him in the pretest).

Perceptions of Alternatives   Respondents were generally hopeful about the impact of alternative sentences and did not perceive them to be lenient. An overwhelming 86 percent agreed that "alternatives give judges the flexibility to make the punishment fit the crime." A man from the Wilkes-Barre area said, "I like these alternatives for a judge because he has a little more right to treat a person as an individual." Respondents also expect alternatives to save taxpayers money: fully 79 percent agree that "alternative sentences are a less expensive way to punish offenders." Another two-thirds (66 percent) thought that "alternatives improve the chances that an offender will be rehabilitated." Only 24 percent agreed with the statement that "alternatives are not harsh enough"; 61 percent disagreed.

Preferred Alternatives   When asked which of the five alternatives should be used most often, respondents preferred three alternatives: boot camp, strict probation plus restitution, and strict probation plus community service. When posttest respondents sentenced the twenty-four hypothetical offenders, they most often relied on those same three alternatives. Strict probation alone and house arrest were the least preferred alternatives.

The Pennsylvanians clearly voiced their preference for sanctions that force convicted offenders to work or pay back the victim or the community for crimes committed. There was a sense that work offered a chance for rehabilitation. One woman said of boot camp, "This teaches self-discipline to some kids that have never had it at home." Restitution was perceived as especially appropriate for property crimes. A woman from the Pittsburgh area said, "Generally the people who embezzle have homes, take vacations. Why should they be able to do that and not pay restitution to the person they embezzled?" However, the public disliked house arrest because the option lacked a work component. A Wilkes-Barre man said, "It's not getting at the cause of the crime. With a white-collar criminal their home is an entertainment capital itself. So they are stuck in their own home with their VCR, three color TVs. Where's the punishment there?"

Perceptions of Prison   It is interesting to note parallels between the public's views of prisons and the least preferred alternatives. A majority (60 percent) thought Pennsylvania's inmates "spend most of their time watching television, playing cards or basketball and not having to work." Almost half (49 percent) thought that "Pennsylvania's prisons turn most offenders into hardened criminals who are more likely to commit crime when they leave prison."

Response to Violations of Alternative Sentences   Respondents' reactions to an offender who violated the terms of his alternative sentence (strict probation) is enlightening in that they reflect a desire for enforcement of sanctions and an overall strategy toward crime—tough but calibrated. Sixty percent wanted more severe alternatives for a man convicted of shoplifting who oversleeps and therefore misses a meeting with his probation officer, with only 5 percent opting for prison. When the offender is described as failing a drug test, 62 percent moved him to more severe alternatives and another 26 percent to prison. When the offender repeats his shoplifting offense, 55 percent moved him to tougher alternatives and 39 percent to prison.

Mandatory Sentences   The issue of mandatory sentences is a vexing one for criminal justice experts. On the one hand, there is public support and even insistence on the adoption of mandatory sentences. On the other hand, such sentencing aggravates fiscal burdens and overcrowded prison conditions. Not surprisingly, our survey found large and stable majorities supporting mandatory sentences. About three-fourths of the public favored mandatory sentences, even after reading two arguments for and three against the policy. The public's attraction to mandatory sentences may be explained by its overwhelming agreement with the statement that "mandatory sentencing laws stop some judges from being too lenient on offenders." For the public, mandatory sentences may represent a kind of insurance policy guaranteeing offenders will receive some level of punishment. It is interesting to speculate on what the public's response would be to mandatory punishments, such as alternative sentences, that do not involve prison.

Conclusion   In Pennsylvania, as in Delaware and Alabama, the public displays a strong willingness to support alternatives to incarceration for nonviolent offenders. This study shows that the public's approach to criminal justice issues is pragmatic and nuanced, not ideological or knee-jerk. Support for alternatives was steady across racial, political, gender, education, and geographic groupings. The boundaries of political permission—the leeway and support to undertake change that citizens will give their governmental leadership—seem to extend to the substantial but delimited use of alternatives.

Oregonians Support Alternatives for Nonviolent Offenders
Jen Kiko Begasse

Oregonians broadly support many kinds of alternative sentences including boot camp, day-reporting centers, restitution, and day fines for nonviolent offenders, according to a recent public opinion study conducted by Doble Research Associates. Alternative sentences that focus merely on monitoring offenders, such as intensively supervised probation and house arrest, were less popular, although they were still favored by a majority of the population.

Public Opinion Study   To explore public opinion about crime and corrections in Oregon, Doble Research Associates, a New York–area research firm with extensive experience studying public opinion, conducted a series of six focus groups in September 1994. Each session, lasting two hours, included a cross section of adults chosen to reflect the adult population in terms of age, gender, and education level. Sessions were held in Portland, Pendleton, Medford, Newport, Salem, and Eugene to ensure geographical representation.

In each session, respondents were first asked general questions about crime and corrections: whether the crime rate was increasing, whether a family member had been a recent victim, whether the state's jails and prisons were overcrowded, and others. The groups were then told to pretend they were citizens' committees advising state officials and policy makers about appropriate sentences for a wide variety of offenders, including an array of nonviolent offenders. People's knowledge of

Table 5.8. Citizens' Understanding of Terms Associated with Crime and Corrections

Terms	People Think of . . .	Not of . . .
Crime	violent crime, murder, rape, armed robbery	shoplifting, bad checks, credit card fraud
Drug dealers	drug kingpins, dealers who target students	addicts dealing to pay for their habit
Drugs	heroin, crack	marijuana
Felony	murder, rape, violent crime	property offenses
Murder	during a robbery or premeditated	crime of passion
Murder victim	total stranger	spouse, friend, family member
Prison inmates	violent, hard core	repeat nonviolent men over 65
Sex offender	predatory rapists, child molesters	family offenders, statutory rapists
Violent crime	rape, armed robbery	bar fight

Source: Doble Research Associates, Inc. (unpublished data provided to author).

terms and key facts (see table 5.8), and their views about selected issues, were also explored. The aim was to develop a series of hypotheses about the boundaries of political permission—the limits the public sets within which policy makers must make decisions and administer Oregon's criminal justice system.

In January, following the analysis of the focus group results, Doble Research conducted telephone interviews with 439 Oregon adults, age eighteen or older, reflecting a cross section of the state's population. Respondents were asked to assess crime and corrections policy and whether the state should make greater use of community-based sanctions or alternative sentences. The sampling error for each item is plus or minus 5 percentage points, with a confidence interval of 95 percent.

Attitudes Toward Prison   Oregonians' attitudes are strongly influenced by mistaken beliefs about crime rates and sentencing practices. First, like most Americans, Oregonians believe violent crime is on the rise, even though data from the FBI and other reliable sources show it has leveled off or decreased. Second, Oregonians believe that about half of those convicted of violent crimes are not incarcerated. Third, they believe that large numbers of dangerous offenders are released early because of prison overcrowding. Data from the Oregon Criminal Justice Council show both of these views to be serious misperceptions: 77 percent of all convicted offenders, violent and nonviolent, are sent to jail or prison, and no offenders at all, especially violent offenders, are released early due to prison overcrowding.

People across the state seemed to be aware that the jails and prisons of Oregon are badly overcrowded. "I know a lot of the prisons pretty well and most of them are badly overcrowded," said a man from Newport. Eighty-six percent of Oregonians believe that, because of overcrowding, violent offenders are regularly set free before serving their whole sentence. "My ex-husband used to work at the jail and it was just a revolving door. They'd do a little crime and go into jail and be out in a couple of days because it was so overcrowded," said a woman from Newport. Such offenders, people think, pose a clear and present danger to the community.

Lastly, Oregonians believe there is a gap between the sentence most offenders receive and the time they actually serve. "They go in for twenty years and get out in three," said a woman from Newport. Such a gap, they feel, means the law does not mean what it says it does and therefore does not act as a deterrent.

Because they think violent offenders are not incarcerated or are released early because of overcrowding, 67 percent of the people of Oregon favor building more prisons *even if that means raising taxes.* While this should not be taken at face value as an absolute expression of people's willingness to raise their own taxes, it stands in stark contrast to what we found in Delaware and Pennsylvania, where fewer than one-third favored raising taxes to build more prisons, and Alabama, where only one in five was in favor.

Although Oregonians want to incarcerate more violent offenders, they are dissatisfied with the prison system. Oregonians believe that instead of making inmates less likely to reoffend, a stint in prison usually makes offenders more dangerous. Sixty-six percent believe that prisons are schools for criminals, turning new inmates into hardened criminals. When inmates are released, people believe they are more likely to commit new crimes, which means prison exacerbates the crime problem. "The problem is that [a prison sentence amounts to] years in training with people who are a lot worse. If you weren't a hardened criminal when you went in, chances are you will be when you come out," said a Salem woman. "Imprisonment in a state facility does not really rehabilitate individuals. It teaches them to be harder criminals," said a Salem man.

Many believe that drugs are readily available in prisons and that inmates have far too much power in controlling day-to-day affairs. "The guards don't run the facility—the inmates do," said a man from Salem.

Many Oregonians think prison life is too soft, saying prisoners enjoy "luxuries" that many law-abiding citizens cannot afford. Many complained that prisoners can watch cable television, or play pool and lift weights at the expense of taxpayers. People were frustrated by what they saw as an injustice, with many calling for a more spartan atmosphere in which inmates' time would be filled with productive work instead of television watching and weight lifting. Sixty-four percent of Oregonians are convinced that instead of working productively, most state inmates sit around all day, watching television or playing cards.

A faulty work ethic, people believe, is at the root of criminal behavior. Therefore, prisons should help inmates develop and internalize a work ethic by making them work. "Most of these people never learned the work ethic. Why not have them work for their keep, on a farm or an industry? They would get a work ethic and see some sense of achievement at the end of the day," said a Newport woman. Even when the moderator suggested that making inmates work could involve higher costs in terms of extra security and transportation, many said they would pay higher taxes if that were the case. Making inmates work was felt to be a forceful policy that combines rehabilitation, restitution, and deterrence.

Whether Oregonians' views on prison are accurate does not alter their political force: these beliefs—no matter how accurate they may be—combine to create a climate that affects corrections policy, drives support for ballot initiatives for ideas like mandatory sentencing, and limits the policy options of state officials.

Views on Alternative Sentences Oregonians want to incarcerate nearly all violent offenders. "Prison is for people who commit murder and rape and violent crimes — not for a guy who swindles people out of money," said a Eugene man. "Let's be sure the ones we put in prison are really a threat," said a Medford man.

However, when it comes to nonviolent offenders, Oregonians strongly favor community-based punishments. People were asked to sentence offenders who committed a variety of offenses, many of which did not involve violence. In addition to jail and prison, people were given a choice of eight community-based punishments: house arrest with and without electronic monitoring; community service/work crew; inpatient/outpatient alcohol and drug treatment; day fines; restitution; boot camp; day-reporting or restitution center; intensively supervised probation (ISP).

Very large majorities wanted to sentence an array of nonviolent offenders — including those convicted of multiple property offenses, possession of illegal drugs, selling small quantities of illegal drugs, and driving under the influence — to one of the community-based punishments.

After learning more, the people of Oregon strongly favor making much greater use of an array of alternatives, especially restitution (96 percent in favor, 87 percent strongly in favor), boot camp (96 percent in favor, 85 percent strongly), and community service (97 percent in favor, 83 percent strongly). Table 5.9 summarizes these findings. Large numbers also favor making greater use of strict probation or ISP (91 percent in favor, 74 percent strongly), work centers (85 percent in favor, 65 percent strongly), and house arrest (72 percent in favor, 47 percent strongly). Oregonians like community-based punishments because they see them as more rehabilitative and as more stringent or demanding. "I'd give [offenders] boot camp so they don't just sit there behind bars," said a Eugene man.

Cost is not a major reason why Oregonians favor community punishments. But when the cost of incarceration is introduced as a factor, it becomes another good reason for alternatives. On the other hand, 90 percent want to use community punishments even if they cost taxpayers more than regular probation, an indication of the depth of public support.

Oregonians like the idea of a three-month stint in boot camp, with many saying it

Table 5.9. Oregonians' Views about Using Alternative Sentences Instead of Prison for Selected Nonviolent Offenders

Should Oregon Make Greater Use Of:	Total Favor (%)	Total Oppose (%)
Community service	97	3
Restitution	96	3
Boot camp	96	4
Strict probation	91	8
Work centers	85	13
House arrest	72	24

Source: Doble Research Associates, Inc. (unpublished data provided to author).

is a tough punishment that would help offenders develop a work ethic and a sense of self-respect—two things that are, Oregonians believe, more crucial to rehabilitation than either education or job training. Many said that young offenders in particular should be sent to boot camp. Some also saw it as a good option for drug addicts.

Oregonians also like community service, seeing it as an effective way to make offenders pay for their crimes and instill the work ethic. Others liked the option because employed offenders could be punished without losing their jobs. People felt that community service, in order to be successful, needed to involve hard work. Physical labor such as cutting brush, for example, was more popular than making offenders work in a library. Many also wanted the work to be visible, arguing that the indignity would deter potential criminals. Some wanted to tailor community service to the crime. For example, making a man convicted of DWI work with Mothers Against Drunk Driving was a popular idea. "A drunk driver should do community service with MADD, where he can really see [the harm] he can potentially do," said a Eugene woman. Finally, some suggested that community service made sense for those who were too poor to pay a fine or make restitution.

Restitution was popular, especially for offenders convicted of theft, writing bad checks, and other nonviolent property crimes. People saw it as a tough and effective punishment that would serve as a deterrent while helping to restore what victims had lost. Some questioned whether the state would enforce it properly. If restitution is to be used, people said, it should be very strictly enforced.

Oregonians have mixed feelings about house arrest, even with electronic monitoring, with some mistakenly thinking the bracelet is easily removed. "Electronic monitoring is a farce. The electronic anklet comes off. There are ways of getting around these things," said a Pendleton man. But their biggest reservation is the sense that offenders do not have to work—do not have to pay back their victims or society. However, people had a much more favorable outlook when house arrest was combined with other punishments such as restitution or community service. "I'd give [a shoplifter] electronic monitoring, restitution, and maybe a day fine. But not electronic monitoring by itself. That's not enough," said a Newport woman.

Similarly, Oregonians feel that probation neither effectively keeps track of offenders nor helps them become rehabilitated. If they had to cut back somewhere, they would cut probation because they see it as ineffective. People believe that probation officers have too large a caseload to monitor effectively, with many saying that once-a-month checkups are so infrequent that offenders can easily deceive officers. "There's just not enough time for the probation officers to watch everyone," said a Portland woman. Most Oregonians favor using intensively supervised probation, which requires offenders to see their probation officer twice a week. However, many people in the focus groups did not believe that probation, even intensively supervised probation, accomplishes much in terms of rehabilitation or monitoring.

The idea of day fines, fining offenders different amounts based on how much they earn in a day, was felt to be equitable and effective. Only a small minority thinks this approach unjust or unfair. "You've got to hit [offenders in their pocketbooks] where it hurts. If they shoplift and pay the current fine, say $500, then it hurts the minimum-wage guy a lot. But the other guy [who has a higher income] doesn't feel it," said a Newport man. But some questioned the effectiveness of any kind of fine, saying fines are often not collected.

People often wanted to combine alternatives, with one following another. For example, people sometimes felt that a fine should accompany community service, or that offenders should go to boot camp, then be put on house arrest. If an offender had drug or alcohol problems, people invariably wanted treatment to be part of the sentence.

Views on Drug Treatment   Oregonians believe that those caught in possession of marijuana or growing small amounts for recreational use should not be incarcerated. Indeed, many do not want to punish such offenders at all, calling such prosecutions a waste of time. "It's a waste of our money to be picking up someone for possession. What has he done to harm society?" said a Portland man. Those who did want to punish such offenders preferred mild sanctions.

Oregonians consider the use and sale of heroin, cocaine, and crack cocaine to be far more serious than marijuana. Cocaine and heroin lead to addiction, people believe, and then to crime to pay for the addiction. Therefore, the use of these drugs cannot be tolerated. While they generally oppose incarcerating drug addicts, Oregonians want addicts to complete treatment programs and perhaps do some kind of structured work like community service. Requiring addicts to complete a drug treatment program was part of nearly everyone's preferred sentence.

Oregonians deeply believe in drug treatment, regardless of cost or success rate. A majority thinks treatment programs rehabilitate those who complete them less than one-third of the time. Yet 88 percent favor greater use of mandatory treatment for offenders or inmates with drug or alcohol problems even if it is more expensive, for two reasons: first, the belief that everyone deserves a second chance; second, the view that recovered addicts are less likely to break the law when released.

Oregonians' belief in rehabilitation, deterrence, and punishment underlies their support for alternatives, especially restitution, community service, and boot camp, where offenders can get a high school diploma or acquire a job skill, as well as work hard in a structured environment. It may also partly explain why they are less enthusiastic about sentences that are used strictly to monitor offenders, such as intensive probation and house arrest.

However, people from across the state said that if day fines and other community-based punishments are to be used, they should be very strictly enforced. Otherwise, people said, they undermine their threefold purpose of simultaneously punishing, deterring, and rehabilitating. If community punishments are not strictly enforced, people said, they end up being counterproductive instead of doing what they promise to do—helping to reduce the ever-increasing rate of crime.

References

Albrecht, H.-J. 1994. *Strafzumessung bei schwerer Kriminalit"t. Eine vergleichende theoretische und empirische Untersuchung zur Herstellung und Darstellung des Strafmasses.* Berlin: Duncker & Humblot.

_____. 1995. "Sentencing Reform in Germany." *Overcrowded Times* 6(1):1, 6–10.

Albrecht, H.-J. and W. Schädler. 1986. *Community Service, Gemeinnützige Arbeit, Dienstverlening, Travail d' Interet General. "A New Option in Punishing Offenders in Europe."* Freiburg: Max-Planck-Institut.

American Bar Association. 1994. *American Bar Association Standards for Criminal Justice, Sentencing Alternatives, and Procedures.* 3d ed. Washington, D.C.: American Bar Association.

American Friends Service Committee. 1971. *Struggle for Justice.* New York: Hill & Wang.

Anttila, Inkeri. 1986. "Trends in Criminal Law." In *Criminal Law in Action: An Overview of Current Issues in Western Societies*, edited by Jan van Dijk, C. Haffmans, and P. Rüter. Arnhem: Gouda Quint bv.

Ashworth, Andrew. 1992. "The Criminal Justice Act 1991." In *Sentencing, Judicial Discretion, and Training*, edited by Colin Munro and Martin Wasik. London: Sweet & Maxwell.

_____. 1995. *Sentencing and Criminal Justice.* 2d ed. London: Butterworths.

Austin, James, and Aaron David McVey. 1989. *The Impact of the War on Drugs.* San Francisco: National Council on Crime and Delinquency.

Australia Law Reform Commission. 1980. *Sentencing of Federal Offenders.* Canberra: Australian Government Publishing Service.

_____. 1988. *Report No. 44: Sentencing.* Canberra: Australia Law Reform Commission.

Australian Institute of Criminology. 1994. *Australian Prison Trends.* Canberra ACT: Australian Institute of Criminology.

Beck, Allen, Darrell Gilliard, Lawrence Greenfeld, Caroline Harlow, Thomas Hester, Louis Jankowski, Tracy Snell, James Stephan, and Danielle Morton. 1993. *Survey of State Prison Inmates, 1991.* Washington, D.C.: Bureau of Justice Statistics.

Beck, Allen J., and Peter M. Brien. 1995. "Trends in U.S. Correctional Populations." In *The Dilemmas of Corrections*, edited by Kenneth C. Haas and Geoffrey P. Alpert. Prospect Heights, Ill.: Waveland Press.

Berk, Richard, and Alex Campbell. 1993. "Preliminary Data on Race and Crack Charging Practices in Los Angeles." *Federal Sentencing Reporter* 6(1):36–38.

Billotte, Roger Griffin. 1981. "The Gorsuch Bill: A Case Study of Determinate Sentencing Legislation in Colorado." M.A. thesis. Fort Collins: Colorado State University.

Bishop, D. M., and C. E. Frazier. 1990. *A Study of Race and Juvenile Processing in Florida.* A report submitted to the Florida Supreme Court Racial and Ethnic Bias Study Commission.

Blumstein, Alfred. 1993. "Making Rationality Relevant—The American Society of Criminology 1992 Presidential Address." *Criminology* 31(1):1–16.

Blumstein, Alfred, and Jacqueline Cohen. 1973. "A Theory of the Stability of Punishment." *The Journal of Criminal Law and Criminology* 64(2):198–206.

Blumstein, Alfred, Jacqueline Cohen, Susan E. Martin, and Michael Tonry, eds. 1983. *Research on Sentencing: The Search for Reform,* 2 vols. Washington, D.C.: National Academy Press.

Blumstein, Alfred, Jacqueline Cohen, and Daniel Nagin. 1978. *Deterrence and Incapacitation: Estimating the Effects of Criminal Sanctions on Crime Rates.* Washington, D.C.: National Academy Press.

Blumstein, Alfred, Jacqueline Cohen, Jeffrey Roth, and Christy Visher. 1986. *Criminal Careers and "Career Criminals."* Washington, D.C.: National Academy Press.

Boerner, David. 1985. *Sentencing in Washington—A Legal Analysis of the Sentencing Reform Act of 1981.* Seattle: Butterworth.

―――. 1993. "The Legislature's Role in Guidelines Sentencing in 'The Other Washington.'" *Wake Forest Law Review* 28:381–420.

Bogan, Kathleen M. 1990. "Constructing Felony Sentencing Guidelines in an Already Crowded State: Oregon Breaks New Ground." *Crime and Delinquency* 36(4):467–487.

―――. 1991. "Sentencing Reform in Oregon." *Overcrowded Times* 2(2):5, 14–15.

Bogan, Kathleen, and David Factor. 1995. "Oregon Voters' Sentencing Initiative Creates Policy Dilemmas." *Overcrowded Times* 6(1):4–5, 20.

Bol, M. W., and J. J. Overwater. 1984. *Dienstverlening, Vervanging van de vrijheidsstraf in het strafrecht voor volwassenen.* 's Gravenhage: Staatsuitgeverij.

―――. 1986. *Recidive van dienstverleners in het strafrecht van volwassenen.* 's Gravenhage: Staatsuitgeverij.

Bundesregierung Österreich. *Sicherheitsericht.* Vienna: Bundesregierung Österreich.

Bureau of Justice Statistics. Various publications. Washington, D.C.: U.S. Department of Justice, Bureau of Justice Statistics.

Bynum, T. C., M. Wordes, and J. C. Corley. 1993. *Disproportionate Representation in Juvenile Justice in Michigan: Examining the Influence of Race and Gender.* A technical report prepared for the Michigan Committee on Juvenile Justice.

Canadian Sentencing Commission. 1987. *Sentencing Reform: A Canadian Approach.* Ottawa: Canadian Government Publishing Centre.

Carter, Dan. 1995. *The Politics of Rage: George Wallace, the Origins of the New Conservatism, and the Transformation of American Politics.* New York: Simon and Schuster.

Chan, J. 1992. "Dangers and Opportunities in the Sentencing Crisis." *Current Issues in Criminal Justice* 3(3):249–250.

Cherner, Philip A. 1989. "Felony Sentencing in Colorado." *Colorado Lawyer* 18:1689–1701.

Clark, Stover. 1992. "Pennsylvania Corrections in Context." *Overcrowded Times* 3(4):4–5.

Clarke, Stevens H. 1992. "North Carolina Prisons Growing." *Overcrowded Times* 3(4):1, 11–13.

Conley, D. 1994. "Adding Color to a Black and White Picture: Using Qualitative Data to Explain Racial Disproportionality in the Juvenile Justice System." *Journal of Research in Crime and Delinquency* 31(2):135–148.

Council of Europe. 1983–1993. *Bulletin d'information pénologique.* No. 1 (1983) to no. 18 (1993). Strasbourg: Council of Europe.

_____. 1995. *Model of the European Sourcebook on Criminal Justice Statistics.* Strasbourg: Council of Europe.

Cox, George H. Jr. 1988. "Building Our Way Out of the Prison Crisis." In *The State of Corrections: Proceedings of the American Correctional Association.* Laurel, Md: American Correctional Association.

Doble, John, Stephen Immerwahr, and Amy Richardson. 1991. *Punishing Criminals: The People of Delaware Consider the Options.* New York: The Edna McConnell Clark Foundation.

Doble, John, and Josh Klein. 1989. *Punishing Criminals: The Public's View. An Alabama Survey.* New York: The Edna McConnell Clark Foundation.

Doob, Anthony. 1995. "The United States Sentencing Commission's Guidelines: If You Don't Know Where You Are Going, You May Not Get There." In *The Politics of Sentencing Reform,* edited by Chris Clarkson and Rod Morgan. Oxford: Oxford University Press.

Doob, A. N., and J. V. Roberts. 1983. *Sentencing: An Analysis of the Public's View of Sentencing.* Ottawa: Department of Justice, Canada.

Edsall, Thomas, and Mary Edsall. 1991. *Chain Reaction: The Impact of Race, Rights, and Taxes on American Politics.* New York: Norton.

English, K., J. Crouch, and S. Pullen. 1989. *Attitudes Toward Crime: A Survey of Colorado Citizens and Criminal Justice Officials.* Colorado: Colorado Department of Public Safety, Division of Criminal Justice.

Eyland, S. 1992. "Truth in Sentencing: A Koori Perspective." Report submitted to the University of Sydney Institute of Criminology in partial fulfillment for the degree of Master of Criminology.

Fängelsestraffkommittén. 1986. *Påföljd för Brott.* SOU (1–3):13–15.

Farkas, Steve. 1993. *Punishing Criminals: Pennsylvanians Consider the Options.* New York: The Edna McConnell Clark Foundation.

Federal Bureau of Investigation. 1993. Various publications. Washington, D.C.: U.S. Government Printing Office.

Feest, J. 1991. "Reducing the Prison Population: Lessons from the West German Experience?" In *Imprisonment, European Perspectives,* edited by J. Muncie and R. Sparks. London: Harvester Wheatshear.

Fichter, J. 1994. "First Experiences with Community Service in the Canton of Vaud." In *Réforme des sanctions pénales,* edited by S. Bauhofer and P. H. Bolle. Chur & Zurich, Switzerland: Rüegger.

Flanagan, T., and K. Maguire, eds. 1990. *Sourcebook of Criminal Justice Statistics—1989.* Washington, D.C.: U.S. Department of Justice, Bureau of Justice Statistics.

_____. 1992. *Sourcebook of Criminal Justice Statistics—1991.* Washington, D.C.: U.S. Department of Justice, Bureau of Justice Statistics.

Flanagan, Timothy J., and Maureen Macleod. 1983. *Sourcebook of Criminal Justice Statistics—1982.* Washington, D.C.: U.S. Department of Justice, Bureau of Justice Statistics.

Folkard, M. S. et al. 1974. *IMPACT Vol. I.: The Design of the Probation Experiment and an Interim Evaluation.* Home Office Research Study no. 24. London: H.M. Stationery Office.

_____. 1976. *IMPACT Vol. II.: The Results of the Experiment.* Home Office Research Study no. 36. London: H.M. Stationery Office.

Frankel, Marvin E. 1972. *Criminal Sentences: Law without Order.* New York: Hill & Wang.

Frase, Richard. 1993. "Implementing Commission-based Sentencing Guidelines: The Lessons of the First Ten Years in Minnesota." *Cornell Journal of Law and Public Policy* 2:279–337.

Freiberg, A. 1992. "Truth in Sentencing?: The Abolition of Remissions in Victoria." *Criminal Law Journal* 16:165.

_____. 1993. "Sentencing Reform in Victoria." *Overcrowded Times* 4(4):7–9.

_____. 1995. "Sentencing Reform in Victoria. A Case Study." In *The Politics of Sentencing Reform*, edited by Chris Clarkson and Rod Morgan. Oxford: Oxford University Press.

Galaway, B. 1984. *Public Acceptance of Restitution as an Alternative to Imprisonment for Property Offenders: A Survey*. Wellington, New Zealand: Department of Justice.

Godefroy, T., and B. Laffargue. 1991. *Changements économiques et répression pénale: plus de chômage, plus d'emprisonnement?* Déviance & Contrôle Social no. 55. Paris: CESDIP.

Gorta, A. 1990. "Truth in Sentencing." Paper presented at the sixth annual conference of the Australian and New Zealand Society of Criminology, Sydney, September.

_____. 1992. "Impact of the *Sentencing Act of 1989* on the NSW Prison Population." *Current Issues in Criminal Justice* 3(3):308–317.

Gorta, A., and S. Eyland. 1990. *Truth in Sentencing: Impact of the Sentencing Act, 1989. Report 1*. NSW Department of Corrective Services Research Publication no. 22.

Gottfredson, Don M., Leslie T. Wilkins, and Peter B. Hoffman. 1978. *Guidelines for Parole and Sentencing*. Lexington, Mass.: Lexington Books.

Gottlieb, D. 1991. "A Review and Analysis of the Kansas Sentencing Guidelines." *Kansas Law Review* 39:65–89.

Graham, J. 1990. "Decarceration in the Federal Republic of Germany. How Practitioners Are Succeeding Where Policy-Makers Have Failed." *The British Journal of Criminology* 30(2):150–170.

Hagedorn, John M. 1994. "Homeboys, Dope Fiends, Legits, and New Jacks." *Criminology* 32(2):197–219.

Hanewinkel, C., and M. Lolkema. 1990. *Dienstverlening in het arrondissement*. Groningen: Assen.

Harvey, L., and K. Pease. 1987. "The Lifetime Prevalence of Custodial Sentences." *British Journal of Criminology* 27:311–315.

Home Office. 1988. *Punishment, Custody, and the Community*. Cm 424. London: H.M. Stationery Office.

_____. 1989. *National Standards for Community Service Orders*. Home Office circular 18/1989.

_____. 1990. *Crime, Justice, and Protecting the Public*. London: Home Office.

_____. 1991. *Custody, Care, and Justice*. London: Home Office.

_____. 1991. *Prison Disturbances*. April 1990. Report of an Inquiry by the Rt. Hon. Lord Justice Woolf, CM. 1456.

_____. 1992. *Projections of Long-Term Trends in the Prison Population to 2000*. Statistical Bulletin 10/92. Home Office Research and Statistics Department, May 28.

_____. 1993. *Projections of Long-Term Trends in the Prison Population to 2001*. Statistical Bulletin 6/93. Home Office Research and Statistics Department, March 30.

Hood, Roger, 1992. *The Death Penalty*. Oxford: Oxford University Press.

Immarigeon, R. 1986. "Surveys Reveal Broad Support for Alternative Sentencing." *Journal of the National Prison Project of the American Civil Liberties Union Foundation* 9:1–4.

Jacobs, James B. 1982. "Sentencing by Prison Personnel." *UCLA Law Review* 30:216–270.

Jareborg, Nils. 1995. "The Swedish Sentencing Reform." In *The Politics of Sentencing Reform*, edited by Chris Clarkson and Rod Morgan. Oxford: Oxford University Press.

Johnston, Denise. 1993. "Report No. 13: Effects of Parental Incarceration." Pasadena, Calif.: Pacific Oaks College, The Center for Children of Incarcerated Parents.

Joutsen, Matti. 1989. *The Criminal Justice System of Finland: A General Introduction*. Helsinki: Finnish Ministry of Justice.

Jung, H. 1992. Sanktionensysteme und Menschenrechte. Bern et al.: Haupt.

Kaiser, G. 1990. *Befinden sich die kriminalrechtlichen Maβregeln in der Krise?* Heidelberg: C. F. Müller Juristischer Verlag.

Kansas Sentencing Commission. 1992. *Kansas Sentencing Guidelines Act: Implementation Manual.* Topeka, Kans.: Kansas Sentencing Commission.

Kempf, C. L., S. H. Decker, and R. L. Bing. 1990. *An Analysis of Apparent Disparities in the Handling of Black Youth within Missouri's Juvenile Justice System.* St. Louis: Department of Administration of Justice, University of Missouri-St. Louis.

Kempf-Leonard, K., C. E. Pope, and W. Feyerherm, eds. 1995. *Minorities in the Juvenile Justice System.* Newbury Park, Calif.: Sage.

Kerner, H. J., and O. Kästner. 1986. *Gemeinnützige Arbeit in der Strafrechtspflege.* Bonn: Schriftenreihe der Deutschen Bewährungshilfe. Neue Folge 5.

Killias, M. 1991. *Introduction to Criminology.* Berne: Stämpfli.

Killias, M., and R. Aeschbacher. 1988. "How Many Swiss Have Experienced Incarceration?" *Bulletin de criminologie* 14:3–14.

Killias, M., and C. Grandjean. 1986. "Unemployment and the Incarceration Rate: The Case of Switzerland between 1890 and 1941." *Déviance et société* 10(4):309–322.

Killias, M., A. Kuhn, and S. Rônez. 1995 "Sentencing in Switzerland." *Overcrowded Times* 6(3):1, 13–17.

Knapp, Kay A. 1984. *The Impact of the Minnesota Sentencing Guidelines: Three-Year Evaluation.* St. Paul, Minn.: Minnesota Sentencing Guidelines Commission.

_____. 1993. "Allocation of Discretion and Accountability within Sentencing Systems." *University of Colorado Law Review* 64:679–705.

Knowles, J. 1987. *Ohio Citizen Attitudes concerning Crime and Criminal Justice.* 5th ed. Columbus, Ohio: Governor's Office of Criminal Justice Services.

Kockelkorn, R., P. H. van der Laan, and C. Meulenberg. 1991. *Knelpunten bij de toepassing van dienstverlening.* 's Gravenhage: Ministry of Justice, Scientific Research and Documentation Center.

Kress, Jack M. 1980. *Prescription for Justice: The Theory and Practice of Sentencing Guidelines.* Cambridge, Mass.: Ballinger.

Kuhn, A. 1987. "The Origins of Prison Overcrowding in Switzerland." *Déviance et société* 11(4):365–379.

_____. 1993. *Punitiveness, Crime Policies, and Prison Overcrowding: Or How to Reduce Prison Populations.* Berne (Switzerland) & Stuttgart (Germany): Haupt.

_____. 1996. "Étude des fluctuations de la population carcérale allemande." *Déviance et société* 20(1):59–83.

La Libération Conditionnelle: Risque ou Chance? 1994. Bâle: Helbling & Lichtenhahn.

Langan, P. A., C. A. Perkins, and J. M. Chaiken. 1994. *Felony Sentences in the United States—1990.* Washington, D.C.: U.S. Department of Justice, Bureau of Justice Statistics.

Lappi-Seppälä, Tapio. 1994. "Alternative Penal Sanctions." In *Finnish National Reports to the Fourteenth Congress of the International Academy of Comparative Law,* edited by A. Suviranta. Helsinki: Finnish Lawyer's Publishing.

Larivee, John J. 1991. "Day Reporting in Massachusetts: Supervision, Sanction, and Treatment." *Overcrowded Times* 2(1):7–8.

Lillie-Blanton, Marsha, James C. Anthony, and Charles R. Schuster. 1993. "Probing the Meaning of Racial/Ethnic Group Comparisons in Crack Cocaine Smoking." *Journal of the American Medical Association* 269(8):993–997.

Lockhart, K. L., P. D. Kurtz, R. Stuphen, and K. Gauger. 1990. *Georgia's Juvenile Justice System: A Retrospective Investigation of Racial Disparity.* Research report submitted to the Georgia Juvenile Justice Council. Part I of the Racial Disparity Investigation. School of Social Work, University of Georgia.

Lubitz, Robin L. 1993. "NC Legislature Considers Sentencing Change." *Overcrowded Times* 4(2):1, 9–10.

Lynch, J. 1993. "A Cross-National Comparison of the Length of Custodial Sentences for Serious Crimes." *Justice Quarterly* 10:639–660.

Maguire, Kathleen, Ann L. Pastore, and Timothy J. Flanagan, eds. 1993. *Sourcebook of Criminal Justice Statistics 1992.* Washington, D.C.: U.S. Department of Justice, Bureau of Justice Statistics.

Mair, George. 1993. "Day Centres in England and Wales." *Overcrowded Times* 4(2):5–7.

Mair, G., Charles Lloyd, Claire Nee, and Rae Subbett. 1994. *Intensive Probation in England and Wales: An Evaluation.* Home Office Research Study no. 133. London: H.M. Stationery Office.

Mande, Mary J., and Kim English. 1989. *The Effect of Public Opinion on Correctional Policy: A Comparison of Opinions and Practice.* Colorado: Colorado Department of Public Safety, Division of Criminal Justice.

Mann, Coramae Richey. 1993. *Unequal Justice—A Question of Color.* Bloomington: Indiana University Press.

Mauer, Marc, and Tracy Huling. 1995. *Young Black Americans and the Criminal Justice System: Five Years Later.* Washington, D.C.: The Sentencing Project.

McDonald, Douglas. 1986. *Punishment without Walls: Community Service Sentences in New York City.* New Brunswick, N.J.: Rutgers University Press.

McDonald, Douglas C., and Kenneth E. Carlson. 1993. *Sentencing in the Federal Courts: Does Race Matter? The Transition to Sentencing Guidelines, 1986–90. Summary.* Washington, D.C.: U.S. Department of Justice, Bureau of Justice Statistics.

Melossi, D. 1995. *The Effects of Economic Circumstances on the Criminal Justice System.* Strasbourg: Council of Europe.

Meyer, Louis B. 1993. "North Carolina's Fair Sentencing Act: An Ineffective Scarecrow." *Wake Forest Law Review* 28:519–570.

Miller, Jerome. 1992*a*. "Hobbling a Generation: Young African American Males in D.C.'s Criminal Justice System." Alexandria, Va: National Center on Institutions and Alternatives.

———. 1992*b*. "Hobbling a Generation: Young African American Males in the Criminal Justice Systems of American Cities—Baltimore, Maryland." Alexandria, Va.: National Center on Institutions and Alternatives.

Ministry of Justice. 1985. *Society and Crime: A Policy Plan for the Netherlands.* [Introduction and main chapters in English.] 's Gravenhage: Dutch Ministry of Justice.

———. 1990. *Law in Motion.* [English version] 's Gravenhage: Dutch Ministry of Justice.

Minnesota Sentencing Guidelines Commission. 1990. *Guidelines.* St. Paul, Minn.: Minnesota Sentencing Guidelines Commission.

Minshell, S. 1991. *Respect in Probation.* Paper given to Conference on Adult Offenders. Lincoln, Nebr. Mimeographed.

Morris, Norval. 1974. *The Future of Imprisonment.* Chicago: University of Chicago Press.

Morris, N., and M. Tonry. 1990. *Between Prison and Probation.* New York: Oxford University Press.

National Council on Crime Prevention. 1977. *Brotts-förebyggande Rådet, Nytt Straffsystem.* Stockholm: Swedish National Council on Crime Prevention.

National Institute on Drug Abuse. 1991. *National Household Survey on Drug Abuse: Population Estimates 1990.* Washington, D.C.: U.S. Government Printing Office.

Nay, G. 1994. "Recent Developments in the Jurisprudence of the Federal Supreme Court's Criminal Law Branch." *Swiss Criminal Law Review* 112(2):170–193.

New York State Committee on Sentencing Guidelines. 1985. *Determinate Sentencing*

Report and Recommendations. New York: New York State Committee on Sentencing Guidelines.

North Carolina Sentencing and Policy Advisory Commission. 1994. *Felony Guidelines.* Raleigh, N.C.: North Carolina Sentencing and Policy Advisory Commission.

OFS (Swiss Bureau of Statistics). 1985. *Swiss Prisons: "No Vacancies!"* Berne: OFS.

————. 1994a. Statistics of Corrections 1993. Berne (unpublished).

————. 1994b. *Drugs and Criminal Law in Switzerland.* Berne: OFS.

————. 1994c. *On the National Origin of Prison Inmates.* Berne: OFS.

Office fédéral de la statistique. *Annuaire statistique de la Suisse (Statistical Yearbook of Switzerland).* Zurich: Neue Zürcher Zeitung.

Oikeustilastollinen vuosikirja 1992. 1993. *Yearbook of Justice Statistics.* Helsinki: Statistics Finland.

Pennsylvania Commission on Sentencing. 1991. *Guidelines Manual.* State College, Pa.: Pennsylvania Commission on Sentencing.

————. 1995. *Sentencing Guidelines Manual.* State College, Pa.: Pennsylvania Commission on Sentencing.

Petersilia, J. 1988. "Probation Reform." In *Controversial Issues in Crime and Justice,* edited by J. E. Scott and T. Hirschi. Beverly Hills, Calif.: Sage.

Pope, C. E., and W. Feyerherm. 1990. "Minority Status and Juvenile Processing." *Criminal Justice Abstracts* 22(2):327–336 (Part I); 22(3):527–542 (Part II).

————. 1993. *Minorities and the Juvenile Justice System: An Executive Summary.* Rockville, Md.: Office of Juvenile Justice and Delinquency Prevention, U.S. Department of Justice, Juvenile Justice Clearing House.

Poupart, L. 1995. "Juvenile Justice Processing of American Indian Youth: Disparity in One Rural County." In *Minorities in the Juvenile Justice System,* edited by K. Kempf-Leonard, C. E. Pope, and W. Feyerherm. Newbury Park, Calif.: Sage.

Prison Service. 1994. *Corporate Plan 1994–97.* London: Prison Service.

Reiss, Albert J., Jr., and Jeffrey Roth, eds. 1993. *Understanding and Controlling Violence.* Washington, D.C.: National Academy Press.

Reitz, Kevin, and Curtis Reitz. 1995. "Building a Sentencing Reform Agenda—The ABA's New Sentencing Standards." *Judicature* 78(4):189–195.

Reuter, Peter, Robert MacCoun, and Patrick Murphy. 1990. "Money from Crime: A Study of the Economics of Drug Dealing in Washington, D.C." Washington, D.C.: RAND.

Roethof Committee. 1984. *Interim rapport van de Commissie kleine criminaliteit.* [In Dutch.] 's Gravenhage: Staatsuitgeverij.

————. 1986. *Eindrapport Commissie kleine criminaliteit.* [In Dutch.] 's Gravenhage: Staatsuitgeverij.

Rothman, David. 1980. *Conscience and Convenience.* Boston: Little, Brown.

Schöch, H. 1992. Empfehlen sich Änderungen und Erg"nzungen bei den strafrechtlichen Sanktionen ohne Freiheitsentzug? Gutachten C zum 59. Deutschen Juristentag, Hannover 1992. München: Beck.

Sourcebooks of Criminal Justice Statistics. Various years. Washington, D.C.: U.S. Department of Justice, Bureau of Justice Statistics.

South African Institute of Race Relations. 1994. *Survey of Race Relations in South Africa, 1993/1994.* Capetown: South African Institute of Race Relations.

South Australia, Criminal Law and Penal Methods Reform Committee (Mitchell Committee). 1973. *First Report: Sentencing and Corrections.* Adelaide: Government Printer.

Spaans, E. C. 1994. *Appels en peren. Een onderzoek naar de recidive van dienstverleners en kortgestraften.* Arnhem: Gouda Quint.

Statistisches Bundesamt. Various years. *Fachserie 10: Rechtspflege. Reihe 4: Strafvollzug. And*

Statistisches Jahrbuch für das vereinte Deutschland (Statistical Yearbook of Unified Germany). Wiesbaden: Statistisches Bundesamt.

Statistisk Årsbok för Sverige 1995. 1995. Stockholm: Statistics Sweden.

Tak, Peter J. P. 1994. "Sentencing and Punishment in the Netherlands." *Overcrowded Times* 5(5):5–8.

Tonry, Michael. 1993. "Sentencing Commissions and Their Guidelines." In *Crime and Justice: A Review of Research*, vol. 17, edited by Michael Tonry. Chicago: University of Chicago Press.

———. 1995. *Malign Neglect: Race, Crime, and Punishment in America*. New York: Oxford University Press.

———. 1996. *Sentencing Matters*. New York: Oxford University Press.

Tonry, Michael, and Kate Hamilton, eds. 1995. *Intermediate Sanctions in Overcrowded Times*. Boston: Northeastern University Press.

Törnudd, Patrik. 1993. *Fifteen Years of Declining Prisoner Rates*. Research Communication no 8. Helsinki: National Research Institute of Legal Policy.

Tournier, P. 1984. "La population carcérale." In *Données sociales*. Paris: INSEE (Institut National de la Statistique et des Études Économiques).

Tubex, H., and S. Snacken. 1995. "L'évolution des longues peines . . . Aperçu international et analyse des causes." *Déviance et société* 19(2):103–126.

United Kingdom. 1990. *Hansard Parliamentary Debates*, House of Commons, November 20, 1990, col. 139 and March 1988, col. 1096.

U.S. Department of Health and Human Services. 1994. "Preliminary Estimates from the 1993 National Household Survey on Drug Abuse." Washington, D.C.: Substance Abuse and Mental Health Services Administration.

U.S. General Accounting Office. 1992. *Sentencing Guidelines: Central Questions Remain Unanswered*. Washington, D.C.: U.S. General Accounting Office.

U.S. Sentencing Commission. 1991. *The Federal Sentencing Guidelines: A Report on the Operation of the Guidelines System and Short-Term Impacts on Disparity in Sentencing, Use of Incarceration, and Prosecutorial Discretion and Plea Bargaining*. Washington, D.C.: U.S. Sentencing Commission.

———. 1995. *Cocaine and Federal Sentencing Policy*. Washington, D.C.: U.S. Sentencing Commission.

van Kalmthout, Anton M., and Peter J. P. Tak. 1992. *Sanctions-Systems in the Member-States of the Council of Europe: Deprivation of Liberty, Community Service, and Other Substitutes*. Boston: Kluwer.

von Hirsch, A., and A. Ashworth, eds. 1992. *Principled Sentencing*. Boston: Northeastern.

von Hirsch, Andrew, and Kathleen Hanrahan. 1979. *The Question of Parole*. Cambridge, Mass.: Ballinger.

von Witzleben, T. 1994. "Intermediate Findings of the Evaluation of the Community Service Pilot Project in the Canton of Berne." In *Réforme des sanctions pénales*, edited by S. Bauhofer and P. H. Bolle. Chur & Zurich, Switzerland: Rüegger.

Walker, J. 1994. "Trends in Crime and Justice." In *The Australian Criminal Justice System: The Mid 1990s*, edited by D. Chappell and P. Wilson. Sydney: Butterworths.

Washington State Office of the Administration for the Courts. 1991. *The Report of the Courts of Washington—1991*. Olympia: Office of the Administration for the Courts.

Washington State Sentencing Guidelines Commission. 1991. *A Statistical Summary of Adult Felony Sentencing*. Olympia, Wash.: Washington State Sentencing Guidelines Commission.

Weikel, Dan. 1995. "War on Crack Targets Minorities Over Whites." *Los Angeles Times*, May 21.

Wilbanks, William. 1987. *The Myth of a Racist Criminal Justice System.* Monterey, Calif.: Brooks/Cole.

Wordes, M., and T. C. Bynum. 1995. "Policing Juveniles: Is There Bias Against Youth of Color?" In *Minorities in the Juvenile Justice System*, edited by K. Kempf-Leonard, C. E. Pope, and W. Feyerherm. Newbury Park, Calif.: Sage.

Wordes, M., T. C. Bynum, and C. J. Corley. 1994. "Locking Up Youth: The Impact of Race on Detention Decisions." *Journal of Research in Crime and Delinquency* 31(2):149–165.

Wright, Ronald F., and Susan Ellis. 1993. "A Progress Report on the North Carolina Sentencing and Policy Advisory Commission." *Wake Forest Law Review* 28:421–461.

Zdenkowski, G. 1994. "Contemporary Sentencing Issues." In *The Australian Criminal Justice System: The Mid 1990s*, edited by D. Chappell and P. Wilson. Sydney: Butterworths.

Acknowledgments

All of the articles in this collection were originally published in the journal *Overcrowded Times* in the issues listed below.

Hans-Jörg Albrecht: "Sentencing and Punishment in Germany" (February 1995)

Andrew Ashworth: "New Sentencing Laws Take Effect in England" (October 1992); "English Sentencing since the Criminal Justice Act 1991" (December 1995)

Jen Kiko Begasse: "Oregonians Support Alternatives for Nonviolent Offenders" (August 1995)

David Boerner: "Sentencing Policy in Washington, 1992–1995" (June 1995)

Kathleen Bogan: "Sentencing Reform in Oregon" (March 1991); "Oregon Guidelines, 1989–1994" (April 1995)

Kenneth E. Carlson: "Drug Policies Causing Racial and Ethnic Differences in Federal Sentencing" (December 1994)

Stevens H. Clarke: "North Carolina Prisons Are Growing" (August 1992)

Debra Dailey: "Minnesota's Sentencing Guidelines—Past and Future" (February 1992); "Minnesota's Sentencing Guidelines—1995 Update" (December 1995)

David Diroll: "Ohio Adopts Determinate Sentencing" (August 1995)

John Doble: "Survey Shows Alabamians Support Alternatives" (January 1991); "Delawareans Favor Prison Alternatives" (September 1991)

Anthony N. Doob: "Sentencing Reform in Canada" (August 1994)

David Factor: "Oregon Guidelines, 1989–1994" (April 1995)

Steve Farkas: "Pennsylvanians Prefer Alternatives to Prison" (April 1993)

Richard S. Frase: "Prison Population Growing under Minnesota Guidelines" (February 1993)

Arie Freiberg: "Sentencing Reform in Victoria" (August 1993); "Sentencing and Punishment in Australia in the 1990s" (February 1995)

Richard Gebelein: "Sentencing Reform in Delaware" (March 1991)

Angela Gorta: "Truth in Sentencing in New South Wales" (April 1993)

David Gottlieb: "Kansas Adopts Sentencing Guidelines" (June 1993)

Judy Greene: "Massachusetts, Missouri, and Oklahoma Establish Sentencing Commissions" (June 1995)

Tracy Huling: "One in Three Young Black Men Is Ensnared in the Justice System" (December 1995)

Stephen Immerwahr: "Delawareans Favor Prison Alternatives" (September 1991)

Cynthia Kempinen: "Pennsylvania's Sentencing Guidelines—The Process of Assessment and Revision" (December 1995)

Martin Killias: "Sentencing in Switzerland" (June 1995)

John Kramer: "The Evolution of Pennsylvania's Sentencing Guidelines" (August 1992); "Pennsylvania's Sentencing Guidelines—The Process of Assessment and Revision" (December 1995)

André Kuhn: "Sentencing in Switzerland" (June 1995); "Prison Populations in Western Europe" (February 1996)

Roxanne Lieb: "Washington State: A Decade of Sentencing Reform" (July 1991); "Washington Prison Population Growth Out of Control" (February 1993)

Robin Lubitz: "North Carolina Legislature Considers Sentencing Change" (April 1993)

Marc Mauer: "Young Black Men and the Criminal Justice System" (January 1991); "One in Three Young Black Men Is Ensnared in the Justice System" (December 1995)

Douglas McDonald: "Drug Policies Causing Racial and Ethnic Differences in Federal Sentencing" (December 1994)

Jerome Miller: "Forty-Two Percent of Black D.C. Males, Ages Eighteen to Thirty-Five, under Criminal Justice System Control" (June 1992); "Fifty-Six Percent of Young Black Males in Baltimore under Justice System Control" (December 1992)

Rod Morgan: "Punitive Policies and Politics Crowding English Prisons" (June 1994)

David Moxon: "England Abandons Unit Fines" (August 1993)

Richard J. Oldroyd: "Utah's Conjoint Guidelines for Sentencing and Parole" (February 1994)

Carl Pope: "Racial Disparities in the Juvenile Justice System" (December 1994)

Stan C. Proband: "North Carolina Legislature Adopts Guidelines" (October 1993); "Success in Finland in Reducing Prison Use" (October 1994)

Thomas J. Quinn: "Voluntary Guidelines Effective in Delaware" (February 1992)

Curtis R. Reitz: "American Bar Association Adopts New Sentencing Standards" (June 1994)

Kevin R. Reitz: "American Bar Association Adopts New Sentencing Standards" (June 1994)

Julian V. Roberts: " American Attitudes about Punishment: Myth and Reality" (April 1992)

Simone Rônez: "Sentencing in Switzerland" (June 1995)

Peter J.P. Tak: "Sentencing and Punishment in the Netherlands" (October 1994);

"Netherlands Successfully Implements Community Service Orders" (April 1995)

Stephan Terblanche: "Sentencing in South Africa" (April 1995)

T. M. Thorp: "Sentencing and Punishment in New Zealand, 1981–1993" (April 1994)

Michael Tonry: "Racial Disparities Getting Worse in U.S. Prisons and Jails" (April 1994); "Drug Policies Increasing Racial Disparities in U.S. Prisons" (June 1994)

Patrik Törnudd: "Sentencing and Punishment in Finland" (December 1994)

Andrew von Hirsch: "Sentencing Reform in Sweden" (August 1995)

Martin Wasik: "England Repeals Key 1991 Sentencing Reforms" (August 1993)

Thomas Weigend: "Germany Reduces Use of Prison Sentences" (April 1992)

Marianne Wesson: " Sentencing Reform in Colorado—Many Changes, Little Progress" (December 1993)

Ronald F. Wright: "North Carolina Prepares for Guidelines Sentencing" (February 1994); "North Carolina Avoids Early Trouble with Guidelines" (February 1995)